T0198163

Get the eBooks FREE!

(PDF, ePub, Kindle, and liveBook all included)

We believe that once you buy a book from us, you should be able to read it in any format we have available. To get electronic versions of this book at no additional cost to you, purchase and then register this book at the Manning website.

Go to https://www.manning.com/freebook and follow the instructions to complete your pBook registration.

That's it!
Thanks from Manning!

Geoprocessing with Python

Geoprocessing with Python

CHRIS GARRARD

MANNING

SHELTER ISLAND

For online information and ordering of this and other Manning books, please visit
www.manning.com. The publisher offers discounts on this book when ordered in quantity.
For more information, please contact

Special Sales Department
Manning Publications Co.
20 Baldwin Road
PO Box 761
Shelter Island, NY 11964
Email: orders@manning.com

Manning Publications Co.
20 Baldwin Road
PO Box 761
Shelter Island, NY 11964

Development editor: Jennifer Stout
Technical development editor: Karsten Strøbæk
Copyeditor: Katie Petito
Proofreader: Katie Tennant
Technical proofreader: Rizwan Bilbul
Typesetter: Marija Tudor
Cover designer: Marija Tudor

ISBN: 9781617292149
Printed in the United States of America
3 4 5 6 7 8 9 10 – SP – 21 20 19

brief contents

contents

preface

Although I'd taken a lot of programming classes in college, I never fully appreciated programming until I had a job that involved a lot of repetitive tasks. After amusing myself by automating much of that job, I decided to return to school and study biology, which is when I took my first GIS course. I was instantly in love, and managed to convince someone to give me a biology degree for writing an extension for ArcView GIS (a precursor to ArcGIS, for you Esri fans out there). After finishing that up, I went to work for the Remote Sensing/Geographic Information Systems Laboratory at Utah State University. One of my first projects involved some web mapping, and I soon became a big fan of the open source UMN MapServer software. That was my introduction to open source geospatial software, including GDAL.

I'm fairly certain that I didn't appreciate the power of the GDAL/OGR library when I first learned about it, but I came to my senses once I started using it in my C++ and C# code. In the College of Natural Resources, there weren't many people around who were interested in coding, but I did get to point people to the GDAL command-line utilities on a regular basis. But then Esri introduced Python as the scripting language of choice for ArcGIS, and things started to change. I don't think I had used Python much before then, but playing with arcgisscripting (the original Esri Python module) made me realize how much I enjoyed working with Python, so naturally I had to start using GDAL with it as well.

More importantly for this book, my coworker John Lowry suggested that we team-teach a Python-for-GIS class. He taught students how to use Python with ArcGIS, and I taught them about GDAL. The class turned out to be popular, so we taught it that way for another few years until John moved away. I took over the entire class and have been teaching it in various configurations ever since. I've never bothered to take the class material from the first two years off the web, however, which is how Manning found me. They asked if I would write a book on using GDAL with Python. I'd never had the desire to write a book, so it took a bit of persuasion to convince me to do it. In the end, it was my love for teaching that won me over. I've discovered over the years that I really enjoy teaching, mostly because I love watching students incorporate what they're learning into the rest of their work. This is especially true of graduate students, some of whom might not have completed their research in a timely manner (or at all) if they hadn't learned how to write code. I know that these skills will continue to assist them throughout their careers, and my hope is that this book will provide the same help to you, no matter if you're a student, professional, or a hobbyist. This is fun stuff, and I hope you enjoy it as much as I do!

acknowledgments

I knew that writing a book would be difficult, but it was even harder than I anticipated, and it ended up taking quite a bit longer than I thought it would. My coworkers, friends, and family were very supportive throughout the process, and definitely deserve my thanks. Chris McGinty and Tommy Thompson provided feedback on some of the text. My neighbors Marybeth and McKay Wilson knew I was busy and seemed to take joy in mowing my lawn or clearing snow from my driveway before I could get to it, and they also dropped treats by occasionally. My friend Gayle Edlin published her own book while I was working on this one, which proved to me that it really could be done!

My Manning editor, Jennifer Stout, was my biggest cheerleader, always encouraging me when I got bogged down. Thanks for being so patient with me!

There were many reviewers who provided invaluable feedback throughout the process: Alban Thomas, Alfredo Alessandrini, Chris Gaschler, Fredric Ragnar, Gonzalo Vazquez, Jackie Wilson, Jiří Fejfar, Marcus Geselle, Nate Ron-Ferguson, Ramesha Murthy, Ryan Stelly, Scott Chaussée, Shaun Langley, and Thorsten Szutzkus. Rizwan Bilbul gave the complete manuscript a technical proofread.

The book never would've been finished without the help of the rest of the team at Manning, including Katie Tennant, Katie Petito, Kevin Sullivan, Chuck Larson, and Marija Tudor.

And last but not least, I'd like to acknowledge my students over the years, because without them this never would've happened. Nobody would've asked me to write a book if I hadn't left old class materials stranded on the web, nor would I have realized how much I enjoyed helping people learn.

about this book

I wrote *Geoprocessing for Python* to help you learn the basics of working with geospatial data, mostly using GDAL/OGR. There are other options, of course, but some of them build on top of GDAL, so if you understand the material in this book, you'll probably be able to pick them up without too much trouble. This is not a book on GIS or remote sensing, although some background theory will be explained. Instead, this book will teach you how to write Python code for manipulating and creating spatial data, along with some simple analyses. You can use these building blocks to implement more-complicated analyses of your own devising.

Who should read this book

This book is for anyone who wants to learn to work with geospatial data. Some basics of GIS and remote sensing are explained so that readers new to geospatial analysis will know why they're learning certain things, but the code starts out simple enough so that people with a geospatial background—but not much coding experience—will also benefit.

How this book is organized

This book is organized into 13 chapters. It starts out with a general introduction to geospatial data and Python and then covers vector data, spatial reference systems, raster data, and visualization.

- Chapter 1 is an introduction to spatial data and analysis. It describes types of analyses you can perform with different types of data, along with the difference between vector and raster data and the uses of each.
- Chapter 2 is a quick Python primer.
- Chapter 3 explains what the OGR library is and teaches you how to read, write, and edit vector data sources.

- Chapter 4 dives into the differences between vector formats. Although various formats can be treated the same in many cases, here you'll learn about specific capabilities.
- Chapter 5 teaches you how to filter and select data based on spatial and attribute relationships.
- Chapter 6 describes the nitty-gritty details of creating and editing point, line, and polygon geometries.
- Chapter 7 shows you how to look at spatial relationships between geometries and how you might use these concepts for simple analyses.
- Chapter 8 includes an introduction to spatial reference systems and then teaches you how to work with them and transform data between them.
- Chapter 9 explains what the GDAL library is and teaches you how to read and write raster datasets. It also shows you how to convert between real-world coordinates and pixel offsets.
- Chapter 10 teaches you how to work with aspects of raster data such as ground control points, color tables, histograms, and attribute tables. It also covers the use of callback functions and error handlers.
- Chapter 11 describes how to use NumPy and SciPy for map algebra, including local, focal, zonal, and global analyses, and covers some methods for resampling data.
- Chapter 12 shows you some techniques for supervised and unsupervised map classification.
- Chapter 13 teaches you how to use matplotlib and Mapnik to visualize your data.

If you're familiar with spatial data and analyses, you can safely skip chapter 1. Similarly, if you're already familiar with Python, then there's no need to read chapter 2. If you've never programmed at all, you might find that you need to read a little more theory than can be provided in one chapter, but chapter 2 should be a good start, at least. If you're only interested in vector data, you can ignore chapters 9-11. Likewise, if you're only interested in raster data, chapters 3-7 can be skipped.

This book also has several appendixes. The first two, included in the pBook, contain installation instructions for the software used in this book and a list of data resources. Three additional appendixes (containing reference material for the three modules included with GDAL: ogr, osr, and gdal) are online-only and can be downloaded from www.manning.com/books/geoprocessing-with-python.

About the code

This book contains many examples of source code both in numbered listings and in line with normal text. In both cases, source code is formatted in a `fixed-width font like this` to separate it from ordinary text.

In many cases, the original source code has been reformatted; I've added line breaks and reworked indentation to accommodate the available page space in the book, and occasionally used line-continuation markers (➡). Additionally, comments in the source code have often been removed from the listings when the code is described in the text. Code annotations accompany many of the listings, highlighting important concepts.

I've tried to make variable names understandable while still keeping them short enough so that the code can fit on a line in the book. You might want to use more-descriptive variable names in your code, however.

Source code for the examples can be downloaded from www.manning.com/books/geoprocessing-with-python or from https://github.com/cgarrard/osgeopy-code. The example datasets are also available from the Manning link or from https://app.box.com/osgeopy.

Author Online

The purchase of *Geoprocessing in Python* includes free access to a private web forum run by Manning Publications, where you can make comments about the book, ask technical questions, and receive help from the author and from other users. To access the forum and subscribe to it, point your web browser to www.manning.com/books/geoprocessing-with-python. This page provides information on how to get on the forum once you're registered, what kind of help is available, and the rules of conduct on the forum.

Manning's commitment to our readers is to provide a venue where a meaningful dialogue between individual readers and between readers and the author can take place. It is not a commitment to any specific amount of participation on the part of the author, whose contribution to the forum remains voluntary (and unpaid). We suggest you try asking her some challenging questions lest her interest stray! The Author Online forum and the archives of previous discussions will be accessible from the publisher's website as long as the book is in print.

Other online resources

If you need help with the Python language itself, there are a lot of tutorials online, such as the one at www.codecademy.com/learn/python.

If you need help with GDAL/OGR, the gdal-dev mailing list is a great place to ask questions and get advice. Sign up or view the archives at http://lists.osgeo.org/listinfo/gdal-dev.

The Python GDAL/OGR Cookbook found at https://pcjericks.github.io/py-gdalogr-cookbook/ contains a lot of useful examples.

After learning how to use OGR, you might also be interested in learning how to use Fiona (http://toblerity.org/fiona/), which is a module designed to read and write vector data and is built on top of OGR. Shapely (http://toblerity.org/shapely/) is a useful module for manipulating geometries.

Rasterio (https://github.com/mapbox/rasterio) is built on top of GDAL and is another good module for working with raster data.

about the author

Chris Garrard has worked as a developer for the Remote Sensing/Geographic Information Systems Laboratory in the Quinney College of Natural Resources at Utah State University for almost 15 years. She has been teaching a Python-for-GIS course for about half of that time, and has also taught workshops on campus and at conferences. She loves showing people that there are open source alternatives for processing spatial data, but her favorite thing about teaching is that "Aha!" moment when someone realizes just how much the ability to code will help them with their work.

about the cover illustration

The illustration on the cover of *Geoprocessing with Python* is captioned "Man from Dalmatia, Croatia." Dalmatia is a historical region of Croatia on the Adriatic coast. It was once a province of the Roman Empire, and over its history has been fought over and controlled by the Goths, the Byzantines, the Venetians, and the Austro-Hungarian Empire. This illustration is taken from a recent reprint of Balthasar Hacquet's *Images and Descriptions of Southwestern and Eastern Wenda, Illyrians, and Slavs* published by the Ethnographic Museum in Split, Croatia, in 2008. Hacquet (1739–1815) was an Austrian physician and scientist who spent many years studying the botany, geology, and ethnography of many parts of the Austro-Hungarian Empire, as well as the Veneto, the Julian Alps, and the western Balkans, inhabited in the past by peoples of many different tribes and nationalities. Hand-drawn illustrations accompany the many scientific papers and books that Hacquet published.

The rich diversity of the drawings in Hacquet's publications speaks vividly of the uniqueness and individuality of Alpine and Balkan regions just 200 years ago. This was a time when the dress codes of two villages separated by a few miles identified people uniquely as belonging to one or the other, and when members of an ethnic tribe, social class, or trade could be easily distinguished by what they were wearing. Dress codes have changed since then and the diversity by region, so rich at the time, has faded away. It is now often hard to tell the inhabitant of one continent from another, and today's inhabitants of the towns and villages on the shores of the Baltic or Mediterranean or Black Seas are not readily distinguishable from residents of other parts of Europe.

We at Manning celebrate the inventiveness, the initiative, and the fun of the computer business with book covers based on costumes from two centuries ago, brought back to life by illustrations such as this one.

Introduction

This chapter covers
- Introducing basic types of spatial data
- What is geoprocessing?
- Using QGIS

Humans have been making maps for far longer than we've been writing, and even the famed Lascaux caves in France have a star map on their walls. We know that ancient peoples all over the world used maps, including the Babylonians, Greeks, and Chinese. The art of cartography has evolved over the millennia, from cave walls as mediums to clay tablets, parchment, paper, and now digital. Maps have also gotten much more detailed, as well as accurate, as technology has been developed and improved. In fact, most of us would probably have a hard time recognizing the most primitive maps as maps at all.

It took mankind a long time to go from cave walls to mass-produced road maps, but the degree of change in the last few decades has been staggering. *Geographic Information Systems* (GISs) became more common and easier to use, giving more people the ability to both analyze spatial data and produce their own high-quality maps. Then came web mapping and services that allow users to make custom maps online and share them with the world. Many of us even carry devices in our pockets

1

that can display a map showing our current location and tell us how to get to a new restaurant that we want to try. Not only that, but the available data has also changed dramatically. Makers of those early maps would be blown away by our roadmaps overlaid on top of aerial photography and our talking GPS units.

Thanks to these recent advances in technology, along with free and open source tools, you have access to powerful software to work with your own data. This book aims to teach you the basic concepts of working with spatial data and how to do so with the Python programming language and a few open source tools. After reading this book, you'll write Python scripts to solve basic data analysis problems and have the background knowledge to answer more-complicated questions.

1.1   *Why use Python and open source?*

Several compelling reasons exist for using Python and open source tools for processing spatial data. First, Python is a powerful programming language that has the advantage of being much easier to learn than some other languages, and it's also easy to read. It's a good language to start with if you've never programmed before, and if you're coming from other languages, you'll probably find Python easy to pick up.

Learning Python is a good move, even if you never again use it for spatial analysis after reading this book. Many different Python modules are available for a wide range of applications, including web development, scientific data analysis, and 3D animation. In fact, geospatial applications are only a small subset of what Python is used for.

In addition, Python is multiplatform, so unless you've used an extra module that's specific to one operating system, a Python script that you write on one machine will run on any other machine, provided the required modules are installed. You can use your Linux box to develop a set of scripts and then give them to a colleague who uses Windows, and everything should work fine. You do need to install a Python interpreter to run the code, but those are freely available for major desktop operating systems.

Python ships with the core language and numerous modules that you can optionally use in your code. In addition, many more modules are available from other sources. For example, the *Python Package Index* (PyPI), available at https://pypi.python .org/pypi, lists more than 60,000 additional modules, all used for different purposes, and all free. That's not to say that everything Python is free, however. Several of you coming from a GIS background are no doubt familiar with *ArcPy*, which is a Python module that comes with *ArcGIS*, and is not useable without an ArcGIS license.

Not only is there an abundance of free Python packages, but many of them are also open source. Although many people associate open source software with software that doesn't cost money, that's only part of it. The real meaning is that the source code is made available for you to use if you wish. The fact that you have access to the source code means that nothing is a "black box" (if you want to take the time to learn what's inside the box), but also that you can modify the code to suit your needs. This is extremely liberating. I've used open source tools that didn't quite do what I wanted, so I tweaked the source code, recompiled, and then had a utility that did exactly what

I needed. This is impossible with proprietary software. These two types of freedom associated with open source software make it an attractive model.

Several different types of open source licenses exist, some of which not only allow you to modify the code as needed, but even allow you to turn around and sell your derived work without providing the source code and your modifications. Other licenses require that if you use the software, then your software must also be open source.

We'll cover a few popular open source Python modules for geospatial data in this book. Several were originally developed in other languages, but became so common and well respected that they were either ported to other languages, or bindings were developed so that they could be used in other languages. For example, the *Geospatial Data Abstraction Library* (GDAL) is an extremely popular C/C++ library for reading and writing spatial data, and bindings have been developed for Python, .NET, Ruby, and other languages. The GDAL library is even used by many proprietary software packages. Because of the library's widespread use, this book concentrates on GDAL/OGR. If you can learn to use this, then moving to other libraries shouldn't be difficult. In fact, several nice libraries are built on top of GDAL/OGR that are probably easier to use, but don't necessarily provide all of the functionality that's present in GDAL. See appendix A for installation instructions for the modules used in this book.

Another advantage to going with open source tools is that active user communities exist for some of these packages, and you may find that bugs and other issues are addressed much more quickly than with many proprietary software packages. You can even discuss the finer points of the libraries with the actual developers via email lists.

1.2 Types of spatial data

You'll learn how to work with the two main types of spatial data, vector and raster. Vector data is made up of points, lines, and polygons, while raster data is a two- or three-dimensional array of data values such as the pixels in a photograph. A dataset containing country boundaries is an example of vector data. In this case, each country is generally represented as a polygon. Datasets that use lines to represent roads or rivers, or points to show the location of weather stations, are other examples. Early primitive maps, such as those drawn on cave walls, only showed the features themselves. Later maps contained labels for features of interest such as cities or seaports; for example, the Portolan map of northwest Africa shown in figure 1.1.

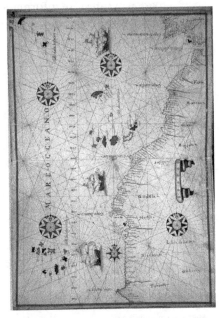

Figure 1.1 A Portolan map of the northwest coast of Africa, circa 1590

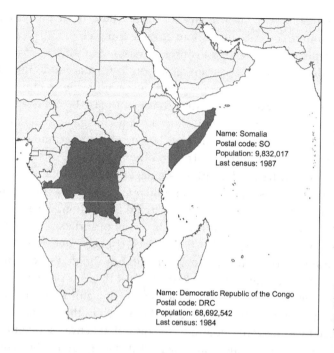

Name: Somalia
Postal code: SO
Population: 9,832,017
Last census: 1987

Name: Democratic Republic of the Congo
Postal code: DRC
Population: 68,692,542
Last census: 1984

Figure 1.2   You can store attributes such as name and population for each geographic feature in a dataset.

Using digital data, you have the advantage of attaching multiple attribute values to each feature, whether you plan to display the information on a map or not. For each road, you can store information such as its name, speed limit, number of lanes, or anything else you can think of. Figure 1.2 shows an example of data you might store with each country in a dataset.

Of the several reasons why this is useful, the obvious one is that you can label features using one of the attributes. For example, figure 1.2 could show country names as well as outlines. All of this data can also help you make more-interesting maps that might even tell a story. The population counts stored for each feature in figure 1.2 could be used to symbolize countries based on population, so it's evident at a glance which countries are most populated (figure 1.3).

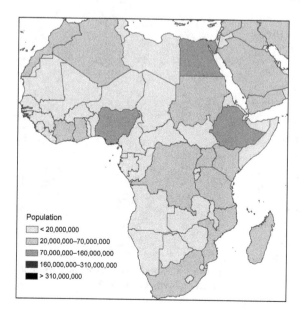

Population
☐ < 20,000,000
☐ 20,000,000–70,000,000
☐ 70,000,000–160,000,000
☐ 160,000,000–310,000,000
■ > 310,000,000

Figure 1.3   Countries symbolized based on population

Figure 1.4 Lake Victoria straddles Uganda, Kenya, and Tanzania. Spatial analysis could help you determine the proportion of the lake that falls in each country.

Spatial overlay analyses are also easy using vector data. Say you wanted to know what percentage of Lake Victoria was in Uganda, Kenya, and Tanzania. You could always guesstimate the answer based on figure 1.4, but you could also use GIS software to get more accurate numbers. You'll do simple analyses like this by the time you finish this book.

Attribute values attached to features can also add to the power of spatial operations. For example, say you had a dataset containing the locations of water wells with attributes that included depth and flow rate. If you also had a dataset for the same area containing geologic landforms or soil types, you could analyze this data to see if flow rate or required well depth was affected by landform or soil type.

Unlike the early mapmakers, you also have access to raster data. Rasters, as the datasets are called, are two- or three-dimensional arrays of values, the way a photograph is a two-dimensional array of pixel values. In fact, aerial photographs such as the one shown in figure 1.5 are a commonly used type of raster data. Satellite images sometimes look similar, although they generally have lower

Figure 1.5 An aerial photograph near Seattle, Washington

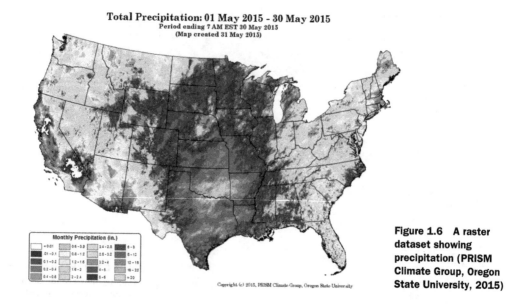

Total Precipitation: 01 May 2015 - 30 May 2015
Period ending 7 AM EST 30 May 2015
(Map created 31 May 2015)

**Figure 1.6   A raster
dataset showing
precipitation (PRISM
Climate Group, Oregon
State University, 2015)**

resolutions. The cool thing about satellite imagery is that much of it is collected using nonvisible light so it can provide information that a simple photograph cannot.

Raster datasets are well suited to any continuous data, not only photographs. Precipitation data like that shown in figure 1.6 is a good example. Rain doesn't usually stop at a sudden boundary, so it's hard to draw a polygon around it. Instead, a grid of precipitation amounts works much better and can capture local variation more easily. The same idea applies to temperature data, and many other variables, as well. Another example is a *digital elevation model* (DEM), in which each pixel contains an elevation value.

Raster data is better suited for different types of analysis than vector data. Satellite imagery and aerial photos are commonly used for tasks such as vegetation mapping. Because water only flows downhill, elevation models can be used to determine watershed boundaries. Even simple math can be used to perform useful analyses with raster data. For example, simple ratios of one wavelength value to another can help identify healthy vegetation or measure soil moisture.

Blocks of adjacent pixels can also be used to calculate useful information. For example, you can use a DEM to calculate slope, which can then be used for runoff analysis, vegetation mapping, or planning a ski resort. But to calculate slope, you need the elevation of surrounding cells. In figure 1.7, you use all of the pixel values shown

54	53	51
53	52	50
50	50	48

**Figure 1.7   All nine elevation values shown here would
be used to calculate the slope for the center pixel.**

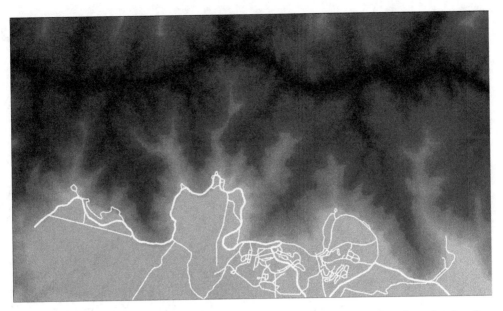

Figure 1.8   Simple map of the Grand Canyon with vector roads layer drawn on top of a raster elevation dataset

to calculate the slope of the center pixel. For any other pixel, you need the surrounding nine cells to calculate slope for it, too. These sets of pixels are called *windows*, and you can do many other kinds of analyses by moving a window around a raster so each pixel is in the center of its own window.

Vector and raster data can also be used together. Think of a hybrid web mapping application that shows a photographic basemap with roads drawn on top of it. The basemap is raster data and the roads shown on top are vectors. Figure 1.8 shows an example of a simple map that uses a raster DEM of the Grand Canyon as a basemap and shows a vector line dataset drawn on top.

1.3   *What is geoprocessing?*

Geoprocessing is a general term for manipulating spatial data, whether raster or vector. As you can imagine, that covers an awful lot of ground. I've always thought of using GIS with geoprocessing as a tool much like statistics in that it can be applied to pretty much everything. You even use geoprocessing in your daily life, whether you realize it or not. For example, I tend to take a different route to work depending on whether I'm driving or riding a bicycle because I prefer to avoid high-traffic roads with no shoulder when riding my bike. Steep hills are also not a concern while driving, but they are when I'm biking. Basing my route selection not only on spatial factors such as the direction of the road and elevation gain, but also on attributes such as the amount of traffic and road width is a type of geoprocessing. You probably make similar decisions every day.

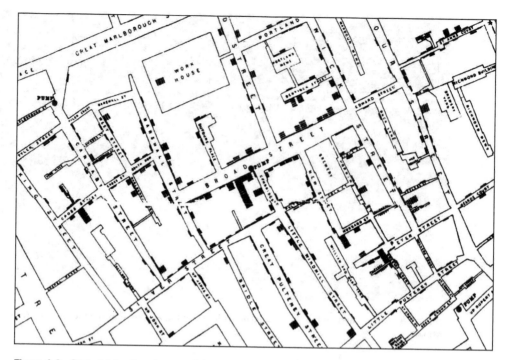

Figure 1.9   Part of John Snow's map of the Soho cholera outbreak of 1854

You have many reasons to be interested in geoprocessing, other than selecting a route to work. Let's look at a few examples of applications. One famous example of early spatial analysis is the story of John Snow, an English physician who lived in the 1800s. Although parts of the story have been disputed, the gist of it is that he used spatial analysis to determine the cause of a cholera outbreak in 1854. A section of his map is shown in figure 1.9, with the Broad Street pump in the middle. You can see that it looks like bar charts are anchored on nearby streets. Each of these bars is made of horizontal lines, with one per cholera victim. Snow realized that most of the victims probably got their water from the pump on Broad Street, because that was the closest one, and he convinced authorities to shut the pump down. This is significant not only because it's an early example of spatial analysis, but also because it wasn't yet known that cholera was contracted from contaminated water. Because of this, Snow is considered one of the fathers of modern epidemiology.

Spatial analysis is still an important part of epidemiology, but it's used for many other things, too. I've worked on projects that include studying the habits of a threatened species, modeling vegetation cover over large areas, comparing data from pre- and post-flood events to see how the river channels changed, and modeling carbon sequestration in forests. You can probably find examples of spatial analysis wherever your interests lie. Let's consider a few more examples.

Chinese researchers Luo et al.[1] used spatial analysis, along with historical records, to pinpoint the locations of missing courier stations along the Silk Road. The historical records contained descriptions of the route, including distance traveled and general direction between stations. The locations of several stations were already known, and the researchers knew that ancient travelers were unlikely to follow a straight line, but instead follow rivers or other landforms. They used all of this information to determine likely geographic areas for the still-missing stations. They then used high-resolution satellite imagery to search these areas for geometric shapes that could be station ruins. After visiting the sites in person, they determined that one, in fact, was an old courier station, and two others were likely military facilities during the Han Dynasty.

For a completely different application, Moody et al.[2] were interested in the potential for using microalgae as a biofuel. They used a microalgae growth model and meteorological data from various locations around the globe to simulate biomass productivity. Because the meteorological data was only from certain sites, the results were then spatially interpolated to provide a global map of productivity potential. It turns out that the most promising locations are in Australia, Brazil, Colombia, Egypt, Ethiopia, India, Kenya, and Saudi Arabia.

This is interesting, but spatial analyses also affect your everyday life. Have you noticed that your automobile insurance premium differs depending on where you live? It's likely that a sort of spatial analysis also affected the location of your favorite coffee shop or grocery store. Several new elementary and high schools are being built in my community, and their locations were determined in part by the spatial distribution of future students, along with the availability of suitable pieces of real estate.

Spatial analysis isn't limited to geography, either. Rose et al.[3] demonstrated that GIS can be used to analyze the distribution of nano- and microstructures in bone. They could use this to see how bone remodeling events corresponded to parts of the bone that experience high levels of compression and tension.

You personally might need to make data more suitable for a map, such as eliminating unwanted features or simplifying complex lines so they display faster on a web map. Or you might analyze demographic data to plan for future transportation needs. Perhaps you're interested in how vegetation responds to different land management practices, such as prescribed burns or mowing. Or maybe it's something else entirely.

Although geoprocessing techniques can be rather complicated, many are fairly simple. It's the simple ones that you'll learn about in this book, but they're the foundation for everything else. By the time you're done, you'll read and write spatial data in many

[1] Luo, L., X. Wang, C. Liu, H. Guo, and X. Du. 2014. Integrated RS, GIS and GPS approaches to archaeological prospecting in the Hexi Corridor, NW China: a case study of the royal road to ancient Dunhuang. Journal of Archaeological Science. 50: 178-190. doi:10.1016/j.jas.2014.07.009.

[2] Moody, J. W., C. M. McGinty, and J. C. Quinn. 2014. Global evaluation of biofuel potential from microalgae. Proceedings of the National Academy of Sciences of the United States of America. 111: 8691-8696. doi: 10.1073/pnas.1321652111.

[3] Rose, D. C., A. M. Agnew, T. P. Gocha, S. D. Stout, and J. S. Field. 2012. Technical note: The use of geographical information systems software for the spatial analysis of bone microstructure. American Journal of Physical Anthropology. 148: 648–654. doi: 10.1002/ajpa.22099.

formats, both vector and raster. You'll subset vector data by attribute value or by spatial location. You'll know how to perform simple vector geoprocessing, including overlay and proximity analyses. In addition, you'll know how to work with raster datasets, including resizing pixels, performing calculations based on multiple datasets, and moving window analyses.

   You'll know how to do all of this with Python rather than by pushing buttons in a software package. The ability to script your processes like this is extremely powerful. Not only does it make it easy to batch process many datasets at once (something I do often), but it gives you the ability to customize your analysis instead of being limited to what the software user interface allows. You can build your own custom toolkits based on your workflow, and use these over and over. Automation is another big one, and it's the reason I fell in love with scripting in the first place. I hate pushing buttons and doing the same thing over and over, but I'll happily spend time figuring out how to automate something so I never have to think about it again. One last advantage that I'll mention here is that you always know exactly what you did, as long as you don't lose your script, because everything is right there.

1.4   *Exploring your data*

You'll see ways to visualize your data as you work with it in Python, but the best way to explore the data is still to use a desktop GIS package. It allows you to easily visualize the data spatially in multiple ways, but also inspect the attributes included with the data. If you don't have access to GIS software already, QGIS is a good open source option and is the one we'll be using when needed in this book. It's available from www.qgis.org, and it runs on Linux, Mac OS X, and Windows.

Downloadable code and sample data

The examples in this book use code and sample data that's available for download from the following links. You'll need to download these if you want to follow along. The code contains examples from the book but also custom utilities used by the examples, and all of the data used in the examples is included.

- Code: https://github.com/cgarrard/osgeopy-code and www.manning.com/books/geoprocessing-with-python
- Data: https://app.box.com/osgeopy and www.manning.com/books/geoprocessing-with-python

---

This isn't a book on QGIS, so I won't talk much about how to use it. Documentation is available on their website, and you can find one or two books published on the topic. However, I'll briefly discuss how to load data and take a look. If you've never used a GIS before, then QGIS might look a bit daunting when you first open it up, but it's not hard to use it to view data. For example, to load up one of the shapefiles in the example data for this book, select Add Vector Layer… from the Layer menu in QGIS. In the dialog that opens, make sure that the File button is selected and then use the Browse

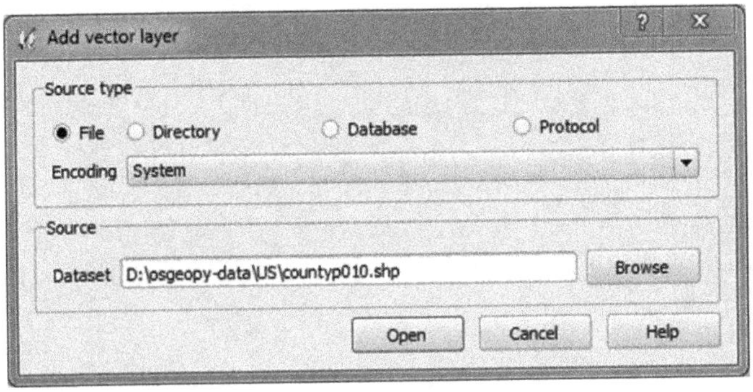

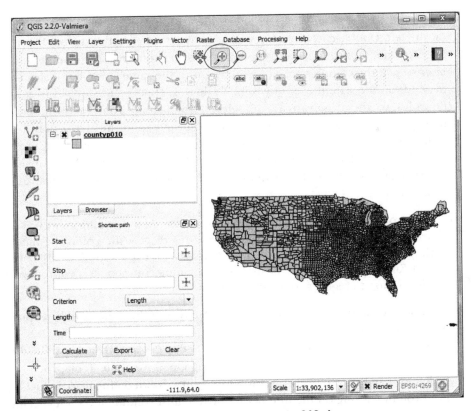

Figure 1.10 The dialog for adding a vector layer to QGIS

button to select a shapefile. A good choice to start out with is the countyp010.shp file in the US folder (figure 1.10).

After selecting a file, click Open in the Add vector layer dialog, and the spatial data will draw in QGIS, as shown in figure 1.11. You can use the magnifying glass tool (circled in figure 1.11) to zoom in on part of the map.

Figure 1.11 QGIS window immediately after loading countyp010.shp

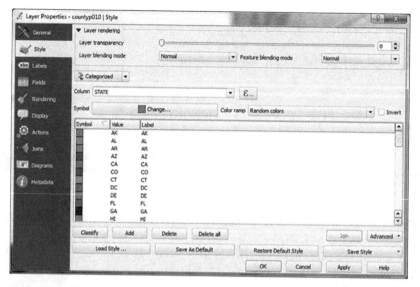

Figure 1.12   QGIS Style dialog configured to draw the counties in each state in a different color

You'll also see the name of the layer, countyp010 in this case, shown in the Layers list on the left. Double-click on a layer and you'll get a Properties dialog. If you click on the Style tab, then you can change how the data is drawn. Let's change the counties layer so that the counties are not all drawn with the same color, but instead the color depends on the state the county is in. To do this, choose Categorized from the drop-down list, set the column to STATE, select a Color ramp from the dropdown list, and then click Classify. You'll see a list of all of the states and the colors they'll be drawn with, as shown in figure 1.12. You can change the color ramp by selecting a new one from the list, clicking Delete All, and then clicking Classify again. You can also change a particular entry in the list by double-clicking on the color swatch next to the state abbreviation.

NOTE TO PRINT BOOK READERS: COLOR GRAPHICS   Many graphics in this book are best viewed in color. The eBook versions display the color graphics, so they should be referred to as you read. To get your free eBook in PDF, ePub, and Kindle formats, go to https://www.manning.com/books/geoprocessing-with-python to register your print book.

Once you're happy with your colors, click Apply, and the colors will be applied in the main QGIS window (figure 1.13).

You can view the attribute data that's attached to the spatial data by right-clicking on the layer name in the Layers list and selecting Open Attribute Table. Each row in the table shown in figure 1.14 corresponds to a county drawn on the map. In fact, try selecting a row by clicking on the number in the left-most column and then clicking on the Zoom map to selected rows button (circled in figure 1.14) and watch what happens.

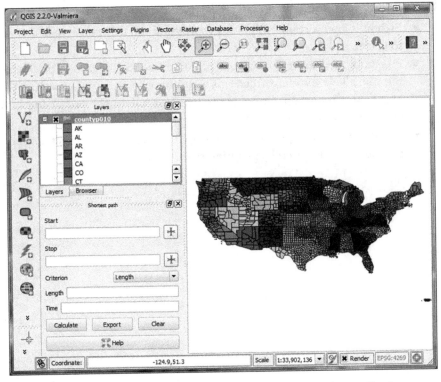

Figure 1.13   Results of applying the symbology from figure 1.12 to the counties layer

	AREA	PERIMETER	COUNTYP010	STATE	COUNTY	FIPS	STATE_FIPS	SQUARE_MIL
0	0.00239414646	0.57037166542	1.00000000000	LA	Jefferson Parish	22051	22	9.932
1	0.00064387335	0.13918850596	2.00000000000	ME	Sagadahoc County	23023	23	2.213
2	0.00029573696	0.11466394609	3.00000000000	NC	Carteret County	37031	37	1.158
3	0.00006176934	0.07273739517	4.00000000000	PA	Delaware County	42045	42	0.227
4	0.87367413872	36.90592850260	5.00000000000	WA	NULL	53000	53	2788.630
5	0.68630400240	6.25083038049	6.00000000000	WA	Whatcom County	53073	53	2163.723
6	1.59122576763	6.45664543886	7.00000000000	MT	Valley County	30105	30	5061.970

Attribute table - countyp010 :: Features total: 3641, filtered: 3641, selected: 0

Show All Features

Figure 1.14   Attribute table for the counties layer

Take time to play with QGIS and read at least part of the documentation on the website. The software is extremely powerful and worth getting to know. I'll talk about it more throughout the book, but not a whole lot. You'll want to use it to inspect the sample data and the results of any data you create, however.

1.5 Summary

- Python is a powerful multiplatform programming language that's relatively easy to learn.
- Free and open source software is not only free with regard to price (free beer), but also allows for many freedoms with how it's used (free speech).
- Many excellent open source Python modules exist for processing both vector and raster geospatial data.
- You don't give up quality by using open source tools. In fact several of these packages are also used by proprietary software.

Python basics

You can do many things with desktop GIS software such as QGIS, but if you work with spatial data for long, you'll inevitably want to do something that isn't available through the software's interface. If you know how to program, and are clever enough, you can write code that does exactly what you need. Another common scenario is the need to automate a repetitive processing task instead of using the point-and-click method over and over again. Not only is coding more fun and intellectually stimulating than pointing and clicking, but it's also much more efficient when it comes to repetitive tasks. You have no shortage of languages you could learn and work with, but because Python is used with many GIS software packages, including QGIS and ArcGIS, it's an excellent language for working with spatial data. It's also powerful, but at the same time a relatively easy-to-learn language, so that makes it a good choice if you're starting out with programming.

Another reason for using Python is that it's an interpreted language, so programs written in Python will run on any computer with an interpreter, and

15

interpreters exist for any operating system you're likely to use. To run a Python script, you need the script and an interpreter, which is different from running an .exe file, for example, where you only need one file. But if you have an .exe file, you can only run it under the Windows operating system, which is a bummer if you want to run it on a Mac or Linux. However, if you have a Python script, you can run it anywhere that has an interpreter, so you're no longer limited to a single operating system.

2.1   *Writing and executing code*

Another advantage of interpreted languages is that you can use them interactively. This is great for playing around and learning a language, because you can type a line of code and see the results instantly. You can run the Python interpreter in a terminal window, but it's probably easier to use *IDLE*, which is a simple development environment installed with Python. Two different types of windows exist in IDLE, *shells* and *edit windows*. A shell is an interactive window in which you can type Python code and get immediate results. You'll know that you're looking at an interactive window if you see a >>> prompt, like that in figure 2.1. You can type code after this prompt and execute it by pressing Enter. Many of the examples in this book are run this way to show results. This is an inefficient way to run more than a few lines of code, and it doesn't save your code for later use. This is where the edit window comes in. You can use the File menu in IDLE to open a new window, which will contain an empty file. You can type your code in there and then execute the script using the Run menu, although you'll need to save it with a .py extension first. The output from the script will be sent to the interactive window. Speaking of output, in many of the interactive examples in this book I type a variable name to see what the variable contains, but this won't work if you're running the code from a script. Instead, you need to use print to explicitly tell it to send information to the output window.

In figure 2.1 the string I typed, 'Hello world!', and the output are color coded. This syntax highlighting is useful because it helps you pick out keywords, built-in functions, strings, and error messages at a glance. It can also help you find spelling mistakes if something doesn't change color when you expect it to. Another useful feature of IDLE is *tab completion*. If you start typing a variable or function name and then press the Tab

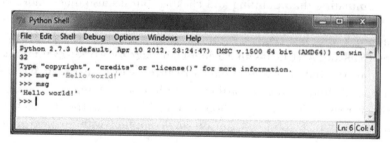

Figure 2.1   An IDLE shell window

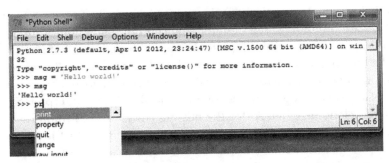

Figure 2.2 Start typing and press the Tab key in order to get a list of possible variables or functions that match what you were typing.

key, a list of options will pop up, as shown in figure 2.2. You can keep typing, and it will narrow the search. You can also use arrow keys to scroll through the list. When the word you want is highlighted, press Tab again, and the word will appear on your screen.

Because Python scripts are plain text files, you aren't forced to use IDLE if you don't want to. You can write scripts in whatever text editor you prefer. Many editors are easy to configure, so you can run a Python script directly without leaving the editor. See the documentation for your favorite editor to learn how to do this. Packages that are designed specifically for working with Python code are Spyder, PyCharm, Wing IDE, and PyScripter. Everybody has their own favorite development environment, and you may need to play with a few different ones before you find an environment that you like.

2.2 Basic structure of a script

Some of the first things you'll see right at the top of most Python scripts are `import` statements. These lines of code load additional modules so that the scripts can use them. A module is basically a library of code that you can access and use from your scripts, and the large ecosystem of specialized modules is another advantage to using Python. You'd have a difficult time working with GIS data in Python without extra modules that are designed for this, similar to the way tools such as GIMP and Photoshop make it easier to work with digital images. The whole point of this book is to teach you how to use these tools for working with GIS data. Along the way, you'll also use several of the modules that come with Python because they're indispensable for tasks such as working with the file system.

Let's look at a simple example that uses one of the built-in modules. The first thing you need to do to use a module is load it using `import`. Then you can access objects in the module by prefixing them with the module name so that Python knows where to find them. This example loads the `random` module and then uses the `gauss` function contained in that module to get a random number from the standard normal distribution:

```
>>> import random
>>> random.gauss(0, 1)
-0.22186423850882403
```

Another thing you might notice in a Python script is the lack of semicolons and curly braces, which are commonly used in other languages for ending lines and setting off blocks of code. Python uses whitespace to do these things. Instead of using a semicolon to end a line, press Enter and start a new line. Sometimes one line of code is too long to fit comfortably on one line in your file, however. In this case, break your line at a sensible place, such as right after a comma, and the Python interpreter will know that the lines belong together. As for the missing curly braces, Python uses indentation to define blocks of code instead. This may seem weird at first if you're used to using braces or end statements, but indentation works as well and forces you to write more readable code. Because of this, you need to be careful with your indentations. In fact, it's common for beginners to run into syntax errors because of wayward indentations. For example, even an extra space at the beginning of a line of code will cause an error. You'll see examples of how indentation is used in section 2.5.

Python is also case sensitive, which means that uppercase and lowercase letters are different from one another. For example, `random.Gauss(0, 1)` wouldn't have worked in the last example because `gauss` needs to be all lowercase. If you get error messages about something being undefined (which means Python doesn't know what it is), but you're sure that it exists, check both your spelling and your capitalization for mistakes.

It's also a good idea to add comments to your code to help you remember what it does or why you did it a certain way. I can guarantee that things that are obvious as you're writing your code will not be so obvious six months later. Comments are ignored by Python when the script is run, but can be invaluable to the real people looking at the code, whether it's you or someone else trying to understand your code. To create a comment, prefix text with a hash sign:

```
# This is a comment
```

In addition to comments, descriptive variable names improve the legibility of your code. For example, if you name a variable `m`, you need to read through the code to figure out what's stored in that variable. If you name it `mean_value` instead, the contents will be obvious.

2.3 Variables

Unless your script is extremely simple, it will need a way to store information as it runs, and this is where variables come in. Think about what happens when you use software to open a file, no matter what kind of file it is. The software displays an Open dialog, you select a file and click OK, and then the file is opened. When you press OK, the name of the selected file is stored as a variable so that the software knows what file to open. Even if you've never programmed anything in your life, you're probably familiar with this concept in the mathematical sense. Think back to algebra class and computing the value of y based on the value of x. The x variable can take on any value, and y changes in response. A similar concept applies in programming. You'll use many different variables, or x's, that will affect the outcome of your script. The outcome can be anything you want it to be and isn't limited to a single y value, however. It might be a number, if your goal is to calculate a statistic on your data, but it could as easily be one or more entirely new datasets.

Creating a variable in Python is easy. Give it a name and a value. For example, this assigns the value of 10 to a variable called n and then prints it out:

```
>>> n = 10
>>> n
10
```

If you've used other programming languages such as C++ or Java, you might be wondering why you didn't need to specify that the variable n was going to hold an integer value. Python is a dynamically typed language, which means that variable types aren't checked until runtime, and you can even change the data type stored in a variable. For example, you can switch n from an integer to a string and nobody will complain:

```
>>> n = 'Hello world'
>>> n
Hello world
```

Although you can store whatever you want in a variable without worrying about data type, you will run into trouble if you try to use the variable in a way that's inconsistent with the kind of data stored in it. Because the data types aren't checked until runtime, the error won't happen until that line of the script is executed, so you won't get any warning beforehand. You'll get the same errors in the Python interactive window that would occur in a script, so you can always test examples there if you're not sure if something will work. For example, you can't add strings and integers together, and this shows what happens if you try:

```
>>> msg = n + 1
Traceback (most recent call last):
  File "<stdin>", line 1, in <module>
TypeError: Can't convert 'int' object to str implicitly
```

Remember that n contains Hello world, which cannot be added to 1. If you're using Python 2.7, the core of the problem is the same, but your error message will look like this instead:

```
TypeError: cannot concatenate 'str' and 'int' objects
```

Notice that you use a single equal sign to assign a value to a variable. To test for equality, always use a double equal sign:

```
>>> n = 10
>>> n == 10
True
```

When you're first starting out, you might be more comfortable hardcoding values into your script instead of using variables when you don't have to. For example, say you need to open a file in the script, maybe on line 37. You'll probably be tempted to type the filename on line 37 when the file is opened. This will certainly work, but you'll find that things are easier to change later if you instead define a variable containing the filename early in the script and then use that variable on line 37. First, this makes it easier to find the values you need to change, but even more importantly, it will be much easier to adapt your code so that you can use it in more situations. Instead of line 37 looking something like this,

```
myfile = open('d:/temp/cities.csv')
```

you'd define a variable early on and then use it when needed:

```
fn = 'd:/temp/cities.csv'
<snip a bunch of code>
myfile = open(fn)
```

It might be hard to remember to do this at first, but you'll be glad you did if you have to adapt your code to use other data.

2.4   *Data types*

As your code becomes more complex, you'll find that it's extremely difficult to store all of the information that your script needs as numbers and strings. Fortunately, you can use many different types of data structures, ranging from simple numbers to complex objects that can contain many different types of data themselves. Although an infinite number of these object types can be used (because you can define your own), only a small number of core data types exist from which the more complex ones are built. I'll briefly discuss several of those here. Please see a more comprehensive set of Python documentation for more details, because this leaves out much information.

2.4.1   *Booleans*

A *Boolean* variable denotes true or false values. Two case-sensitive keywords, `True` and `False`, are used to denote these values. They can be used in standard Boolean operations, like these:

```
>>> True or False
True
>>> not False
True
>>> True and False
False
>>> True and not False
True
```

Other values can also resolve to `True` or `False` when value testing and performing Boolean operations. For example, `0`, the `None` keyword, blank strings, and empty lists, tuples, sets, and dictionaries all resolve to `False` when used in Boolean expressions. Anything else resolves to `True`. You'll see examples of this in section 2.5.

2.4.2   *Numeric types*

As you'd expect, you can use Python to work with numbers. What you might not expect, however, is that distinct kinds of numbers exist. *Integers* are whole numbers, such as 5, 27, or 592. *Floating-point numbers*, on the other hand, are numbers with decimal points, such as 5.3, 27.0, or 592.8. Would it surprise you to know that 27 and 27.0 are different? For one, they might take up different amounts of memory, although the details depend on your operating system and version of Python. If you're using Python 2.7 there's a major difference in how the two numbers are used for mathematical operations, because integers don't take decimal places into account. Take a look at this Python 2.7 example:

```
>>> 27 / 7
3
>>> 27.0 / 7.0
3.857142857142857
>>> 27 / 7.0
3.857142857142857
```

As you can see, if you divide an integer by another integer, you still end up with an integer, even if there's a remainder. You get the correct answer if one or both of the numbers being used in the operation is floating-point. This behavior has changed in Python 3.x, however. Now you get floating-point math either way, but you can still force integer math using the // floor division operator:

```
>>> 27 / 7
3.857142857142857
>>> 27 // 7
3
```

> **WARNING** Python 3.x performs floating-point math by default, even on integers, but older versions of Python perform integer math if all inputs are integers. This integer math often leads to undesirable results, such as 2 instead of 2.4, in which case you must ensure that at least one input is floating-point.

Fortunately, you have a simple way to convert one numeric data type to the other, although be aware that converting floating-point to integer this way truncates the number instead of rounding it:

```
>>> float(27)
27.0
>>> int(27.9)
27
```

If you want to round the number instead, you must use the round function:

```
>>> round(27.9)
28
```

Python also supports complex numbers, which contain real and imaginary parts. As you might recall, these values result when you take the square root of a negative number. We won't use complex numbers in this book, but you can read more about them at python.org if you're interested.

2.4.3 *Strings*

Strings are text values, such as 'Hello world'. You create a string by surrounding the text with either single or double quotes—it doesn't matter which, although if you start a string with one type, you can't end it with the other because Python won't recognize it as the end of the string. The fact that either one works makes it easy to include quotes as part of your string. For example, if you need single quotes inside your string, as you would in a SQL statement, surround the entire string with double quotes, like this:

```
sql = "SELECT * FROM cities WHERE country = 'Canada'"
```

If you need to include the same type of quote in your string that you're using to delineate it, you can use a backslash before the quote. The first example here results in an error because the single quote in "don't" ends the string, which isn't what you want. The second one works, thanks to the backslash:

```
>>> 'Don't panic!'
  File "<stdin>", line 1
    'Don't panic!'
          ^
SyntaxError: invalid syntax
>>> 'Don\'t panic!'
"Don't panic!"
```

Notice the caret symbol (^) under the spot where Python ran into trouble. This can help you narrow down where your syntax error is. The double quotes that surround the string when it's printed aren't part of the string. They show that it's a string, which is obvious in this case, but wouldn't be if the string was "42" instead. If you use the print function, the quotes aren't shown:

```
>>> print('Don\'t panic!')
Don't panic!
```

> **TIP** Although most of these examples from the interactive window don't use print to send output to the screen, you must use it to send output to the screen from a script. If you don't, it won't show up. In Python 3, print is a function and like all functions, you must pass the parameters inside parentheses. In Python 2, print is a statement and the parentheses aren't required, but they won't break anything, either.

JOINING STRINGS

You have several ways to join strings together. If you're only concatenating two strings, then the simplest and fastest is to use the + operator:

```
>>> 'Beam me up ' + 'Scotty'
'Beam me up Scotty'
```

If you're joining multiple strings, the format method is a better choice. It can also join values together that aren't all strings, something the + operator can't do. To use it, you create a template string that uses curly braces as placeholders, and then pass values to take the place of the placeholders. You can read the Python documentation online to see the many ways you can use this for sophisticated formatting, but we'll look at the basic method of specifying order. Here, the first item passed to `format` replaces the {0} placeholder, the second replaces {1}, and so on:

```
>>> 'I wish I were as smart as {0} {1}'.format('Albert', 'Einstein')
'I wish I were as smart as Albert Einstein'
```

To see that the numeric placeholders make a difference, try switching them around but leaving everything else the same:

```
>>> 'I wish I were as smart as {1}, {0}'.format('Albert', 'Einstein')
'I wish I were as smart as Einstein, Albert'
```

The fact that the placeholders reference specific values means that you can use the same placeholder in multiple locations if you need to insert an item in the string more than once. This way you don't have to repeat anything in the list of values passed to `format`.

ESCAPE CHARACTERS

Remember the backslash that you used to include a quote inside a string earlier? That's called an *escape character* and can also be used to include nonprintable characters in strings. For example, "\n" includes a new line, and "\t" represents a tab:

```
>>> print('Title:\tMoby Dick\nAuthor:\tHerman Melville')
Title:  Moby Dick
Author: Herman Melville
```

The fact that Windows uses backslashes as path separators causes angst for beginning programmers who use Windows, because they tend to forget that a single backslash isn't a backslash. For example, pretend you have a file called cities.csv in your d:\temp folder. Try asking Python if it exists:

```
>>> import os
>>> os.path.exists('d:\temp\cities.csv')
False
```

To get an idea of why that fails, when you know that the file does indeed exist, try printing the string instead:

```
>>> print('d:\temp\cities.csv')
d:      emp\cities.csv
```

The "\t" was treated as a tab character! You have three ways to solve this problem. Either use forward slashes or double backslashes, or prefix the string with an r to tell Python to ignore escape characters:

```
>>> os.path.exists('d:/temp/cities.csv')
True
>>> os.path.exists('d:\\temp\\cities.csv')
True
>>> os.path.exists(r'd:\temp\cities.csv')
True
```

I prefer the latter method if I'm copying and pasting paths, because it's much easier to add one character at the beginning than to add multiple backslashes.

2.4.4 *Lists and tuples*

A *list* is an ordered collection of items that are accessed via their index. The first item in the list has index 0, the second has index 1, and so on. The items don't even have to all be the same data type. You can create an empty list with a set of square brackets, [], or you can populate it right off the bat. For example, this creates a list with a mixture of numbers and strings and then accesses some of them:

```
>>> data = [5, 'Bob', 'yellow', -43, 'cat']
>>> data[0]
5
>>> data[2]
'yellow'
```

You can also use offsets from the end of the list, with the last item having index -1:

```
>>> data[-1]
'cat'
>>> data[-3]
'yellow'
```

You're not limited to retrieving one item at a time, either. You can provide a starting and ending index to extract a slice, or sublist. The item at the ending index isn't included in the returned value, however:

```
>>> data[1:3]
['Bob', 'yellow']
>>> data[-4:-1]
['Bob', 'yellow', -43]
```

You can change single values in the list, or even slices, using indices:

```
>>> data[2] = 'red'
>>> data
[5, 'Bob', 'red', -43, 'cat']
>>> data[0:2] = [2, 'Mary']
>>> data
[2, 'Mary', 'red', -43, 'cat']
```

Use append to add an item to the end of the list, and del to remove an item:

```
>>> data.append('dog')
>>> data
[2, 'Mary', 'red', -43, 'cat', 'dog']
>>> del data[1]
>>> data
[2, 'red', -43, 'cat', 'dog']
```

It's also easy to find out how many items are in a list or if it contains a specific value:

```
>>> len(data)
5
>>> 2 in data
True
>>> 'Mary' in data
False
```

Tuples are also ordered collections of items, but they can't be changed once created. Instead of brackets, tuples are surrounded by parentheses. You can access items and test for existence the same as with lists:

```
>>> data = (5, 'Bob', 'yellow', -43, 'cat')
>>> data[1:3]
('Bob', 'yellow')
>>> len(data)
5
>>> 'Bob' in data
True
```

Like I said, you're not allowed to change a tuple once it has been created:

```
>>> data[0] = 10
Traceback (most recent call last):
  File "<stdin>", line 1, in <module>
TypeError: 'tuple' object does not support item assignment
```

Because of this, use lists instead of tuples when it's possible that the data will change.

Error messages are your friend

When you get an error message, be sure to look carefully at the information it provides because this can save you time figuring out the problem. The last line is a message giving you a general idea of what the problem is, as seen here:

```
>>> data[0] = 10
Traceback (most recent call last):
  File "<stdin>", line 1, in <module>
TypeError: 'tuple' object does not support item assignment
```

You could deduce from this error message that your code tried to edit a tuple object somehow. Before the error message, you'll see a list of the lines of code that were executed before it ran into a problem. This is called a *stack trace*. In this example, <stdin> means the interactive window, so the line number isn't as helpful. But look at the following, which traces through two lines of code:

```
Traceback (most recent call last):
  File "D:\Temp\trace_example.py", line 7, in <module>      ←── Code on line 7
    y = add(x, '1')
  File "D:\Temp\trace_example.py", line 2, in add           ←── Code on line 2
    return n1 + n2
TypeError: unsupported operand type(s) for +: 'int' and 'str'
```

The last line tells you the error is from trying to add an integer and a string together. The trace tells you that the problem started with line 7 of the file trace_example.py. Line 7 calls a function called add, and the error happens on line 2 inside of that function. You can use the information from the stack trace to determine where the error occurred, and where the original line of code that triggered it is. In this example, you know that either you passed bad data to the add function on line 7, or else an error exists in the add function on line 2. That gives you two specific places to look for a mistake.

2.4.5   Sets

Sets are unordered collections of items, but each value can only occur once, which makes it an easy way to remove duplicates from a list. For example, this set is created using a list that contains two instances of the number 13, but only one is in the resulting set:

```
>>> data = set(['book', 6, 13, 13, 'movie'])
>>> data
{'movie', 6, 'book', 13}
```

You can add new values, but they'll be ignored if they're already in the set, such as 'movie' in this example:

```
>>> data.add('movie')
>>> data.add('game')
>>> data
{'movie', 'game', 6, 'book', 13}
```

Sets aren't ordered, so you can't access specific elements. You can check if items are in the set, however:

```
>>> 13 in data
True
```

Sets also make it easy to do things such as combine collections (union) or find out which items are contained in both sets (intersection):

```
>>> info = set(['6', 6, 'game', 42])
>>> data.union(info)                         ◄── All values from
{6, 'movie', 13, 'game', 'book', '6', 42}          both sets
>>> data.intersection(info)
{'game', 6}                             ◄── Only values
                                            contained in both
```

You've already seen that you can use sets to remove duplicates from a list. An easy way to determine if a list contains duplicate values is to create a set from the list and check to see if the set and list have the same length. If they don't, then you know duplicates were in the list.

2.4.6   Dictionaries

Dictionaries are indexed collections, like lists and tuples, except that the indices aren't offsets like they are in lists. Instead, you get to choose the index value, called a *key*. Keys can be numbers, strings, or other data types, as can the values they reference. Use curly braces to create a new dictionary:

```
>>> data = {'color': 'red', 'lucky number': 42, 1: 'one'}
>>> data
{1: 'one', 'lucky number': 42, 'color': 'red'}
>>> data[1]
'one'
```

```
>>> data['lucky number']
42
```

As with lists, you can add, change, and remove items:

```
>>> data[5] = 'candy'
>>> data
{1: 'one', 'lucky number': 42, 5: 'candy', 'color': 'red'}
>>> data['color'] = 'green'
>>> data
{1: 'one', 'lucky number': 42, 5: 'candy', 'color': 'green'}
>>> del data[1]
>>> data
{'lucky number': 42, 5: 'candy', 'color': 'green'}
```

You can also test to see if a key exists in the dictionary:

```
>>> 'color' in data
True
```

This is a powerful way to store data when you don't know beforehand what it will be. For example, say you needed to remember the spatial extent for each file in a collection of geographic datasets, but the list of datasets changed each time you ran your script. You could create a dictionary and use the filenames as keys and the spatial extents as values, and then this information would be readily available later in your script.

2.5 Control flow

The first script you write will probably consist of a sequence of statements that are executed in order, like all of the examples we have looked at so far. The real power of programming, however, is the ability to change what happens based on different conditions. Similar to the way you might use sale prices to decide which veggies to buy at the supermarket, your code should use data, such as whether it's working with a point or a line, to determine exactly what needs to be done. *Control flow* is the concept of changing this order of code execution.

2.5.1 If statements

Perhaps the simplest way to change execution order is to test a condition and do something different depending on the outcome of the test. This can be done with an if statement. Here's a simple example:

```
if n == 1:
    print('n equals 1')
else:
    print('n does not equal 1')
```

If the value of the n variable is 1, then the string "n equals 1" will be printed. Otherwise, the string "n does not equal 1" will be printed. Notice that the if and else lines end with a colon and that the code depending on a condition is indented under the condition. This is a requirement. Once you quit indenting code, then the code quits depending on the condition. What do you think the following code will print?

```
n = 1
if n == 1:
    print('n equals 1')
else:
    print('n does not equal 1')
print('This is not part of the condition')
```

Well, n is equal to 1, so the equality message prints out, and then control is transferred to the first line of code that isn't indented, so this is the result:

```
n equals 1
This is not part of the condition
```

You can also test multiple conditions like this:

```
if n == 1:
    print('n equals 1')
elif n == 3:
    print('n equals 3')
elif n > 5:
    print('n is greater than 5')
else:
    print('what is n?')
```

In this case, n is first compared to 1. If it's not equal to 1, then it's compared to 3. If it's not equal to that, either, then it checks to see if n is greater than 5. If none of those conditions are true, then the code under the else statement is executed. You can have as many elif statements as you want, but only one if and no more than one else. Similar to the way the elif statements aren't required, neither is an else statement. You can use an if statement all by itself if you'd like.

This is a good place to illustrate the idea that different values can evaluate to True or False while testing conditions. Remember that strings resolve to True unless they're blank. Let's test this with an if statement:

```
>>> if '':
...     print('a blank string acts like True')
... else:
...     print('a blank string acts like false')
...
a blank string acts like false
```

If you'd used a string containing any characters at all, including a single space, then the preceding example would have resolved to True instead of False. If you have a Python console open, go ahead and try it and see for yourself. Let's look at one more example that resolves to True because the list isn't empty:

```
>>> if [1]:
...     print('a non-empty list acts like True')
... else:
```

```
...      print('a non-empty list acts like False')
...
a non-empty list acts like True
```

You can use this same idea to test that a number isn't equal to zero, because zero is the same as False, but any other number, positive or negative, will be treated as True.

2.5.2 While statements

A while statement executes a block of code as long as a condition is True. The condition is evaluated, and if it's True, then the code is executed. Then the condition is checked again, and if it's still True, then the code executes again. This continues until the condition is False. If the condition never becomes False, then the code will run forever, which is called an *infinite loop* and is a scenario you definitely want to avoid. Here's an example of a while loop:

```
>>> n = 0
>>> while n < 5:
...      print(n)
...      n += 1
...
0
1
2
3
4
```

The += syntax means "increment the value on the left by the value on the right," so n is incremented by 1. Once n is equal to 5, it's no longer less than 5, so the condition becomes False and the indented code isn't executed again.

2.5.3 For statements

A for statement allows you to iterate over a sequence of values and do something for each one. When you write a for statement, you not only provide the sequence to iterate over, but you also provide a variable name. Each time through the loop, this variable contains a different value from the sequence. This example iterates through a list of names and prints a message for each one:

```
>>> names = ['Chris', 'Janet', 'Tami']
>>> for name in names:
...      print('Hello {}!'.format(name))
...
Hello Chris!
Hello Janet!
Hello Tami!
```

The first time through the loop, the name variable is equal to 'Chris', the second time it holds 'Janet', and the last time it is equal to 'Tami'. I called this variable name, but it can be called anything you want.

THE RANGE FUNCTION

The range function makes it easy to iterate over a sequence of numbers. Although this function has more parameters, the simplest way to use it is to provide a number n, and it will create a sequence from 0 to n-1. For example, this will count how many times the loop was executed:

```
>>> n = 0
>>> for i in range(20):
...     n += 1
...
>>> print(n)
20
```

The variable i wasn't used in this code, but nothing is stopping you from using it. Let's use it to calculate the factorial of 20, although this time we'll provide a starting value of 1 for the sequence, and have it go up to but not include the number 21:

```
>>> n = 1
>>> for i in range(1, 21):
...     n = n * i
...
>>> print(n)
2432902008176640000
```

You'll see in later chapters that this variable is also useful for accessing individual items in a dataset when they aren't directly iterable.

2.5.4   *break, continue, and else*

A few statements apply to while and for loops. The first one, break, will kick execution completely out of the loop, as in this example that stops the loop when i is equal to 3:

```
>>> for i in range(5):
...     if i == 3:
...         break
...     print(i)
...
0
1
2
```

Without the break statement, this loop would have printed the numbers 0 through 4.

The continue statement jumps back up to the top of the loop and starts the next iteration, skipping the rest of the code that would normally be executed during the current loop iteration. In this example, continue is used to skip the code that prints i if it's equal to 3:

```
>>> for i in range(5):
...     if i == 3:
...         continue
...     print(i)
...
0
```

```
1
2
4
```

Loops can also have an `else` statement. Code inside of this clause is executed when the loop is done executing, unless the loop was stopped with `break`. Here we'll check to see if the number 2 is in a list of numbers. If it is, we'll break out of the loop. Otherwise, the `else` clause is used to notify us that the number wasn't found. In the first case, the number is found, `break` is used to exit the loop, and the `else` statement is ignored:

```
>>> for i in [0, 5, 7, 2, 3]:
...     if i == 2:
...         print('Found it!')
...         break
... else:
...     print('Could not find 2')
...
Found it!
```

But if the number isn't found, so `break` is never called, then the `else` clause is executed:

```
>>> for i in [0, 5, 7, 3]:
...     if i == 2:
...         print('Found it!')
...         break
... else:
...     print('Could not find 2')
...
Could not find 2
```

You could use this pattern to set a default value for something if an appropriate value wasn't found in a list. For example, say you needed to find and edit a file with a specific format in a folder. If you can't find a file with the correct format, you need to create one. You could loop through the files in the folder, and if you found an appropriate one you could break out of the loop. You could create a new file inside the `else` clause, and that code would only run if no suitable existing file had been found.

2.6   *Functions*

If you find that you reuse the same bits of code over and over, you can create your own function and call that instead of repeating the same code. This makes things much easier and also less error-prone, because you won't have nearly as many places to make typos. When you create a function, you need to give it a name and tell it what parameters the user needs to provide to use it. Let's create a simple function to calculate a factorial:

```
def factorial(n):
    answer = 1
    for i in range(1, n + 1):
        answer = answer * i
    return answer
```

The name of this function is `factorial`, and it takes one parameter, n. It uses the same algorithm you used earlier to calculate a factorial and then uses a `return` statement to send the answer back to the caller. You could use this function like this:

```
>>> fact5 = factorial(5)
```

Functions can also have optional parameters that the user doesn't need to provide. To create one of these, you must provide a default value for it when you create the function. For example, you could modify factorial to optionally print out the answer:

```
def factorial(n, print_it=False):
    answer = 1
    for i in range(1, n + 1):
        answer = answer * i
    if print_it:
        print('{0}! = {1}'.format(n, answer))
    return answer
```

If you were to call this function with only a number, nothing would get printed because the default value of `print_it` is False. But if you pass True as the second parameter, then a message will print before the answer is returned:

```
>>> fact5 = factorial(5, True)
5! = 120
```

It's easy to reuse your functions by saving them in a .py file and then importing them the way you would any other module. The one hitch is that your file needs to be in a location where Python can find it. One way to do this is to put it in the same folder as the script that you're running. For example, if the factorial function was saved in a file called myfuncs.py, you could import `myfuncs` (notice there's no .py extension) and then call the function inside of it:

```
import myfuncs
fact5 = myfuncs.factorial(5)
```

Because certain characters aren't allowed in module names, and module names are only filenames without the extension, you need to be careful when naming your files. For example, underscores are allowed in module names, but hyphens aren't.

2.7 Classes

As you work through this book, you'll come across variables that have other data and functions attached to them. These are objects created from classes. Although we won't cover how to create your own classes in this book, you need to be aware of them because you'll still use ones defined by someone else. Classes are an extremely powerful concept, but all you need to understand for the purposes of this book are that they're data types that can contain their own internal data and functions. An object or variable that is of this type contains these data and functions, and the functions operate on that particular object. You saw this with several of the data types we looked at earlier, such as lists. You can have a variable of type list, and that variable contains all of the functions, such as append, that come with being a list. When you call append on a list, it only appends data to that particular list and not to any other list variables you might have.

Classes can also have methods that don't apply to a particular object, but to the data type itself. For example, the Python datetime module contains a class, or type, called date. Let's get that data type out of the module and then use it to create a new date object, which we can then ask which day of the week it is, where Monday is 0 and Sunday is 6:

```
>>> import datetime
>>> datetype = datetime.date
>>> mydate = datetype.today()
>>> mydate
datetime.date(2014, 5, 18)
>>> mydate.weekday()
6
```

The datetype variable holds a reference to the date type itself, not to a particular date object. The data type has a method, today, that creates a new date object. The date object stored in the mydate variable stores date information internally and uses that to determine what day of the week the date refers to, Sunday in this case. You couldn't ask the datetype variable what weekday it was, because it doesn't contain any information about a particular date. You don't need to get a reference to the data type and could have created mydate with datetime.date.today(). Now suppose you want to find out what day of the week May 18 was in 2010. You can create a new date object based on the existing one, but with the year changed, and then you can ask the new one what day of the week it represents:

```
>>> newdate = mydate.replace(year=2010)
>>> newdate
datetime.date(2010, 5, 18)
>>> newdate.weekday()
1
```

Apparently May 18, 2010, was a Tuesday. The original mydate variable hasn't changed, and will still report that it refers to a Sunday.

You'll use objects created from classes throughout this book. For example, whenever you open a dataset, you'll get an object that represents that dataset. Depending on the type of data, that object will have different information and functions associated with it. Obviously, you need to know about the classes being used to create these objects, so that you know what data and functions they contain. The GDAL modules contain fairly extensive classes, which are documented in appendixes B, C, and D. (Appendixes C through E are available online on the Manning Publications website at www.manning.com/books/geoprocessing-with-python.)

2.8   *Summary*

- The Python interpreter is useful for learning how things work or trying out small bits of code, but writing scripts is more efficient for running multiple lines of code. Plus, you can save scripts and use them later, which is one of the main reasons for programming.
- Modules are libraries of code that you can load into your script and use. If you need to do something with Python, chances are good that somewhere a module exists that will help you out, no matter what it is you're trying to do.
- Get used to storing data in variables, because it will make your code much easier to adapt later.
- Python has a few core data types, all of which are extremely useful for different types of data and different situations.
- You can use control flow statements to change which lines of code execute based on various conditions or to repeat the same code multiple times.
- Use functions to make your code reusable.

Reading and writing vector data

This chapter covers

- Understanding vector data
- Introducing OGR
- Reading vector data
- Creating new vector datasets
- Updating existing datasets

They seem to be rare these days, but you've probably seen a paper roadmap designed to be folded up and kept in your car. Unlike the more recent web maps that we're used to using, these maps don't use aerial imagery. Instead, features on the maps are all drawn as geometric objects—namely, points, lines, and polygons. These types of data, where the geographic features are all distinct objects, are called *vector datasets*.

Unless you only plan to look at maps that someone else has made, you'll need to know how to read and write these types of data. If you want to work with existing data in any way, whether you're summarizing, editing, deriving new data, or performing sophisticated spatial analyses, you need to read it in from a file first. You also need to write any new or modified data back out to a disk. For example, if you had a nationwide city dataset but needed to analyze only data from cities with 100,000 people or more, you could extract those cities out of your original dataset and run your analysis on them while ignoring the smaller towns. Optionally, you could also save the smaller dataset to a new file for later use.

In this chapter you'll learn basic ideas behind vector data and how to use the OGR library to read, write, and edit these types of datasets.

3.1   *Introduction to vector data*

At its most basic, *vector data* are data in which geographic features are represented as discrete geometries—specifically, points, lines, and polygons. Geographic features that have distinct boundaries, such as cities, work well as vector data, but continuous data, such as elevation, don't. It would be difficult to draw a single polygon around all areas with the same elevation, at least if you were in a mountainous area. You could, however, use polygons to differentiate between different elevation ranges. For example, polygons showing subalpine zones for a region would be a good proxy for an elevation range, but you'd lose much of the detailed elevation data within those polygons. Many types of data are excellent candidates for a vector representation, though, such as features in the roadmap mentioned earlier. Roads are represented as lines, counties and states are polygons, and depending on the scale of the map, cities are drawn as either points or polygons. In fact, all of the features on the map are probably represented as points, lines, or polygons.

The type of geometry used to draw a feature can be dependent on scale, however. Figure 3.1 shows an example of this. On the map of New York State, cities are shown as points, major roads as lines, and counties as polygons. A map of a smaller area, such as New York City, will symbolize features differently. In this case, roads are still lines, but the city and its boroughs are polygons instead of points. Now points would be used to represent features such as libraries or police stations.

You can imagine many other examples of geographic data that lend themselves to being represented this way. Anything that can be described with a single set of coordinates, such as latitude and longitude, can be represented as a point. This includes cities, restaurants, mountain peaks, weather stations, and geocache locations. In addition to their x and y coordinates (such as latitude and longitude), points can have a third z coordinate that represents elevation.

Geographic areas with closed boundaries can be represented as polygons. Examples are states, lakes, congressional districts, zip codes, and land ownership, along with many of the same features that can be symbolized as points such as cities and parks. Other features that could be represented as polygons, but probably not as points, include countries, continents, and oceans.

Linear features, such roads, rivers, power lines, and bus routes, all lend themselves to being characterized as lines. Once again, however, scale can make a difference. For example, a map of New Orleans could show the Mississippi River as a polygon rather than a line because it's so wide. This would also allow the map to show the irregular banks of the river, rather than just a smooth line, as shown in figure 3.2.

Figure 3.1 An example of how scale changes the geometries used to draw certain features. New York City is a point on the state map, but is made of several polygons on the city map.

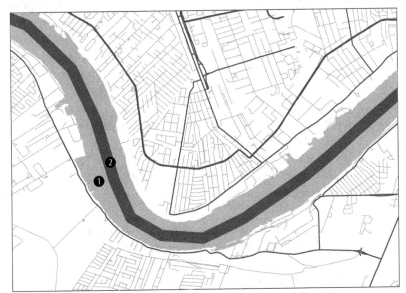

Figure 3.2 The difference between using polygon ❶ and line ❷ geometries to represent the Mississippi River. The polygon shows the details along the banks, while the line doesn't.

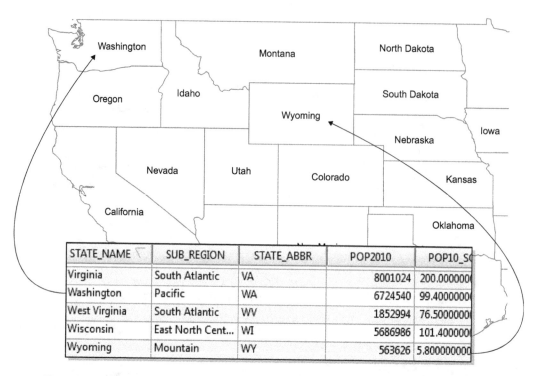

Figure 3.3   An attribute table for a dataset containing state boundaries within the United States. Each state polygon has an associated row in the data table with several attributes, including state name and population in 2010.

Vector data is more than geometries, however. Each one of these features also has associated attributes. These attributes can relate directly to the geometry itself, such as the area or perimeter of a polygon, or length of a line, but other attributes may be present as well. Figure 3.3 shows a simple example of a states dataset that stores the state name, abbreviation, population, and other data along with each feature. As you can see from the figure, these attributes can be of various types. They can be numeric, such as the city population or road speed limit, strings like city or road names, or dates such as the date the land parcel was purchased or last appraised. Certain types of vector data also support BLOBs (binary large objects), which can be used to store binary data such as photographs.

It should be clear by now that this type of data is well suited for making maps, but some reasons might not be so obvious. One example is how well it scales when drawing. If you're familiar with web graphics, you probably know that vector graphics such as SVG (scalable vector graphics) work much better than raster graphics such as PNG when displayed at different scales. Even if you know nothing about SVG, you've surely seen an image on a website that's pixelated and ugly. That's a raster graphic displayed at a higher resolution than it was designed for. This doesn't happen with vector graphics, and the exact same principle applies to vector GIS data. It always looks smooth, no matter the scale.

That doesn't mean that scale is irrelevant, though. As you saw earlier, scale affects the type of geometry used to represent a geographic feature, but it also affects the resolution you should use for a feature. A simple way to think of resolution is to equate it to detail. The higher the resolution, the more detail can be shown. For example, a map of the United States wouldn't show all of the individual San Juan Islands off the coast of Washington State, and in fact, the dataset wouldn't even need to include them. A map of only Washington State, however, would definitely need a higher-resolution dataset that includes the islands, as seen in figure 3.4. Keep in mind that resolution isn't important only for display, but also for analysis. For example, the two maps of Washington would provide extremely different measurements for coastline length.

Figure 3.4 An example showing the difference that resolution makes. The dataset shown with the thick outline has a lower resolution than the one shown with shading. Notice the difference in the amount of detail available in the two datasets.

The coastline paradox

Have you ever thought about how to measure the coastline of a landmass? As first pointed out by the English mathematician Lewis Fry Richardson, this isn't as easy as you might think, because the final measurement depends totally on scale. For example, think about a wild section of coastline with multiple headlands, with a road running along beside it. Imagine that you drive along that road and use your car's odometer to measure the distance, and then you get out of the car and walk back the way you came. But when on foot, you walk out along the edges of the headlands and follow other curves in the coast that the road doesn't. It should be easy to imagine that you'd walk farther than you drove because you took more detours. The same principle applies when measuring the entire coastline, because you can measure more variation if you measure in smaller increments. In fact, measuring the coast of Great Britain in 50-km increments instead of 100-km increments increases the final measurement by about 600 km. You can see another example of this, using part of Washington State,

(continued)

in figure 3.3. If you were to measure all of the twists and turns in the higher-resolution dataset, you'd get a longer coastline measurement than if you measured the lower-resolution coastline shown by the dark line, which doesn't even include many of the islands.

As mentioned previously, vector data isn't only for making maps. In fact, I couldn't make a pretty map if my life depended on it, but I do know a little bit more about data analysis. One common type of vector data analysis is to measure relationships between geographic features, typically by overlaying them on one another to determine their spatial relationship. For example, you could determine if two features overlap spatially and what that area of overlap is. Figure 3.5 shows the New Orleans city boundaries overlaid on a wetlands dataset. You could use this information to determine where wetlands exist within the city of New Orleans and how much of the city's area is or isn't wetland.

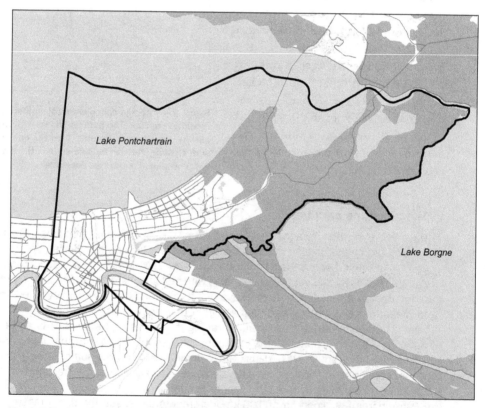

Figure 3.5   An example of a vector overlay operation. The dark outline is the City of New Orleans boundary, and the darker land areas are wetlands. These two datasets could be used to determine the percentage of land area within the New Orleans boundary that is wetlands.

Another aspect of spatial relationships is the distance between two features. You could find the distance between two weather stations, or all of the sandwich shops within one mile of your office. I helped out with a study a few years ago in which the researchers needed both distances and spatial relationships. They needed to know how far GPS-collared deer traveled between readings, but also the direction of travel and how they interacted with man-made features such as roads. One question in particular was if they crossed the roads, and if so, how often.

Speaking of roads, vector datasets also do a good job of representing networks, such as road networks. A properly configured road network can be used to find routes and drive times between two locations, similar to the results you see on various web-mapping sites. Businesses can also use information like this to provide services. For example, a pizza joint might use network analysis to determine which parts of town they can reach within a 15-minute drive to set their delivery area.

As with other types of data, you have multiple ways to store vector data. Similar to the way you can store a photograph as a JPEG, PNG, TIFF, bitmap, or one of many other file types, many different file formats can be used for storing vector data. I'll talk more about the possibilities in the next chapter, but for now I'll briefly mention a few common formats, several of which we'll use in this chapter.

Shapefiles are a popular format for storing vector data. A shapefile isn't made of a single file, however. In fact, this format requires a minimum of three binary files, each of which serves a different purpose. Geometry information is stored in .shp and .shx files, and attribute values are stored in a .dbf file. Additionally, other data, such as indexes or spatial reference information, can be stored in even more files. Generally you don't need to know anything about these files, but you do need to make sure that they're all kept together in the same folder.

Another widely used format, especially for web-mapping applications, is GeoJSON. These are plain text files that you can open up and look at in any text editor. Unlike a shapefile, a GeoJSON dataset consists of one file that stores all required information.

Vector data can also be stored in relational databases, which allows for multiuser access as well as various types of indexing. Two of the most common options for this are spatial extensions built for widely used database systems. The PostGIS extension runs on top of PostgreSQL, and SpatiaLite works with SQLite databases. Another popular database format is the Esri file geodatabase, which is completely different in that it isn't part of an existing database system.

3.2   *Introduction to OGR*

The OGR Simple Features Library is part of the Geospatial Data Abstraction Library (GDAL), an extremely popular open source library for reading and writing spatial data. The OGR portion of GDAL is the part that provides the ability to read and write many different vector data formats. OGR also allows you to create and manipulate geometries; edit attribute values; filter vector data based on attribute values or spatial location; and it also offers data analysis capabilities. In short, if you want to use GDAL to work with vector data, OGR is what you need to learn about, and you will, in the next four chapters.

The GDAL library was originally written in C and C++, but it has bindings for several other languages, including Python, so there's an interface to the GDAL/OGR library from Python, not that the code was rewritten in Python. Therefore, to use GDAL with Python, you need to install both the GDAL library and the Python bindings for it. If you haven't yet done this, please see appendix A for detailed installation instructions.

> **NOTE** What does the OGR acronym stand for, anyway? It used to stand for OpenGIS Simple Features Reference Implementation, but because OGR isn't fully compliant with the OpenGIS Simple Features specification, the name was changed and now the OGR part of it doesn't stand for anything and is only historical in nature.

Several functions used in this chapter are from the ospybook Python module available for download at www.manning.com/books/geoprocessing-with-python. You'll want to install this module, too. The sample datasets are available from the same site.

Before you start working with OGR, it's useful to look at how various objects in the OGR universe are related to each other, as shown in figure 3.6. If you don't understand this hierarchy, then the steps required to read and write data won't make much sense. When you use OGR to open a data source, such as a shapefile, GeoJSON file, SpatiaLite, or PostGIS database, you'll have a `DataSource` object. This data source can have one or more child `Layer` objects, one for each dataset contained in the data source. Many vector data formats, such as the shapefile examples used in this chapter, can only contain one dataset. But others, such as SpatiaLite, can contain multiple datasets, and you'll see examples of this in the next chapter. Regardless of how many datasets are in a data source, each one is considered a layer by OGR. Even several of my students, who use GIS regularly for their classes and research, get confused by this if they mostly use shapefiles, because it's counterintuitive to them that something called a layer sits between the data source and the actual data.

And speaking of the actual data, each layer contains a collection of `Feature` objects that holds the geometries and their attributes. If you load vector data into a GIS, such as QGIS, and then look at the attribute table, you'll see something similar to

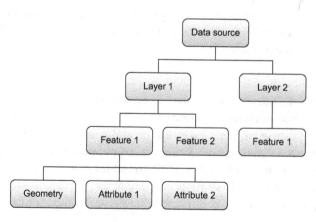

Figure 3.6 The OGR class structure. Each data source can have multiple layers, each layer can have multiple features, and each feature contains a geometry and one or more attributes.

	scalerank	featurecla	LABELRANK	SOVEREIGNT	SOV_A3	ADM0_DIF	LEVEL	TYPE	
0	3	Admin-0 country	5.00	Netherlands	NL1	1.00	2.00	Country	Arub
1	0	Admin-0 country	3.00	Afghanistan	AFG	0.00	2.00	Sovereign coun...	Afgh
2	0	Admin-0 country	3.00	Angola	AGO	0.00	2.00	Sovereign coun...	Angc
3	3	Admin-0 country	6.00	United Kingd...	GB1	1.00	2.00	Dependency	Angu
4	0	Admin-0 country	6.00	Albania	ALB	0.00	2.00	Sovereign coun...	Albai

Figure 3.7 An example of an attribute table shown in QGIS. Each row corresponds to a feature, and each column is an attribute field.

figure 3.7. Each row in the table corresponds to a feature, such as the feature representing Afghanistan. Each column corresponds to an attribute field, and in this case two of the attributes are SOVEREIGNT and TYPE. Although you can open data tables that don't have any spatial information or geometries associated with the features, we'll work with datasets that do have geometries. As you can see in figure 3.7, the geometries don't show up in the attribute table in QGIS, although other GIS software packages, such as ArcGIS, do show a shape column in the attribute table.

The first step to accessing any vector data is to open the data source. For this, you need to have an appropriate driver that tells OGR how to work with your data format. The GDAL/OGR website lists more than 70 vector formats that OGR is capable of reading, although it can't write to all of them. Each one of these has its own driver. It's likely that your version of OGR doesn't support all of those listed, but you can always compile it yourself if you need something that's missing (note that this is easier said than done in many cases). See www.gdal.org/ogr_formats.html for the list of all available formats and specific details pertaining to each one.

> **DEFINITION** A driver is a translator for a specific data format, such as GeoJSON or shapefile. It tells OGR how to read and write that particular format. If no driver for a format is compiled into OGR, then OGR can't work with it.

If you aren't sure if your installation of GDAL/OGR supports a particular data format, you can use the ogrinfo command-line utility to find out which drivers are available. The location of this utility on your computer depends on your operating system and how you installed GDAL, so you might need to refer back to appendix A. If you aren't used to using a command line, you may be tempted to double-click the ogrinfo executable file, but that won't get you anywhere useful. Instead, you need to run ogrinfo from a terminal window or Windows command prompt. At any rate, once you find the executable, you'll want to run it with the --formats option. Figure 3.8 shows an example of running it on my Windows 7 machine, although I've cut off most of the output.

Figure 3.8 An example of running the ogrinfo utility from a GDAL command prompt on a Windows computer

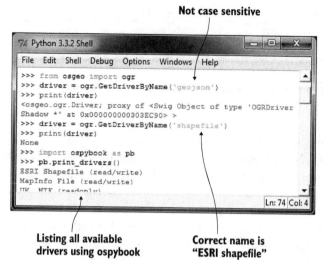

Figure 3.9   **Sample Python interactive session showing how to get drivers**

As you can see, ogrinfo not only tells you which drivers are included with your version of OGR, but also whether it can write to each one as well as read from it.

> **TIP**   Information about vector formats supported by OGR can be found at www.gdal.org/ogr_formats.html.

You can also determine which drivers are available using Python. In fact, let's try it. Start by opening up your favorite Python interactive environment. I'll use IDLE (figure 3.9) because it's the one that's packaged with Python, but you can use whichever one you're comfortable with. The first thing you need to do is import the ogr module so that you can use it. This module lives inside the osgeo package, which was installed when you installed the Python bindings for GDAL. All of the modules in this package are named with lowercase letters, which is how you need to refer to them in Python. Once you've imported ogr, then you can use ogr.GetDriverByName to find a specific driver:

```
from osgeo import ogr
driver = ogr.GetDriverByName('GeoJSON')
```

Use the name from the Code column on the OGR Vector Formats webpage. If you get a valid driver and print it out, you'll see information about where the object is stored in memory. The important thing is that there was something for it to print out because it means you successfully found a driver. If you pass an invalid name, or the name of a missing driver, the function will return None. See figure 3.9 for examples.

A function called print_drivers in the ospybook module will also print out a list of available drivers. This is shown in figure 3.9.

3.3   *Reading vector data*

Now that you know what formats are available to work with, it's time to read data. You'll start with a cities shapefile, the ne_50m_populated_places.shp dataset in the global

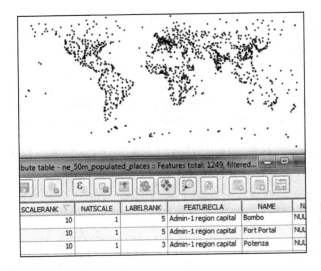

Figure 3.10 The geometries and attributes from ne_50m_populated_places.shp as seen in QGIS

subfolder of your osgeopy-data folder. Feel free to open it up in QGIS and look. Not only will you see the cities shown in figure 3.10, but you'll also see that the attribute table contains a collection of fields, most of which aren't visible in the screenshot.

Listing 3.1 shows a little script that prints out the names, populations, and coordinates for the first 10 features in this dataset. Don't worry if it doesn't make much sense at first glance because we'll go over it in excruciating detail in a moment. The file is included with the source code for this chapter, so if you want to try it out, you can open it in IDLE, change the filename in the third line of code to match your setup, and then choose Run Module under the Run menu.

Listing 3.1 Printing data from the first ten features in a shapefile

```
import sys
from osgeo import ogr                                          ◄─┐ Don't forget
                                                                  │ to import ogr
fn = r'D:\osgeopy-data\global\ne_50m_populated_places.shp'
ds = ogr.Open(fn, 0)                                           Open the data
if ds is None:                                                 source
    sys.exit('Could not open {0}.'.format(fn))
lyr = ds.GetLayer(0)

i = 0
for feat in lyr:
    pt = feat.geometry()
    x = pt.GetX()                                              Get x, y coordinates
    y = pt.GetY()
    name = feat.GetField('NAME')
    pop = feat.GetField('POP_MAX')                             Get attribute values
    print(name, pop, x, y)
    i += 1
    if i == 10:
        break
del ds
```

The basic outline is simple. The first thing you do is open the shapefile and make sure that the result of that operation isn't equal to None, because that would mean the data source couldn't be opened. I tend to call this variable ds, short for data source. After making sure the file is opened, you retrieve the first layer from the data source. Then you iterate through the first 10 features in the layer and for each one, get the geometry object, its coordinates, and the NAME and POP_MAX attribute values. Then you print the information about the feature before moving on to the next one. When done, you delete the ds variable to force the file to close.

If you successfully ran the code, you should have 10 lines of output that look something like this, although you won't have the parentheses if using Python 3:

```
('Bombo', 75000, 32.533299524864844, 0.5832991056146284)
('Fort Portal', 42670, 30.27500161597942, 0.671004121125236)
<snip>
('Clermont-Ferrand', 233050, 3.080008095928406, 45.779982115759424)
```

Let's look at this in a little more detail. You open a data source by passing the filename and an optional update flag to the Open function. This is a standalone function in the OGR module, so you prefix the function name with the module name so that Python can find it. If the second parameter isn't provided it defaults to 0, which will open the file in read-only mode. You could have passed 1 or True to open it in update, or edit, mode instead.

If the file can't be opened, then the Open function returns None, so the next thing you do is check for this and print out an error message and quit if needed. I like to check for this so I can solve the problem immediately and in the manner of my choosing (quitting, in this case) instead of waiting for the script to crash when it tries to use the nonexistent data source. Change the filename in listing 3.1 to a bogus one and run the script if you want to see this behavior in action:

```
fn = r'D:\osgeopy-data\global\ne_50m_populated_places.shp'
ds = ogr.Open(fn, 0)
if ds is None:
    sys.exit('Could not open {0}.'.format(fn))
lyr = ds.GetLayer(0)
```

Remember that data sources are made of one or more layers that hold the data, so after opening the data source you need to get the layer from it. Data sources have a function called GetLayer that takes either a layer index or a layer name and returns the corresponding Layer object inside that particular data source. Layer indexes start at 0, so the first layer has index 0, the second has index 1, and so on. If you don't provide any parameters to GetLayer, then it returns the first layer in the data source. The shapefile only has one layer, so the index isn't technically needed in this case.

Now you want to get the data out of your layer. Recall that each layer is made of one or more features, with each feature representing a geographic object. The geometries and attribute values are attached to these features, so you need to look at them to get your data. The second half of the code in listing 3.1 loops through the first 10 features in the layer and prints information about each one. Here's the interesting part of it again:

```
for feat in lyr:
    pt = feat.geometry()
    x = pt.GetX()
    y = pt.GetY()
    name = feat.GetField('NAME')
    pop = feat.GetField('POP_MAX')
    print(name, pop, x, y)
```

The layer is a collection of features that you can iterate over with a for loop. Each time through the loop, the feat variable will be the next feature in the layer, and the loop will iterate over all features in the layer before stopping. You don't want to print out all 1,249 features, though, so you force it to stop after the first 10.

The first thing you do inside the loop is get the geometry from the feature and stick it in a variable called pt. Once you have the geometry, you grab its x and y coordinates and store them in variables to use later.

Next you retrieve the values from the NAME and POP_MAX fields and store those in variables as well. The GetField function takes either an attribute name or index and returns the value of that field. Once you have the attributes, you print out all of the information you gathered about the current feature.

One thing you should be aware of is that the GetField function returns data that's the same data type as that in the underlying dataset. In this example, the value in the name variable is a string, but the value stored in pop is a number. If you want the data in another format, check out appendix B to see a list of functions that return values as a specific type. For example, if you wanted pop to be a string so that you could concatenate it to another string, you could use GetFieldAsString.

```
pop = feat.GetFieldAsString('POP_MAX')
```

Note that not all data formats support all field types, and not all data can successfully be converted between types, so you should test things thoroughly before relying on these automatic conversions. Not only are these functions useful for converting data between types, but you can also use them to make data types more evident in your code. For example, if you use GetFieldAsInteger, then it's obvious to anyone reading your code that the value is an integer.

3.3.1 Accessing specific features

Sometimes you don't need every feature, so you have no reason to iterate through all of them as you've done so far. One powerful method of limiting features to a subset is to select them by attribute value or spatial extent, and you'll do that in chapter 5. Another way is to look at features with specific offsets, also called *feature IDs* (FIDs). The offset is the position that the feature is at in the dataset, starting with zero. It depends entirely on the position of the feature in the file and has nothing to do with the sort order in memory. For example, if you open the ne_50m_populated _places shapefile in

A. Native sort order

	NAME
0	Bombo
1	Fort Portal
2	Potenza
3	Campobasso
4	Aosta
5	Mariehamn
6	Ramallah
7	Vatican City

B. Sorted by name

	NAME ╱
346	Abakan
908	Abeche
1163	Abidjan
83	Aboa Station
772	Abu Dhabi
878	Abuja
859	Acapulco
1147	Accra

Figure 3.11   The attribute table for the ne_50m_populated_places shapefile. Table A shows the native sort order, with the FIDs in order. Table B has been sorted by city name, and the FIDs are no longer ordered sequentially.

QGIS and look at the attribute table, it would show Bombo as the first record in the table, as in figure 3.11A. See the numbers in the left-most column? Those are the offset values. Now try sorting the table by name by clicking on the NAME column header, as shown in figure 3.11B. Now the first record shown in the table is the one for Abakan, but it has an offset of 346. As you can see, that left-most column isn't a row number like you see in spreadsheets, where the row numbers are always in the right order no matter how you sort the data. These numbers represent the order in the file instead.

If you know the offset of the feature you want, you can ask for that feature by FID. To get the feature for Vatican City, you use Get-Feature(7).

You can also get the total number of features with GetFeatureCount, so you could grab the last feature in the layer like this:

```
>>> num_features = lyr.GetFeatureCount()
>>> last_feature = lyr.GetFeature(num_features - 1)
>>> last_feature.NAME
'Hong Kong'
```

You have to subtract one from the total number of features because the first index is zero. If you had tried to get the feature at index num_features, you'd have gotten an error message saying that the feature ID was out of the available range. This snippet also shows an alternate way of retrieving an attribute value from a feature, instead of using GetField, but it only works if you know the names beforehand so that you can hardcode them into your script.

THE CURRENT FEATURE

Another important point is that the functions that return features keep track of which feature was last accessed; this is the *current feature.* When you first get the layer object, it has no current feature. But if you start iterating through features, the first time through the loop, the current feature is the one with an FID of zero. The second time through the loop, the current feature is the one with offset 1, and so on. If you use GetFeature to get the one with an FID of 5, that's now the current feature, and if you then call GetNextFeature or start a loop, the next feature returned will be the one with offset 6. Yes, you read that right. If you iterate through the features in the layer, it doesn't start at the first one if you've already set the current feature.

Based on what you've learned so far, what do you think would happen if you iterated through all of the features and printed out their names and populations, but then later tried to iterate through a second time to print out their names and coordinates? If you guessed that no coordinates would print out, you were right. The first loop stops when it runs out of features, so the current feature is pointing past the last one and isn't reset to the beginning (see figure 3.12). No next feature is there when the second loop starts, so nothing happens. How do you get the current feature to point to the beginning again? You wouldn't want to use a FID of zero, because if you tried to iterate through them all, the first feature would be skipped. To solve this problem, use

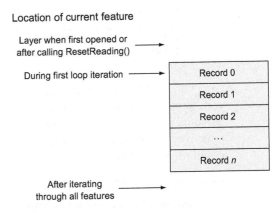

Figure 3.12 The location of the current feature pointer at various times

the `layer.ResetReading()` function, which sets the current feature pointer to a location before the first feature, similar to when you first opened the layer.

3.3.2 Viewing your data

Before we continue, you might find it useful to know about functions in the ospybook module that will help you visualize your data without opening it in another software program. These don't allow the level of interaction with the data that a GIS does, so opening it in QGIS is still a much better option for exploring the data in any depth.

VIEWING ATTRIBUTES

You can print out attribute values to your screen using the `print_attributes` function, which looks like this:

```
print_attributes(lyr_or_fn, [n], [fields], [geom], [reset])
```

- `lyr_or_fn` is either a layer object or the path to a data source. If it's a data source, the first layer will be used.
- `n` is an optional number of records to print. The default is to print them all.
- `fields` is an optional list of attribute fields to include in the printout. The default is to include them all.
- `geom` is an optional Boolean flag indicating whether the geometry type is printed. The default is `True`.
- `reset` is an optional Boolean flag indicating whether the layer should be reset to the first record before printing. The default is `True`.

For example, to print out the name and population for the first three cities in the populated places shapefile, you could do something like this from a Python interactive window:

```
>>> import ospybook as pb
>>> fn = r'D:\osgeopy-data\global\ne_50m_populated_places.shp'
>>> pb.print_attributes(fn, 3, ['NAME', 'POP_MAX'])
```

```
FID     Geometry                        NAME            POP_MAX
0       POINT (32.533, 0.583)           Bombo           75000
1       POINT (30.275, 0.671)           Fort Portal     42670
2       POINT (15.799, 40.642)          Potenza         69060
3 of 1249 features
```

Normally, you must provide arguments to functions in the order they're listed, but if you want to provide an optional argument without specifying values for earlier optional parameters, you can use keywords to specify which parameter you mean. For example, If you wanted to set geom to False without specifying a list of fields, you could do it like this:

```
pb.print_attributes(fn, 3, geom=False)
```

This function works well for viewing small numbers of attributes, but you'll probably regret using it to print all attributes of a large file.

PLOTTING SPATIAL DATA

The ospybook module also contains convenience classes to help you visualize your data spatially, although you'll learn how to do it yourself in the last chapter. To use these, you must have the matplotlib Python module installed. To plot your data, you need to create a new instance of the VectorPlotter class and pass a Boolean parameter to the constructor indicating if you want to use interactive mode. If interactive, the data will be drawn immediately when you plot it. If not interactive, you'll need to call draw after plotting the data, and everything will be drawn at once. Either way, once you've created this object, you can use it to plot your data with the plot method:

```
plot(self, geom_or_lyr, [symbol], [name], [kwargs])
```

- geom_or_lyr is a geometry, layer, or path to a data source. If a data source, the first layer will be drawn.
- symbol is an optional pyplot symbol to draw the geometries with.
- name is an optional name to assign to the data so it can be accessed later.
- kwargs are optional pyplot drawing parameters that are specified by keyword (you'll see the abbreviation kwargs used often for an indeterminate number of keyword arguments).

The plot function can optionally use parameters from the pyplot interface in matplotlib. You'll see a few used in this book, but to see more you can read the pyplot documentation at http://matplotlib.org/1.5.0/api/pyplot_summary.html. Let's start with an example that plots the populated places shapefile on top of country outlines:

```
>>> import os
>>> os.chdir(r'D:\osgeopy-data\global')
>>> from ospybook.vectorplotter import VectorPlotter
>>> vp = VectorPlotter(True)
>>> vp.plot('ne_50m_admin_0_countries.shp', fill=False)
>>> vp.plot('ne_50m_populated_places.shp', 'bo')
```

The first thing you do is use the built-in os module to change your working directory, which allows you to use filenames later instead of typing the entire path. Then you pass

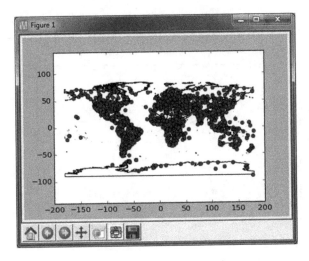

Figure 3.13 The output from plotting the global populated places shapefile on top of the country outlines

True to `VectorPlotter` to create an interactive plotter. The `fill` pyplot parameter causes the countries shapefile to be drawn as hollow polygons, and the `'bo'` symbol for populated places means blue circles. This results in a plot that looks like figure 3.13.

You don't need to do anything special if you want to use this in a script, but you should know that when the plotter isn't created with interactive mode, it will stop script execution until you close the window that pops up. I've also discovered that depending on the environment I'm running the script from, sometimes it closes itself automatically if I created it with interactive mode, so I never get the chance to view it. Because of this, if I'm using a `VectorPlotter` in a script instead of a Python interactive window, I usually create it using non-interactive mode and call `draw` at the end of the script. The source code for this chapter has examples of this.

3.4 Getting metadata about the data

Sometimes you also need to know general information about a dataset, such as the number of features, spatial extent, geometry type, spatial reference system, or the names and types of attribute fields. For example, say you want to display your data on top of a Google map. You need to make sure that your data use the same spatial reference system as Google, and you need to know the spatial extent so that you can have your map zoom to the correct part of the world. Because different geometry types have different drawing options, you also need to know geometry types to define the symbology for your features.

You've already seen how to get some of these, such as the number of features in a layer with `GetFeatureCount`. Remember that this applies to the layer and not the data source, because each layer in a data source can have a different number of features, geometry type, spatial extent, or attributes.

The spatial extent of a layer is the rectangle constructed from the minimum and maximum bounding coordinates in all directions. Figure 3.14 shows the Washington large_cities file and its extent. You can get these bounding coordinates from a layer

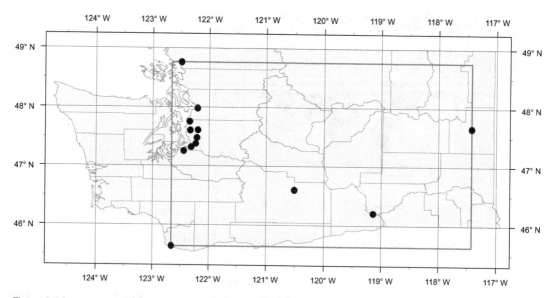

Figure 3.14   Here you can see the spatial extent of the large_cities dataset. The minimum and maximum longitude (x) values are approximately -122.7 and -117.4, respectively. The minimum and maximum latitude (y) values are approximately 45.6 and 48.8.

object with the GetExtent function, which returns a tuple of numbers as (min_x, max_x, min_y, max_y). Here's an example:

```
>>> ds = ogr.Open(r'D:\osgeopy-data\Washington\large_cities.geojson')
>>> lyr = ds.GetLayer(0)
>>> extent = lyr.GetExtent()
>>> print(extent)
(-122.66148376464844, -117.4260482788086, 45.638729095458984,
 48.759552001953125)
```

Compare these numbers to those in figure 3.14 to better understand what's returned in the extent tuple.

You can also get the geometry type from the layer object, but there's a catch. The GetGeomType function returns an integer instead of a human-readable string. But how is that useful? The OGR module has a number of constants, shown in table 3.1, which are basically unchangeable variables with descriptive names and numeric values. You can compare the value you get with GetGeomType to one of these constants in order to check if it's that geometry type. For example, the constant for point geometries is wkbPoint and the one for polygons is wkbPolygon, so continuing with the previous example, you could find out if large_cities.shp is a point or polygon shapefile like this:

```
>>> print(lyr.GetGeomType())          │  Returns a number
1
>>> print(lyr.GetGeomType() == ogr.wkbPoint)
True                                          │  Compare with
>>> print(lyr.GetGeomType() == ogr.wkbPolygon)│  constants to
False                                         │  get type
```

Table 3.1 Common geometry type constants. You can find more in appendix B.

Geometry type	OGR constant
Point	`wkbPoint`
Mulitpoint	`wkbMultiPoint`
Line	`wkbLineString`
Multiline	`wkbMultiLineString`
Polygon	`wkbPolygon`
Multipolygon	`wkbMultiPolygon`
Unknown geometry type	`wkbUnknown`
No geometry	`wkbNone`

If the layer has geometries of varying types, such as a mixture of points and polygons, `GetGeomType` will return `wkbUnknown`.

> **NOTE** The wkb prefix on the OGR geometry constants stands for well-known binary (WKB), which is a standard binary representation used to exchange geometries between different software packages. Because it's binary, it isn't human-readable, but a well-known text (WKT) format does exist that is readable.

Sometimes you'd rather have a human-readable string, however, and you can get this from one of the feature geometries. The following example grabs the first feature in the layer, gets the geometry object from that feature, and then prints the name of the geometry:

```
>>> feat = lyr.GetFeature(0)
>>> print(feat.geometry().GetGeometryName())
POINT
```

Another useful piece of data you can get from the layer object is the spatial reference system, which describes the coordinate system that the dataset uses. Your GPS unit probably shows unprojected, or geographic, coordinates by default. These are the latitude and longitude coordinates that we're all familiar with. These geographic coordinates can be converted to many other types of coordinate systems, however, and if you don't know which of these systems a dataset uses, then you have no way of knowing where on the earth the coordinates refer to. Obviously, this is a crucial bit of metadata, and I'll talk more about it in chapter 8. For now, you only need to know that you can get this information. If you print it out, you'll get a string that describes the reference system in WKT format, like that shown in listing 3.2.

Listing 3.2 Example of well-known text representation of a spatial reference system

```
>>> print(lyr.GetSpatialRef())
GEOGCS["NAD83",
    DATUM["North_American_Datum_1983",
        SPHEROID["GRS 1980",6378137,298.257222101,
            AUTHORITY["EPSG","7019"]],
```

```
    TOWGS84[0,0,0,0,0,0,0],
    AUTHORITY["EPSG","6269"]],
  PRIMEM["Greenwich",0,
    AUTHORITY["EPSG","8901"]],
  UNIT["degree",0.0174532925199433,
    AUTHORITY["EPSG","9122"]],
  AUTHORITY["EPSG","4269"]]
```

Depending on your GIS experience, this output may or may not mean much to you. Don't worry if it makes no sense now, because you'll learn all about it later.

Last, you can also get information about the attribute fields attached to the layer. The easiest way to do this is to use the schema property on the layer object to get a list of FieldDefn objects. Each of these contains information such as the attribute column name and data type. Here's an example of printing out the name and data type of each field:

```
>>> for field in lyr.schema:
...     print(field.name, field.GetTypeName())
...
CITIESX020 Integer
FEATURE String
NAME String
<snip>
```

Part of this output was left out in the interest of space, but you can run the code yourself to see the rest of the fields in the layer. You'll learn more about working with FieldDefn objects in section 3.5.2.

3.5 *Writing vector data*

Reading data is definitely useful, but you'll probably need to edit existing or create new datasets. Listing 3.3 shows how to create a new shapefile that contains only the features corresponding to capital cities in the global populated places shapefile. The output will look like the cities in figure 3.15.

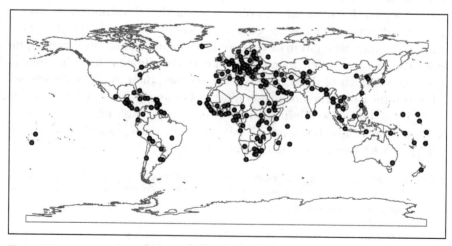

Figure 3.15 Capital cities with country outlines for reference

Listing 3.3 Exporting capital cities to a new shapefile

```
import sys
from osgeo import ogr

ds = ogr.Open(r'D:\osgeopy-data\global', 1)        ◀── Open folder data
if ds is None:                                           source for writing
    sys.exit('Could not open folder.')
in_lyr = ds.GetLayer('ne_50m_populated_places')    ◀── Get input
                                                        shapefile

if ds.GetLayer('capital_cities'):                  │ Delete layer if it exists
    ds.DeleteLayer('capital_cities')
out_lyr = ds.CreateLayer('capital_cities',
                         in_lyr.GetSpatialRef(),   │ Create a point layer
                         ogr.wkbPoint)
out_lyr.CreateFields(in_lyr.schema)

out_defn = out_lyr.GetLayerDefn()                  │ Create a blank feature
out_feat = ogr.Feature(out_defn)
for in_feat in in_lyr:
    if in_feat.GetField('FEATURECLA') == 'Admin-0 capital':
        geom = in_feat.geometry()
        out_feat.SetGeometry(geom)                 │ Copy geometry
        for i in range(in_feat.GetFieldCount()):   │ and attributes
            value = in_feat.GetField(i)
            out_feat.SetField(i, value)
        out_lyr.CreateFeature(out_feat)            ◀── Insert the feature

del ds                                             ◀── Close files
```

In this example you open up a folder instead of a shapefile as the data source. A nice feature of the shapefile driver is that it will treat a folder as a data source if a majority of the files in the folder are shapefiles, and each shapefile is treated as a layer. Notice that you pass 1 as the second parameter to Open, which will allow you to create a new layer (shapefile) in the folder. You pass the shapefile name, without the extension, to GetLayer to get the populated places shapefile as a layer. Even though you open it differently here than in listing 3.1, you can use it in exactly the same way.

Because OGR won't overwrite existing layers, you check to see if the output layer already exists, and delete it if it did. Obviously you wouldn't want to do this if you didn't want the layer overwritten, but in this case you can overwrite data as you test different things.

Then you create a new layer to store your output data in. The only required parameter for CreateLayer is a name for the layer, which should be unique within the data source. You do have, however, several optional parameters that you should set when possible:

```
CreateLayer(name, [srs], [geom_type], [options])
```

- name is the name of the layer to create.
- srs is the spatial reference system that the layer will use. The default is None, meaning that no spatial reference system will be assigned.

- geom_type is a geometry type constant from table 3.1 that specifies the type of geometry the layer will hold. The default is wkbUnknown.
- options is an optional list of layer-creation options, which only applies to certain vector format types.

The first of these optional parameters is the spatial reference, which defaults to None if not provided. Remember that without spatial reference information, it's extremely difficult to figure out where the features are on the planet. Sometimes the spatial reference is implicit in the data; for example, KML only supports unprojected coordinates using the WGS 84 datum, but you should set this if possible. In this case, you copy the spatial reference information from the original shapefile to the new one. We'll discuss spatial reference systems and how to use them in more detail in chapter 8.

The second optional parameter is one of the OGR geometry type constants from either table 3.1 or appendix B. This specifies the type of geometries that the layer will contain. If not provided, it defaults to ogr.wkbUnknown, although in many cases this will be updated to the correct value after you add features to the layer and it can be determined from them.

The last optional parameter is a list of layer-creation option strings in the form of *option=value*. These are documented for each driver on the OGR formats webpage. Not all vector data formats have layer-creation options, and even if a format does have options, you're under no obligation to use them.

You use the following code to create a new point shapefile called capital_cities.shp that uses the same spatial reference system as the populated places shapefile. You do one more thing, though. The schema property on the input layer returns a list of attribute field definitions for that layer, and you pass that list to CreateFields to create the exact same set of attribute fields in the new layer:

```
out_lyr = ds.CreateLayer('capital_cities',
                         in_lyr.GetSpatialRef(),
                         ogr.wkbPoint)
out_lyr.CreateFields(in_lyr.schema)
```

Now, to add a feature to a layer, you need to create a dummy feature that you add the geometry and attributes to, and then you insert that into the layer. The next step is to create this blank feature. Creating a feature requires a feature definition that contains information about the geometry type and all of the attribute fields, and this is used to create an empty feature with the same fields and geometry type. You need to get the feature definition from the layer you plan to add features to, but you must do it after you've added, deleted, or updated any fields. If you get the feature definition first, and then change the fields in any way, the definition will be out of date. This means that a feature you try to insert based on this outdated definition will not match reality, as seen in figure 3.16. This will cause Python to die a horrible death, and you definitely don't want that.

```
out_defn = out_lyr.GetLayerDefn()
out_feat = ogr.Feature(out_defn)
```

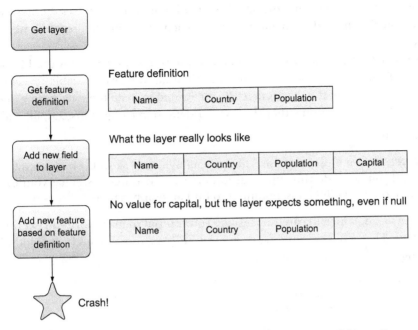

Feature definition

Name	Country	Population

What the layer really looks like

Name	Country	Population	Capital

No value for capital, but the layer expects something, even if null

Name	Country	Population	

Figure 3.16 **Always get feature definitions after making changes to fields, or the definition will not match reality.**

Now that you have a feature to put information into, it's time to start looping through the input dataset. For each feature, you check to see if its FEATURECLA attribute was equal to 'Admin-0 capital', which means it's a capital city. If it is, then you copy the geometry from it into the dummy feature. Then you loop through all of the fields in the attribute table and copy the values from the input feature into the output feature. This works because you create the fields in the new shapefile based on the fields in the original, so they're in the same order in both shapefiles. If they were in different orders, you'd have to use their names to access them, but you can use indexes here because you know that they match:

```
for in_feat in in_lyr:
    if in_feat.GetField('FEATURECLA') == 'Admin-0 capital':
        geom = in_feat.geometry()
        out_feat.SetGeometry(geom)
        for i in range(in_feat.GetFieldCount()):
            value = in_feat.GetField(i)
            out_feat.SetField(i, value)
        out_lyr.CreateFeature(out_feat)
```

Once you copy all of the attribute fields over, you insert the feature into the layer using CreateFeature. This function saves a copy of the feature, including all of the information you add to it, to the layer. The feature object can then be reused, and whatever you do to it won't affect the data that have already been added to the layer.

This way you don't have the overhead of creating multiple features, because you can create a single one and keep editing its data each time you want to add a new feature to the layer.

You delete the ds variable at the end of the script, which forces the files to close and all of your edits to be written to disk. Deleting the layer variable doesn't do the trick; you must close the data source. If you wanted to keep the data source open, you could call SyncToDisk on either the layer or data source object instead, like this:

```
ds.SyncToDisk()
```

> **WARNING** You must close your files or call SyncToDisk to flush your edits to disk. If you don't do this, and your interactive environment still has your data source open, you'll be disappointed to find an empty dataset.

It's always a good idea to carefully inspect your output to make sure you get the results you want. The best way would be to open it in QGIS, or you could get a good idea by plotting it from Python (figure 3.17):

```
>>> vp = VectorPlotter(True)
>>> vp.plot('ne_50m_admin_0_countries.shp', fill=False)
>>> vp.plot('capital_cities.shp', 'bo')
```

Let's return to the topic of adding attribute values for a moment. You might be wondering if multiple functions exist for setting attribute field values as with retrieving values. The answer is generally no. Most data will be converted to the correct type for you, but you may not like the results if a conversion isn't possible. For example, pretend for a minute that you made a mistake and inserted the population into the Name field, and the name into the Population field. Do you think that the population could be converted to a string and successfully inserted into the Name field? How about converting the country name to a number so it could go in the Population field?

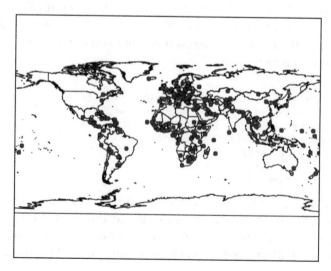

Figure 3.17   The result of plotting the new capital cities shapefile on top of country outlines

Well, converting a number to a string works fine, but converting a string to a number is problematic. The string "3578" can be translated into the number 3578, but what about the string "Russia"? If you try it in a Python interactive window by typing int('Russia'), you'll get an error, but OGR will insert a zero into the Population field instead of crashing. Sometimes this behavior is to your advantage because you don't need to convert data before inserting it in a feature, but it can also be a problem if you mistakenly try to insert the wrong type of data into a field.

3.5.1 *Creating new data sources*

You used an existing data source in listing 3.3, but sometimes you'll need to create new ones. Fortunately, it's not difficult. Perhaps the most important part is that you use the correct driver. It's the driver that does the work here, and each driver only knows how to work with one vector format, so using the correct one is essential. For example, the GeoJSON driver won't create a shapefile, even if you ask it to create a file with an .shp extension. As shown in figure 3.18, the output will have an .shp extension, but it will still be a GeoJSON file at heart.

You have a couple of ways to get the required driver. The first is to get the driver from a dataset that you've already opened, which will allow you to create a new data source using the same vector data format as the existing data source. In this example, the driver variable will hold the ESRI shapefile driver:

```
ds = ogr.Open(r'D:\osgeopy-data\global\ne_50m_admin_0_countries.shp')
driver = ds.GetDriver()
```

The second way to get a driver object is to use the OGR function GetDriverByName and pass it the short name of the driver. Remember that these names are available on the OGR website, by using the ogrinfo utility that comes with GDAL/OGR, or the print_drivers function available in the code accompanying this book. This example will get the GeoJSON driver:

```
json_driver = ogr.GetDriverByName('GeoJSON')
```

Once you have a driver object, you can use it to create an empty data source by providing the data source name. This new data source is automatically open for writing, and

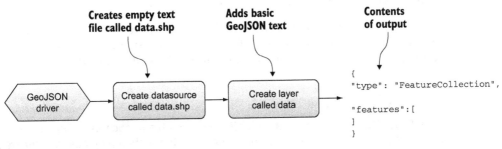

Figure 3.18 Using the GeoJSON driver to create a file with an .shp extension will still create a GeoJSON file, not a shapefile.

you can add layers to it the way you did in listing 3.3. If the data source can't be created, then `CreateDataSource` returns `None`, so you need to check for this condition:

```
json_ds = json_driver.CreateDataSource(json_fn)
if json_ds is None:
    sys.exit('Could not create {0}.'.format(json_fn))
```

A few data formats have creation options that you can use when creating a data source, although these aren't required. Like layer-creation options, these parameters are documented on the OGR website. Don't confuse the two, because data source and layer-creation options are two different things. Both types are passed as a list of strings, however. Let's see how you'd use a data source–creation option to create a full-fledged SpatiaLite data source instead of SQLite. This will fail if your version of OGR wasn't built with SpatiaLite support, though:

```
driver = ogr.GetDriverByName('SQLite')
ds = driver.CreateDataSource(r'D:\ osgeopy-data\global\earth.sqlite',
                    ['SPATIALITE=yes'])
```

Another thing to be aware of when creating new data sources is that you can't overwrite an existing data source. If a chance exists that your code might legitimately try to overwrite a dataset, then you'll need to delete the old one before attempting to create the new one. One way to deal with this would be to use the Python `os.path.exists` function to see if a file already exists before you attempt to create a data source; or you could wait and deal with it if your original attempt fails, either after checking for `None` or by using a try/except block. Either way, you should use the driver to delete the existing source instead of using a Python built-in function. Why? Because the driver will make sure that all required files are deleted. For example, if you're deleting a shapefile, the shapefile driver will delete the .shp, .dbf, .shx, and any other optional files that may be present. If you were using the Python built-in module to delete the shapefile, you'd have to make sure your code checked for all of these files. Here's an example of one way to deal with an existing data source:

```
if os.path.exists(json_fn):
    json_driver.DeleteDataSource(json_fn)
json_ds = json_driver.CreateDataSource(json_fn)
if json_ds is None:
    sys.exit('Could not create {0}.'.format(json_fn))
```

> **TIP** If you try to create a shapefile as a data source rather than a layer (where the data source is the containing folder), and the shapefile already exists, you'll get an odd error message saying that the shapefile isn't a directory.

Using OGR exceptions

By default, OGR doesn't raise an error if it has a problem, such as failing to create a new data source. This is why you check for `None`, but Python programmers generally expect an error to be raised instead. You can enable this behavior if you'd like, by calling `ogr.UseExceptions()` at the beginning of your code. Although most of the time this works as anticipated, I've discovered that it doesn't always raise an error when I expect. For example, no error is raised if OGR fails to open a data source. However, in instances where it does raise an error, you don't need to check for `None` before continuing. Using exceptions also gives you flexibility with handling errors.

For example, here's a contrived situation where I'm pretending to process data, then I want to save some temporary data to a GeoJSON file, and then I want to keep processing something else. If I can't create the temporary file, I want to skip that step and go on to the next bit of data processing rather than crashing. Here's the sample code:

```
ogr.UseExceptions()                                    ◄─┐ Turn on exceptions
fn = r'D:\ osgeopy-data\global\africa.geojson'
driver = ogr.GetDriverByName('GeoJSON')
print('Doing some preliminary analysis...')


try:
    ds = driver.CreateDataSource(fn)                   ◄─┐ Attempt to save
    lyr = ds.CreateLayer('layer')                        │ some data
    # Do more stuff, like create fields and save data


except RuntimeError as e:          ◄── Print error message and continue
    print(e)


print('Doing some more analysis...')
```

Suppose that the africa.geojson file already exists. This code doesn't check for that, so you know it will fail when you call `CreateDataSource`. If you weren't using OGR exceptions, this script would fail at that point and never get to the last `print` statement. But because you're using exceptions, you'll get an error message saying that the file couldn't be created, and then it will continue on to the last `print` statement, and the output will look like this:

```
Doing some preliminary analysis...
The GeoJSON driver does not overwrite existing files.
Doing some more analysis...
```

Try it out yourself and comment out the first line, and watch how the behavior changes.

3.5.2 Creating new fields

You saw in listing 3.3 how to copy attribute field definitions from one layer to another, but you can also define your own custom fields. Several different field types are available, but not all are supported by all data formats. This is another situation when the online documentation for the various formats will come in handy, so hopefully you've bookmarked that page.

To add a field to a layer, you need a `FieldDefn` object that contains the important information about the field, such as name, data type, width, and precision. The `schema` property you used in listing 3.3 returns a list of these, one for each field in the layer. You can create your own, however, by providing the name and data type for the new field to the `FieldDefn` constructor. The data type is one of the constants from table 3.2.

Table 3.2 Field type constants. There are more shown in appendix B, but I have been unable to make them work in Python.

Field data type	OGR constant
Integer	`OFTInteger`
List of integers	`OFTIntegerList`
Floating point number	`OFTReal`
List of floating point numbers	`OFTRealList`
String	`OFTString`
List of strings	`OFTStringList`
Date	`OFTDate`
Time of day	`OFTTime`
Date and time	`OFTDateTime`

After you create a basic field definition, but before you use it to add a field to the layer, you can add other constraints such as floating-point precision or field width, although I've noticed that these don't always have an effect, depending on the driver being used. For example, I haven't been able to set a precision in a GeoJSON file, and I've also discovered that you must set a field width if you want to set field precision in a shapefile. This example would create two fields to hold x and y coordinates with a precision of 3:

```
coord_fld = ogr.FieldDefn('X', ogr.OFTReal)
coord_fld.SetWidth(8)
coord_fld.SetPrecision(3)
out_lyr.CreateField(coord_fld)
coord_fld.SetName('Y')
out_lyr.CreateField(coord_fld)
```

Create and add the first field

Reuse the FieldDefn to create a second field

You might have noticed that you don't create two different field definition objects here. Once you've used the field definition to create a field in the layer, you can change the definition's attributes and reuse it to create another field, which makes this easier because you want two fields that were identical except in name.

Also, sometimes the field width will be ignored if it's too small for the data provided. For example, if you create a string field with a width of 6, but then try to insert a value that's 11 characters long, in certain cases the width of the field would increase to hold the entire string. This isn't always possible, however, and it's best to be specific about what you want rather than hope something like this will conveniently happen.

3.6 *Updating existing data*

Sometimes you need to update existing data rather than create an entirely new dataset. Whether this is possible, and which edits are supported, depends on the format of the data. For example, you can't edit GeoJSON files, but many different edits are allowed on shapefiles. We'll discuss getting information about what's supported in the next chapter.

3.6.1 *Changing the layer definition*

Depending on the type of data you're working with, you can edit the layer definition by adding new fields, deleting existing ones, or changing field properties such as name. As with adding new fields, you need a field definition to change a field. Once you have a field definition that you're happy with, you use the AlterFieldDefn function to replace the existing field with the new one:

```
AlterFieldDefn(iField, field_def, nFlags)
```

- iField is the index of the field you want to change. A field name won't work in this case.
- field_def is the new field definition object.
- nFlags is an integer that is the sum of one or more of the constants shown in table 3.3.

Table 3.3 **Flags used to specify which properties of a field definition can be changed. To use more than one, simply add them together**

Field properties that need to change	OGR constant
Field name only	ALTER_NAME_FLAG
Field type only	ALTER_TYPE_FLAG
Field width and/or precision only	ALTER_WIDTH_PRECISION_FLAG
All of the above	ALTER_ALL_FLAG

To change a field's properties, you need to create a field definition containing the new properties, find the index of the existing field, and decide which constants from table 3.3 to use to ensure your changes take effect. To change the name of a field from 'Name' to 'City_Name', you might do something like this:

```
i = lyr.GetLayerDefn().GetFieldIndex('Name')
fld_defn = ogr.FieldDefn('City_Name', ogr.OFTString)
lyr.AlterFieldDefn(i, fld_defn, ogr.ALTER_NAME_FLAG)
```

If you needed to change multiple properties, such as both the name and the precision of a floating-point attribute field, you'd pass the sum of ALTER_NAME_FLAG and ALTER_WIDTH_PRECISION_FLAG, like this:

```
lyr_defn = lyr.GetLayerDefn()
i = lyr_defn.GetFieldIndex('X')
```

```
width = lyr_defn.GetFieldDefn(i).GetWidth()
fld_defn = ogr.FieldDefn('X_coord', ogr.OFTReal)
fld_defn.SetWidth(width)
fld_defn.SetPrecision(4)
flag = ogr.ALTER_NAME_FLAG + ogr.ALTER_WIDTH_PRECISION_FLAG
lyr.AlterFieldDefn(i, fld_defn, flag)
```

Notice that you use the original field width when creating the new field definition. I found out the hard way that if you don't set the width large enough to hold the original data, then the results will be incorrect. To get around the problem, use the original width. For the precision change to take effect, all records must be rewritten. Making the precision larger than it was won't give you more precision, however, because data can't be created from thin air. The precision can be decreased, however.

Instead of summing up flag values, you could cheat and just use ALTER_ALL_FLAG. Only do this if your new field definition is exactly what you want the field to look like after editing, however. The other flags limit what can change, but this one doesn't. For example, if your field definition has a different data type than the original field but you pass ALTER_NAME_FLAG, then the data type will not change, but it will if you pass ALTER_ALL_FLAG.

3.6.2 *Adding, updating, and deleting features*

Adding new features to existing layers is exactly the same as adding them to brand-new layers. Create an empty feature based on the layer definition, populate it, and insert it into the layer. Updating features is much the same, except you work with features that already exist in the layer instead of blank ones. Find the feature you want to edit, make the desired changes, and then update the information in the layer by passing the updated feature to SetFeature instead of CreateFeature. For example, you could do something like this to add a unique ID value to each feature in a layer:

```
lyr.CreateField(ogr.FieldDefn('ID', ogr.OFTInteger))
n = 1
for feat in lyr:
    feat.SetField('ID', n)
    lyr.SetFeature(feat)
    n += 1
```

First you add an ID field, and then you iterate through the features and set the ID equal to the value of the n variable. Because you increment n each time through the loop, each feature has a unique ID value. Last, you update the feature in the layer by passing it to SetFeature.

Deleting features is even easier. All you need to know is the FID of the feature you want to get rid of. If you don't know that number off the top of your head, or through another means, you can get it from the feature itself, like this:

```
for feat in lyr:
    if feat.GetField('City_Name') == 'Seattle':
        lyr.DeleteFeature(feat.GetFID())
```

FID	Name
0	Lima
1	Paris
Deleted	Toronto
3	Geneva
4	Oslo
5	Moscow
Deleted	Tokyo
7	Cape Town

Vacuum/repack →

FID	Name
0	Lima
1	Paris
2	Geneva
3	Oslo
4	Moscow
5	Cape Town

Figure 3.19 The effect of vacuuming or repacking a database. Notice that the FID values change.

For each feature in the layer, you check to see if its `City_Name` attribute is equal to `Seattle`, and if it is, you retrieve the FID from the feature itself and then pass that number to `DeleteFeature`.

Certain formats don't completely kill the feature at this point, however. You may not see it, but sometimes the feature has only been marked for deletion instead of totally thrown out, so it's still lurking in the shadows. Because of this, you won't see any other features get assigned that FID, and it also means that if you've deleted many features, there may be a lot of needlessly used space in your file. See figure 3.19 for a simple example. Deleting these features will reclaim this space. If you have much experience with relational databases, you should be familiar with this idea. It's similar to running Compact and Repair on a Microsoft Access database or using VACUUM on a PostgreSQL database.

How to go about reclaiming this space, or determining if it needs to be done, is dependent on the vector data format being used. Here are examples for doing it for shapefiles and SQLite:

```
ds.ExecuteSQL('REPACK ' + lyr.GetName())          ← Shapefile
ds.ExecuteSQL('VACUUM')              ← SQLite
```

In both cases, you need to open the data source and then execute a SQL statement on it that compacts the database. For shapefiles you need to know the name of the layer, so if the layer is called "cities", then the SQL would be "REPACK cities".

Another issue with shapefiles is that they don't update their metadata for spatial extent when existing features are modified or deleted. If you edit existing geometries or delete features, you can ensure that the spatial extent gets updated by calling this:

```
ds.ExecuteSQL('RECOMPUTE EXTENT ON ' + lyr.GetName())
```

This isn't necessary if you insert features, however, because those extent changes are tracked. It's also not necessary if there's no chance that your edits change the layer's extent.

3.7   *Summary*

- Vector data formats are most appropriate for features that can be characterized as a point, line, or polygon.
- Each geographic feature in a vector dataset can have attribute data, such as name or population, attached to it.
- The type of geometry used to model a given feature may change depending on scale. A city could be represented as a point on a map of an entire country, but as a polygon on a map of a smaller area, such as a county.
- Vector datasets excel for measuring relationships between geographic features such as distances or overlaps.
- You can use OGR to read and write many different types of vector data, but which ones depend on which drivers have been compiled into your version of GDAL/OGR.
- Data sources can contain one or more layers (depending on data format), and in turn, layers can contain one or more features. Each feature has a geometry and a variable number of attribute fields.
- Newly created data sources are automatically opened for writing. If you want to edit existing data, remember to open the data source for writing.
- Remember to make changes to the layer, such as adding or deleting fields, before getting the layer definition and creating a feature for adding or updating data.

Working with different vector file formats

As mentioned in the previous chapter, there are many different vector file formats, and they're not always interchangeable, at least in a practical sense. Certain formats are more appropriate for certain uses than others. In this chapter you'll learn several of the differences and their strengths and weaknesses.

Another consideration with format is what you can and can't do with the data using OGR. In general, working with one type is the same as working with another, but sometimes how you open the data source is different. The larger issue is the difference in capabilities of each driver. For example, certain formats can be read from but not written to, and others can be created but existing data can't be edited. You'll also learn how to determine what you can and can't do with a dataset.

4.1   *Vector file formats*

Up to this point, you've only worked with shapefiles, but many more vector file formats are available. Chances are that you'll probably only use a handful of them on a regular basis, but you need to have an idea of the available options. Several formats have open specifications and are supported by many different software programs, while others are used more sparingly. Certain formats also support more capabilities than others. Most of these formats allow for easy transfer from one user to another, much like you can give someone else your spreadsheet file. A few use database servers, however, which allows for many users to access and edit the same dataset at a central location, but sometimes makes it more difficult to move the data from one place to another.

4.1.1   *File-based formats such as shapefiles and geoJSON*

What I call *file-based formats* are made up of one or more files that live on a disk drive and can be easily transferred from one location to another, such as from your hard drive to another computer or an external drive. Several of these are relational databases, but are designed to be easily moved around (think of Microsoft Access relational databases), so they're considered file-based for the purposes of this discussion. Several of these formats have open standards so anyone can create software to use them, while others are proprietary and limited to smaller numbers of software. Examples of open formats are GeoJSON, KML, GML, shapefiles, and SpatiaLite.

Spatial data can also be stored in Excel spreadsheets, comma- or tab-delimited files, or other similar formats, although this is most common for point data when only x and y coordinates are required. Most spatial data, however, is stored using formats designed specifically for GIS data. Several of these formats are plain text, meaning that you can open them in any text editor and look at them, and others are binary files that require software capable of understanding them.

As mentioned previously, one advantage of plain text files is that you can open them in a text editor and inspect their contents. You can even edit them by hand, rather than using GIS software, if you're so inclined. Listing 4.1 shows an example of a GeoJSON file that contains two cities in Switzerland, Geneva and Lausanne, both represented as points.

Listing 4.1   **An example GeoJSON file with two features**

```
{
  "type": "FeatureCollection",
  "features": [
    {
      "type": "Feature",
      "properties": { "NAME": "Geneva", "PLACE": "city" },
      "geometry": {
        "type": "Point",
        "coordinates": [ 6.1465886, 46.2017589 ]
      }
```

```
    },
    {
      "type": "Feature",
      "properties": { "NAME": "Lausanne", "PLACE": "city" },
      "geometry": {
        "type": "Point",
        "coordinates": [ 6.6327025, 46.5218269 ]
      }
    },
  ]
}
```

It's okay if you don't understand everything in this example. The point here is that you can open and edit the file in a text editor instead of using GIS software. For example, you could easily fix the spelling of a city name or tweak one of the point coordinates. While we're on the subject, it's worth mentioning that small GeoJSON files are automatically rendered as interactive maps when uploaded to GitHub. The example shown here is saved as a gist at https://gist.github.com/cgarrard/8049400. If you have a GitHub account, you can copy this gist to your own account, make changes, and instantly see the result.

Plain text formats such as GeoJSON, KML, and GML are popular for transferring small amounts of data and for web applications, but they don't work so well for data analysis. For one thing, all three of these formats allow different geometry types to be present in the same dataset, which GIS software doesn't really appreciate. For example, data in the popular shapefile format contains all points, all lines, or all polygons, but not a mixture. Therefore, a shapefile could contain roads (lines) or city boundaries (polygons), but not both. A GeoJSON file, on the other hand, can contain a combination of all three geometries in the same dataset, such as the roads and city boundaries mentioned previously that would have to live in two different shapefiles. Because you have only one file to download and process, this is an excellent solution for passing data to a web browser so it can render it on a map. However, most GIS software expects only points, only lines, or only polygons, and won't read the data correctly if it has a mixture. If you need to load the data into GIS software, don't combine multiple geometry types into one dataset, even when allowed.

Perhaps a more serious problem with plain text formats when it comes to data analysis is that they don't have the same indexing capabilities as many binary formats. Indexes are used for searching and accessing data quickly. Attribute indexes allow for searching on values in the attribute fields for the features, such as searching for all cities in a dataset with a population over 100,000. Spatial indexes store information about the spatial location of features in the dataset so that searching can be limited to features in a certain geographic area, for example, when you overlay a small watershed polygon on a larger dataset of water-monitoring stations. A spatial index would be used to quickly find the monitoring stations that fall within the watershed boundary. Both of these operations, finding large cities and finding water-monitoring stations, would be slow on large datasets if the appropriate attribute or spatial index didn't exist. In addition, spatial indexes can help a dataset be drawn more quickly because they help find the features that fall within the viewport. For example, if you're looking at Asian cities and zoom in on Japan, the spatial index helps find Japanese cities faster while ignoring cities in western China.

These issues aren't as important with small datasets, but they're extremely important with large ones. Certain formats have ways around these problems, though. For example, although the KML format doesn't have true spatial indexes, it does allow for datasets to be broken up into different files for different spatial locations. This allows for smaller datasets to be loaded as a user zooms and pans around the map, which increases rendering speed.

Several vector data formats use familiar desktop-based, or personal, relational database software under the hood. This is true for Esri personal geodatabases and GeoMedia .mdb files, which use Microsoft Access databases to store data. Another example of a vector format based on an existing database format is SpatiaLite, a spatial extension for the SQLite database management system. These vector data formats can take advantage of the capabilities built into the database software, such as indexes. The underlying database also imposes much stricter rules for storing data. For example, all geographic features in a dataset must have the same geometry type and the same set of attribute fields. Similar to the way nonspatial databases can contain multiple tables, a spatial database can contain multiple datasets. Although an individual dataset is limited to a single geometry type, a solitary database file can contain multiple datasets, each with different geometry types and attribute fields. This is convenient for keeping related datasets together and for moving them from disk to disk. Figure 4.1 shows a schematic of a single SpatiaLite database file that contains multiple datasets with different geometries.

Other vector formats consist of several files, such as the ever popular shapefile. These datasets store geometries, attribute values, and indexes in separate files. If you move a shapefile from one location to another, you need to ensure that you move all of the required files. Other format types that require multiple files make it a bit easier by using dedicated folders that contain the necessary files. As with shapefiles, you don't need to know anything about the individual files, but you shouldn't change anything in the folder. Two examples of formats that use this system are Esri grids and file geodatabases.

Many other vector data formats haven't been mentioned here, but you should now have an idea of the types of formats and their strengths and weaknesses.

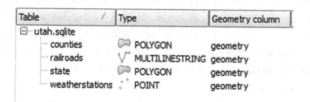

Figure 4.1 A sample SpatiaLite database containing multiple layers with different geometry types. All of these various datasets are contained within one easily transportable file.

4.1.2 Multi-user database formats such as PostGIS

You've seen that file-based formats come in many shapes and sizes, including desktop relational database models such as SpatiaLite. One limitation of these formats is that they don't allow multiple people to edit, or sometimes even use, a specific dataset at the same time. This is where the multi-user client-server database architecture comes in, because the data are stored in a database that is accessible by multiple clients across the network. Users access data from the server rather than opening a file on a local disk. Although this is certainly not for everyone, it's a great choice for making data available to many users from a central location. This is especially useful if the data are updated frequently or are used by many different users, because all users will instantly have access to the updated data. It also allows multiple people to edit a dataset at once, which isn't usually possible with file-based formats. In addition, in many cases the indexing and querying capabilities of these database systems provide faster performance when accessing data.

The most popular client-server database solutions for spatial data include PostgreSQL with the PostGIS spatial extension, ArcSDE, SQL Server, and Oracle Spatial and Graph. If you want to host the data on your own computer, you need to invest in a system like these. My favorite is PostGIS (www.postgis.net) because it's open source and provides a feature-rich environment with many functions, operators, and indexes that are specific to spatial data. Even with huge amounts of data, you can still get good performance. Although you can't zip up a PostGIS dataset and email it to a colleague, it comes with utilities to import and export several popular file-based formats, and it's straightforward to run a query and export the data to a portable format. Not only does PostGIS store the data, but you can use it for many types of analyses as well, without the need for other GIS software. PostGIS also works with raster data.

If you're not familiar with relational databases, then it might take effort to set one of these systems up and learn how to use it. But it's extremely powerful and worth the investment in brain cells if you need to give multiple users simultaneous access to data.

4.2 Working with more data formats

Until now we've only worked with one data format out of many. The basics don't change between formats, though. Once you open the data source, reading the data is pretty much the same. But for kicks, let's look at several formats that support more than one layer, because we haven't done that yet. Until now, we've used the first and only layer in a data source, but if multiple layers exist, you need to know either the name or the index of the one you're interested in. Generally, I'd use ogrinfo to get this information, but because this is a book on Python, let's write a simple function that opens a data source, loops through the layers, and prints their names and indexes:

```python
def print_layers(fn):
    ds = ogr.Open(fn, 0)
    if ds is None:
        raise OSError('Could not open {}'.format(fn))
    for i in range(ds.GetLayerCount()):
        lyr = ds.GetLayer(i)
        print('{0}: {1}'.format(i, lyr.GetName()))
```

This function takes the filename of the data source as a parameter, and the first thing it does is open the file. Then it uses `GetLayerCount` to find out how many layers the data source contains, and iterates through a loop that many times. Each time through the loop, it uses the `i` variable to get the layer at the index corresponding to that iteration. Then it prints the name of the layer and its index. This function is included in the ospybook module, and you'll use it to inspect other data sources in the following examples.

4.2.1 SpatiaLite

Let's start with a SpatiaLite database. This type of data source can contain many different layers, all with unique (and hopefully descriptive) names. To see this, list the layers in the natural_earth_50m.sqlite file in the data download:

```
>>> import ospybook as pb
>>> pb.print_layers(r'D:\osgeopy-data\global\natural_earth_50m.sqlite')
0: countries
1: populated_places
```

As you can see, the dataset has two layers. How would you get a handle to the populated_places layer? Well, you could use either the index or the layer name, so both `ds.GetLayer(1)` and `ds.GetLayer('populated_places')` would do the trick. It's probably better to use the name rather than the index, however, because the index might change if other layers are added to the data source. To prove that this works, try plotting the layer, which will be dots representing cities around the world, as shown in figure 4.2.

```
>>> ds = ogr.Open(r'D:\osgeopy-data\global\natural_earth_50m.sqlite')
>>> lyr = ds.GetLayer('populated_places')
>>> vp = VectorPlotter(True)
>>> vp.plot(lyr, 'bo')
```

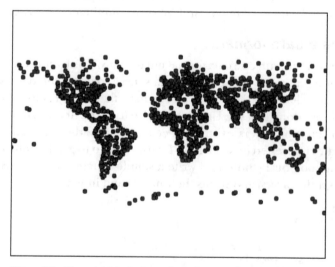

Figure 4.2 The populated_places layer in natural_earth_50m.sqlite

Ogrinfo

GDAL comes with several extremely useful command-line utilities, and in fact, you've already seen how to use ogrinfo to find out which vector data formats your version of OGR supports. You can also use ogrinfo to get information about specific data sources and layers. If you pass it a data source name, it will print a list of layers contained in that data source:

```
D:\osgeopy-data\global>ogrinfo natural_earth_50m.sqlite
INFO: Open of `natural_earth_50m.sqlite'
      using driver `SQLite' successful.
1: countries (Multi Polygon)
2: populated_places (Point)
```

You can also use ogrinfo to see metadata about a layer and even all of the attribute data. This example will show a summary only (-so) of the countries layer in the natural earth SQLite database. This includes metadata such as the extent, spatial reference, and a list of attribute fields and their data types. The second will show all attribute values for the first feature in the layer.

```
ogrinfo -so natural_earth_50m.sqlite countries
```

To display all of the attribute values for the feature with an FID of 1, you could do something like this, where -q means *don't print the metadata* and –geom=NO means *don't print out a text representation of the geometry* (which would be long).

```
ogrinfo -fid 1 -q -geom=NO natural_earth_50m.sqlite countries
```

See http://www.gdal.org/ogrinfo.html for full ogrinfo documentation.

4.2.2 *PostGIS*

What about connecting to a database server such as the PostGIS spatial extension for PostgreSQL? Note a couple of extra considerations that you don't need to worry about with local files. You need to know the connection string to use, which involves host, port, database name, username, and password. You also need permission to connect to the database and tables in question. If you're not managing your own database server, then you might need to talk to the database administrator to set all of this up. The following example connects to the geodata database being served by a PostgreSQL instance running on my local machine. It won't work for you unless you go to the trouble to install PostgreSQL and PostGIS, and then set up a database.

```
>>> pb.print_layers('PG:user=chris password=mypass dbname=geodata')
0: us.counties
1: global.countries
2: global.populated_places
3: time_zones
```

You see four layers here, but they're divided up into three different groups, or schemas. The time zones layer is in the default schema, counties is in the *us* schema, and the remaining two are in the *global* schema. Every user of the database could have access to different schemas, and even different layers within a schema, depending on how the database administrator has set up the security.

As you can see, you can access PostGIS databases with OGR, but you can do many things with a PostGIS database that aren't covered in this book. If you're interested in learning more about it, take a look at *PostGIS in Action*, also published by Manning.

4.2.3   *Folders as data sources (shapefiles and CSV)*

In certain cases OGR will treat entire folders as data sources. Two examples of this are the shapefile and comma-delimited text file (.csv) drivers, which can be used to open either individual files or entire folders as data sources. If you use a folder, then each file inside of the folder is treated as a layer. If a folder contains a variety of file types, then the shapefile driver is used. For example, try listing the layers in the US folder:

```
>>> pb.print_layers(r'D:\osgeopy-data\US')
0: citiesx020 (Point)
1: cities_48 (Point)
2: countyp010 (Polygon)
3: roadtrl020 (LineString)
4: statep010 (Polygon)
5: states_48 (Polygon)
6: volcanx020 (Point)
```

Compare this list to the contents of the folder, and you'll see that it listed each of the shapefiles, but none of the others. The CSV driver is a little pickier, however, and wants all of the files in the folder to be CSV files. Although it won't work with the US folder, it works fine with the csv subfolder. Does this mean that you can't open a CSV file that's in a folder with a bunch of other files? Fortunately, no. All you have to do is treat the CSV file itself as a data source with only one layer. You can do the exact same thing with a shapefile by providing the name of the .shp file.

4.2.4   *Esri file geodatabases*

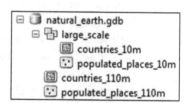

You Esri users out there might expect to see feature datasets inside file geodatabases treated like the schemas in PostGIS. If so, you'll be disappointed, because all you see are feature class names. Figure 4.3 shows what the natural_earth file geodatabase looks like in ArcCatalog, but the large_scale feature dataset name isn't included in the layer names that OGR uses.

Figure 4.3   The natural_earth file geodatabase as seen in ArcCatalog

```
>>> pb.print_layers(r'D:\osgeopy-data\global\natural_earth.gdb')
0: countries_10m
1: populated_places_10m
2: countries_110m
3: populated_places_110m
```

Fortunately, you don't need the feature dataset name to access the layer, though; the feature class name works fine:

```
>>> ds = ogr.Open(r'D:\osgeopy-data\global\natural_earth.gdb')
>>> lyr = ds.GetLayer('countries_10m')
```

File geodatabases have two different drivers. You can read more about the differences on the OGR website, but one huge difference is that the read-only OpenFileGDB driver is compiled into OGR by default and the read/write FileGDB driver isn't because it requires a third-party library from Esri. If somebody gave you a file geodatabase that you needed to change but you didn't have access to the FileGDB driver, you could still use the OpenFileGDB driver to open the geodatabase and copy the data to a format that you could edit. This may not be ideal, but at least you have the option. For example, you could copy the countries_110m feature class in the natural earth geodatabase to a shapefile like this:

```
gdb_ds = ogr.Open(r'D:\osgeopy-data\global\natural_earth.gdb')
gdb_lyr = gdb_ds.GetLayerByName('countries_110m')
shp_ds = ogr.Open(r'D:\Temp', 1)
shp_ds.CopyLayer(gdb_lyr, 'countries_110m')
del shp_ds, gdb_ds
```

You haven't seen the CopyLayer method before. This allows you to easily copy the contents of an entire layer into a new data source or to the same data source but with a different layer name. To use it, you need to get the layer that you want to make a copy of and open the data source that you want to save the copy into. Then call CopyLayer on the data source that will get the copy, and pass it the original layer and a name for the new layer that will be created.

If you do have the Esri FileGDB driver, you can create new file geodatabases, and even feature datasets even though OGR doesn't show you feature dataset names. Listing 4.2 shows a function that imports all of the layers from a data source into a feature dataset within a file geodatabase, but note that this only works if you have the FileGDB driver. If you try to use this function without that driver installed, you'll get an error message that says AttributeError: 'NoneType' object has no attribute 'CreateDataSource'.

Listing 4.2 Function to import layers to a file geodatabase

```
def layers_to_feature_dataset(ds_name, gdb_fn, dataset_name):
    """Copy layers to a feature dataset in a file geodatabase."""
    in_ds = ogr.Open(ds_name)
    if in_ds is None:
        raise RuntimeError('Could not open datasource')
    gdb_driver = ogr.GetDriverByName('FileGDB')
    if os.path.exists(gdb_fn):                              # Open the geodatabase
        gdb_ds = gdb_driver.Open(gdb_fn, 1)                 # if it exists
    else:                                          ← Create the
        gdb_ds = gdb_driver.CreateDataSource(gdb_fn)   geodatabase
    if gdb_ds is None:                                 if needed
        raise RuntimeError('Could not open file geodatabase')
    options = ['FEATURE_DATASET=' + dataset_name]      ← Set the feature
    for i in range(in_ds.GetLayerCount()):                dataset name
        lyr = in_ds.GetLayer(i)
        lyr_name = lyr.GetName()
        print('Copying ' + lyr_name + '...')              # Copy each layer
        gdb_ds.CopyLayer(lyr, lyr_name, options)
```

This function requires three parameters: the path to the original data source, the path to the file geodatabase, and the name of the feature dataset to copy the layers into. After opening the original data source, it checks to see if the file geodatabase exists. If it does, then the geodatabase is opened for writing. If it doesn't exist, it's created. Feature datasets are specified using layer-creation options, so then a list containing a single option for FEATURE_DATASET is created. After that, all of the layers in the original data source are looped over and copied into the geodatabase while keeping the same layer name (although they'll be renamed if naming conflicts arise in the geodatabase). If the FEATURE_DATASET layer-creation option wasn't provided, then the layer will be added to the file geodatabase, but it will be at the top level instead of in a feature dataset.

Now that you have this function, you could copy all of the shapefiles in a folder into a geodatabase like this:

```
layers_to_feature_dataset(
    r'D:\osgeopy-data\global', r'D:\Temp\osgeopy-data.gdb', 'global')
```

If you wanted to have the option of saving the feature classes to the top level of the geodatabase instead of in a feature dataset, you could modify this function so it doesn't pass the option list to CopyLayer if the dataset_name parameter is None or an empty string.

4.2.5   *Web feature services*

You can also access online services, such as *Web Feature Services* (WFS). Let's try this using a WFS hosted by the United States National Oceanic and Atmospheric Administration (NOAA) that serves out hazardous weather watches and advisories. Start with getting the list of available layers:

```
>>> url = 'WFS:http://gis.srh.noaa.gov/arcgis/services/watchWarn/' + \
...        'MapServer/WFSServer'
>>> pb.print_layers(url)
0: watchWarn:WatchesWarnings (MultiPolygon)
1: watchWarn:CurrentWarnings (MultiPolygon)
```

You can loop through these layers like the layers from other data sources, but all of the data are fetched immediately, so there could be quite a lag if the list has lots of features. It looks like the second layer only contains warnings, which are more severe than watches, so it should have less data. Let's find out what type of warning the first feature represents. I've discovered that things crash if I try to use GetFeature with an FID, but you can do it using GetNextFeature:

```
>>> ds = ogr.Open(url)
>>> lyr = ds.GetLayer(1)
>>> feat = lyr.GetNextFeature()
>>> print(feat.GetField('prod_type'))
Tornado Warning
```

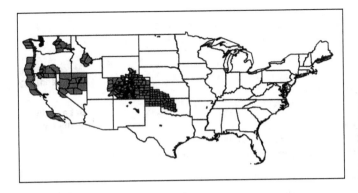

Figure 4.4
The WatchesWarnings layer from the NOAA web feature service. If you plot it, your results will differ because this layer shows real-time data.

I can recommend an easier and faster way to get only the first few features if that's all you want, however. Tack a MAXFEATURES parameter onto your URL, like this:

```
>>> url += '?MAXFEATURES=1'
>>> ds = ogr.Open(url)
>>> lyr = ds.GetLayer(1)
>>> lyr.GetFeatureCount()
1
```

You can also work with the geometries from a WFS. Figure 4.4 shows my results when I used VectorPlotter to draw the watchWarn:WatchesWarnings layer on top of states.

Let's do something a little different—save real-time data from a WFS and use it to build a simple web map using Folium, which is a Python module that creates Leaflet maps. If you have no idea what Leaflet is, that's okay, because you don't have to know anything about web mapping to work through this example. First you need to install Folium, though. On my Windows computer, I opened up a command prompt and used pip to install Folium and Jinja2 (another module that Folium requires in order to work) for Python 3.3 like this:

```
C:\Python33\Scripts\pip install Jinja2
C:\Python33\Scripts\pip install folium
```

If you're not familiar with installing Python modules via pip, please refer to the installation instructions in appendix A. Now let's look at the example script, which breaks things out into functions so code can be easily reused. Listing 4.3 contains a function to retrieve stream gauge data from a WFS and save it as GeoJSON; a function to make the web map showing these stream gauges; a function to get a geometry so that the map focuses on a single state instead of the whole country; and a couple of helper functions to format data for the WFS request and the map.

Listing 4.3   Create a web map from WFS data

```
import os
import urllib
from osgeo import ogr
import folium
```

Get bounding
box from
geometry

```python
def get_bbox(geom):
    """Return the bbox based on a geometry envelope."""
    return '{0},{2},{1},{3}'.format(*geom.GetEnvelope())

def get_center(geom):
    """Return the center point of a geometry."""
    centroid = geom.Centroid()
    return [centroid.GetY(), centroid.GetX()]

def get_state_geom(state_name):
    """Return the geometry for a state."""
    ds = ogr.Open(r'D:\osgeopy-data\US\states.geojson')
    if ds is None:
        raise RuntimeError(
            'Could not open the states dataset. Is the path correct?')
    lyr = ds.GetLayer()
    lyr.SetAttributeFilter('state = "{0}"'.format(state_name))
    feat = next(lyr)
    return feat.geometry().Clone()

def save_state_gauges(out_fn, bbox=None):
    """Save stream gauge data to a geojson file."""
    url = 'http://gis.srh.noaa.gov/arcgis/services/ahps_gauges/' + \
        'MapServer/WFSServer'
    parms = {
        'version': '1.1.0',
        'typeNames': 'ahps_gauges:Observed_River_Stages',
        'srsName': 'urn:ogc:def:crs:EPSG:6.9:4326',
    }
    if bbox:
        parms['bbox'] = bbox
    try:
        request = 'WFS:{0}?{1}'.format(url, urllib.urlencode(parms))
    except:
        request = 'WFS:{0}?{1}'.format(url, urllib.parse.urlencode(parms))
    wfs_ds = ogr.Open(request)
    if wfs_ds is None:
        raise RuntimeError('Could not open WFS.')
    wfs_lyr = wfs_ds.GetLayer(0)

    driver = ogr.GetDriverByName('GeoJSON')
    if os.path.exists(out_fn):
        driver.DeleteDataSource(out_fn)
    json_ds = driver.CreateDataSource(out_fn)
    json_ds.CopyLayer(wfs_lyr, '')

def make_map(state_name, json_fn, html_fn, **kwargs):
    """Make a folium map."""
    geom = get_state_geom(state_name)
    save_state_gauges(json_fn, get_bbox(geom))
    fmap = folium.Map(location=get_center(geom), **kwargs)
    fmap.geo_json(geo_path=json_fn)
    fmap.create_map(path=html_fn)

os.chdir(r'D:\Dropbox\Public\webmaps')
make_map('Oklahoma', 'ok.json', 'ok.html',
         zoom_start=7)
```

Get center point
from geometry

Get a state
geometry

Save gauge WFS
data to GeoJSON

Make the web map

Top-level code

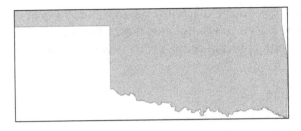

Figure 4.5   The line is the bounding box for the state of Oklahoma.

You can probably understand what the get_state_geom function does and how it does it, because you've seen the same process before. It takes a state name as a parameter, finds the corresponding feature in a layer, and returns the cloned geometry. The filename is hardcoded because you assume that the location of this state boundary file won't change.

The two helper functions are also simple. The get_center function takes a geometry, gets its centroid, and then returns the coordinates as a [y, x] list. The order might seem weird to you, but that's the order that Folium wants them in for the map.

The get_bbox function takes a geometry and returns its bounding coordinates as a string formatted like min_x,min_y,max_x,max_y. This is the format that a WFS uses to spatially subset results, and it's how you'll limit your gauge results to the bounding box of a state. This function takes advantage of the string formatting rules to rearrange the results of GetEnvelope, which returns a geometry's bounding box (figure 4.5) as a [min_x, max_x, min_y, max_y] list.

Now let's look at the slightly more complicated save_state_gauges function. Here you hardcode in the URL for a WFS that returns the observed river stages data from the Advanced Hydrologic Prediction Service. You also create a dictionary containing the parameters to be passed to the WFS. As you already know, the typeNames parameter is the name of the layer to retrieve data from. The version is the WFS version to use, and srsName specifies which coordinate system you'd like your data to be returned in. You can see the available options for this in the WFS's capabilities output, which you can get by tacking ?request=GetCapabilities onto the end of the service URL and visiting it in a web browser. For example, part of the output from http://gis.srh.noaa.gov/arcgis/services/ahps_gauges/MapServer/WFSServer?request=GetCapabilities looks like this:

```
<wfs:FeatureType>
    <wfs:Name>ahps_gauges:Observed_River_Stages</wfs:Name>
    <wfs:Title>Observed_River_Stages</wfs:Title>
    <wfs:DefaultSRS>urn:ogc:def:crs:EPSG:6.9:4269</wfs:DefaultSRS>
    <wfs:OtherSRS>urn:ogc:def:crs:EPSG:6.9:4326</wfs:OtherSRS>
    <snip>
</wfs:FeatureType>
```

From this you can see that the default spatial reference system (DefaultSRS) is EPSG 4269, which happens to be unprojected data using the NAD83 datum. If that doesn't make much sense, don't worry about it for now, because you'll learn all about it in chapter 8. All you need to know now is that web-mapping libraries generally want coordinates that use WGS84, which corresponds to EPSG 4326. Fortunately, that's listed as an `OtherSRS` option in the capabilities output, so you insert it into your parameters dictionary:

```
parms = {
    'version': '1.1.0',
    'typeNames': 'ahps_gauges:Observed_River_Stages',
    'srsName': 'urn:ogc:def:crs:EPSG:6.9:4326',
}
if bbox:
    parms['bbox'] = bbox
```

If the user provided a bbox parameter to the function, you also insert that into your dictionary. If a bbox parameter is provided to the WFS, it returns features that fall in that box instead of returning all of them. Remember that your `get_bbox` function creates a string in the correct format for this based on a geometry's bounding box.

Creating this dictionary wasn't absolutely necessary, because you could have built your query string the same way you did in earlier examples, but I think that using a dictionary makes it easier to see what parameters are being passed. It's easy to create the query string from the dictionary by using the `urlencode` function, which formats everything for you. In Python 2, this function lives in the urllib module, but in Python 3 it lives in urllib.parse, which is why you have the next step in a try/except block. You try to create the query string using the Python 2 function, but if that fails because the script was run with Python 3, then you do it the Python 3 way instead:

```
try:
    request = 'WFS:{0}?{1}'.format(url, urllib.urlencode(parms))
except:
    request = 'WFS:{0}?{1}'.format(url, urllib.parse.urlencode(parms))
```

After creating your query string, you use it to open a connection to the WFS and get the layer. You want to save the output to a local file this time, though, so then you create an empty GeoJSON data source. Data sources have a `CopyLayer` function that copies an existing layer into the data source; this existing layer can be from another data source altogether. You use that function to copy the data from the WFS into your new GeoJSON file:

```
json_ds.CopyLayer(wfs_lyr, '')
```

The second parameter to `CopyLayer` is the name for the new layer, but GeoJSON layers don't have names, so you pass a blank string. You could pass a real layer name, but it wouldn't do much good. When your function returns after creating the layer, the data sources go out of scope, so the files get closed automatically, which is why you don't bother to close them inside the function.

The last function you write is called `make_map`. It wants a state name along with filenames for the output GeoJSON and HTML files. It can also take other named arguments that get passed to Folium, which allows you to pass optional Folium parameters without having to worry about them in your make_map function:

```
def make_map(state_name, json_fn, html_fn, **kwargs):
    """Make a folium map."""
    geom = get_state_geom(state_name)
    save_state_gauges(json_fn, get_bbox(geom))
    fmap = folium.Map(location=get_center(geom), **kwargs)
    fmap.geo_json(geo_path=json_fn)
    fmap.create_map(path=html_fn)
```

The basic outline is shown in figure 4.6, but the first thing this function does is get the geometry for the state of interest. Then it gets the bbox for the geometry and passes that, along with the output GeoJSON filename, to the function that saves the WFS data to file. Then it creates a Folium map centered on the geometry, and also uses any named arguments that the user might have passed in. Remember that ** explodes a dictionary into key/value pairs, so all of the arguments are treated as if they're an exploded dictionary called kwargs. You can read about the optional parameters at http://folium.readthedocs.org/en/latest/. This map uses OpenStreetMap tiles as the basemap by default, but that's one of the things you can change.

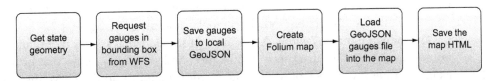

Figure 4.6   Tasks in the make_map function

After creating the basic map, the contents of the GeoJSON file are added and the map is saved to the HTML filename provided by the user. All that's left is to use it.

```
os.chdir(r'D:\Dropbox\Public\webmaps')
make_map('Oklahoma', 'ok.json', 'ok.html',
         zoom_start=7)
```

I used a Dropbox folder so that I could view the output on the web using the Dropbox public link functionality. You probably won't have much luck viewing the output straight from your local drive without using a web server. If you don't have something like Dropbox you can use, check out the sidebar to learn how to start up a simple Python web server on your local machine instead. I wanted to make a map of Oklahoma, and I also passed one of those optional parameters, zoom_start, through to Folium. By default, Folium maps start with a zoom level of 10, which is zoomed in too far to see the entire state. A start level of 7 works much better for this example.

Python SimpleHTTPServer

Python ships with a simple web server that you can use for testing things out, although you probably shouldn't use it for production websites. The easiest way to use it is to open up a terminal window or command prompt, change to the directory that contains the files you want to serve, and then invoke the server from the command line.

For Python 2:

```
D:\>cd dropbox\public\webmaps
D:\Dropbox\Public\webmaps>c:\python27\python -m SimpleHTTPServer
```

For Python 3:

```
D:\>cd dropbox\public\webmaps
D:\Dropbox\Public\webmaps>c:\python33\python -m http.server
```

This will start up a web server running on your local port 8000, so you can get to it in a web browser at http://localhost:8000/. If a file called index.html is in the folder you started the server from (d:\dropbox\public\webmaps, in this case), then that page will automatically be displayed. Otherwise, a list of files in the folder will display, and you can click on one to see it. The URL for the Oklahoma example would be http://localhost:8000/ok.html.

Once you've run the script, you can get the Dropbox public link for ok.html and view it in a web browser. If all went well, it will look something like figure 4.7.

The map in figure 4.7 shows the location of stream gauges, but other than that, it's not too useful. Smaller markers would be nice, and so would popups that provide the gauge reading if you click on the marker. Unfortunately, I don't believe there's a way to do this by adding a GeoJSON file to the map directly, but it's not hard to do manually. Let's add a function to make custom markers, along with a couple of helper

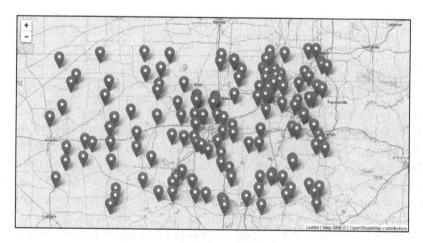

Figure 4.7 A simple Folium map made with a GeoJSON file

functions, and then change the `make_map` function to use those instead of adding the GeoJSON straight to the map.

Listing 4.4 Custom markers for a Folium map

```
colors = {                              ◄─── Colors based
    'action': '#FFFF00',                     on flood status
    'low_threshold': '#734C00',
    'major': '#FF00C5',
    'minor': '#FFAA00',
    'moderate': '#FF0000',
    'no_flooding': '#55FF00',
    'not_defined': '#B2B2B2',
    'obs_not_current': '#B2B2B2',
    'out_of_service': '#4E4E4E'
}

def get_popup(attributes):              ◄─── Create popup text
    """Return popup text for a feature."""   for a feature
    template = '''{location}, {waterbody}</br>
                  {observed} {units}</br>
                  {status}'''
    return template.format(**attributes)

def add_markers(fmap, json_fn):         ◄─── Add markers
    ds = ogr.Open(json_fn)                   to the map
    lyr = ds.GetLayer()
    for row in lyr:
        geom = row.geometry()
        color = colors[row.GetField('status')]
        fmap.circle_marker([geom.GetY(), geom.GetX()],
                           line_color=color,
                           fill_color=color,
                           radius=5000,
                           popup=get_popup(row.items()))

def make_map(state_name, json_fn, html_fn, **kwargs):
    """Make a folium map."""
    geom = get_state_geom(state_name)
    save_state_gauges(json_fn, get_bbox(geom))
    fmap = folium.Map(location=get_center(geom), **kwargs)
    add_markers(fmap, json_fn)          ◄─── Use your
    fmap.create_map(path=html_fn)            new function

os.chdir(r'D:\Dropbox\Public\webmaps')
make_map('Oklahoma', 'ok2.json', 'ok2.html',
         zoom_start=7, tiles='Stamen Toner')
```

The first thing you do here is set up colors to use. These come from the online legend for this map service, which is available at http://gis.srh.noaa.gov/arcgis/rest/services/ahps_gauges/MapServer/0. The keys in the `colors` dictionary are possible values in the `Status` attribute field, and the values are hex strings that describe a color.

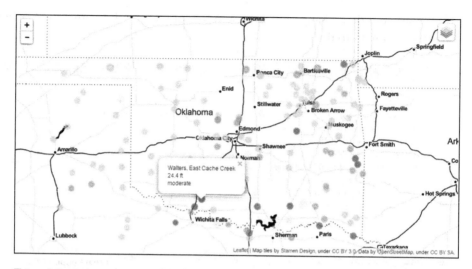

Figure 4.8   A nicer map created by manually constructing colored markers with popups

The get_popup function creates an HTML string by exploding the attributes dictionary for a feature and inserting the values in the corresponding placeholders in a template string. For example, the value from the Location field would get inserted in place of "{location}" in the template string.

The markers are created in the add_markers function, which loops through the GeoJSON layer and creates a marker for each point in the layer. This uses the Folium circle_marker function, which wants a [y, x] list as its first argument. This is where the marker will be placed on the map. You used a different color based on the flood status at that location, and also added a popup to go along with the marker. The radius parameter is the marker radius in pixels. Yours are a little larger than the default.

The last steps are to change the make_map function so that it calls add_markers instead of geo_json, and then to create a new map. This time you use Stamen Toner tiles instead of OpenStreetMap, mostly because the markers are easier to see that way. Your output should look like figure 4.8, and if you click on a marker, you'll see a popup containing the relevant information.

Although it isn't the subject of this book, I hope you enjoyed the short foray into web mapping. If you didn't know anything on the subject and are anything like me, you now have another item on your "to learn" list.

4.3   Testing format capabilities

As mentioned earlier, not all operations are available with all data formats and drivers. How do you find out what's allowed on your data, other than trying it and crossing your fingers that your code doesn't crash? Fortunately, drivers, data sources, and layers are all willing to convey that information if you ask. Table 4.1 shows which capabilities you can check for each of these data types.

Table 4.1   Constants used for testing capabilities

Driver capabilities	OGR constant
Create new data sources	`ODrCCreateDataSource`
Delete existing data sources	`ODrCDeleteDataSource`
DataSource capabilities	**OGR constant**
Create new layers	`ODsCCreateLayer`
Delete existing layers	`ODsCDeleteLayer`
Layer capabilities	**OGR constant**
Read random features using `GetFeature`	`OLCRandomRead`
Add new features	`OLCSequentialWrite`
Update existing features	`OLCRandomWrite`
Supports efficient spatial filtering	`OLCFastSpatialFilter`
Has an efficient implementation of `GetFeatureCount`	`OLCFastFeatureCount`
Has an efficient implementation of `GetExtent`	`OLCFastGetExtent`
Create new fields	`OLCCreateField`
Delete existing fields	`OLCDeleteField`
Reorder fields in the attribute table	`OLCReorderFields`
Alter properties of existing fields	`OLCAlterFieldDefn`
Supports transactions	`OLCTransactions`
Delete existing features	`OLCDeleteFeature`
Has an efficient implementation of `SetNextByIndex`	`OLCFastSetNextByIndex`
Values of string fields are guaranteed to be UTF-8 encoding	`OLCStringsAsUTF8`
Supports ignoring fields when fetching feature data, which can speed up data access	`OLCIgnoreFields`

To check for a given capability, all you have to do is call the `TestCapability` function on a driver, data source, or layer, and pass a constant from table 4.1 as a parameter. The function will return `True` if that operation is allowed and `False` if it isn't. Try using this to determine if you can add new shapefiles to a folder:

```
>>> dirname = r'D:\ osgeopy-data\global'
>>> ds = ogr.Open(dirname)
>>> ds.TestCapability(ogr.ODsCCreateLayer)
False
>>> ds = ogr.Open(dirname, 1)
>>> ds.TestCapability(ogr.ODsCCreateLayer)
True
```

Data source is opened read-only

Data source is opened for writing

As you probably could've guessed, you're allowed to create new layers when the folder has been opened for writing, but not when it has been opened read-only. How could you use this information to make sure you didn't attempt to do something that would cause an error? You can modify your code to add checks before you try to do any editing:

```
ds = ogr.Open(new_fn, 1)
if ds is None:
    sys.exit('Could not open {0}.'.format(new_fn))
lyr = ds.GetLayer(0)

if not lyr.TestCapability(ogr.OLCCreateField):          Check that fields can be added
    raise RuntimeError('Cannot create fields.')

lyr.CreateField(ogr.FieldDefn('ID', ogr.OFTInteger))
```

This snippet will raise an error and not continue if you aren't allowed to add fields to the layer. You could catch and handle this error if you needed to, or let it bail out. If you don't want to handle the errors, the biggest reason for checking beforehand is to make sure that all edits are possible before you start.

For example, what if a layer supported editing fields but not deleting features, and you wanted to do both? If you edited the fields before deleting the features, then part of your changes would take place (the field edits) before your code crashed when trying to delete features. Obviously, this is a problem if you want all or none when it comes to your edits. If partial edits don't bother you, then you may not want to worry about this issue, but you can avoid the problem by checking capabilities beforehand and not proceeding if you're not allowed to make all of your changes.

Another option, if partial edits are okay in your book but you still want to handle errors instead of letting the script crash, is to use OGR exceptions. You wouldn't need to add any code to test capabilities, but you'd need to remember to add `ogr.Use-Exceptions()` somewhere early in your script. Using this approach, the attempt to delete a feature would still fail, but it then throws a `RuntimeError` that you could catch.

A function in the ospybook module called `print_capabilities` will print what capabilities a driver, data source, or layer supports. Here's how to use it from the Python interactive window:

```
>>> driver = ogr.GetDriverByName('ESRI Shapefile')
>>> pb.print_capabilities(driver)
*** Driver Capabilities ***
ODrCCreateDataSource: True
ODrCDeleteDataSource: True
```

Because this function only prints out information, you can't use it in your code to determine what action to take based on available capabilities. You can use it in an interactive window to determine what actions were allowed on an object, though.

4.4   *Summary*

- The vector file format you choose to use might depend on the application. You might go with GeoJSON for making a web map, but use shapefiles or PostGIS for data analysis.
- Perhaps the most popular data transfer format is the shapefile because it's simple, the specifications are public, and it has been around for a long time.
- Formats based on databases, such as SpatiaLite, PostGIS, and Esri geodatabases, tend to be more efficient and support more features than other vector formats.
- Although the syntax for opening various data source types differs, once you have the data source open, you can access the layers and features the same way no matter the source.
- Multiple layers in a data source can be different from one another. For example, they can have different geometry types, attribute fields, spatial extents, and spatial reference systems.
- You can use `TestCapability` to determine which edits are allowed on your dataset.

Filtering data with OGR

This chapter covers

- Efficiently selecting features using attribute values
- Using spatial location to select features
- Joining attribute tables from different layers

Back in chapter 3, you learned how to iterate through all of the features in a layer and use attribute values for each one to determine if it was interesting. You've got easier ways to throw out features that you don't want, however, and that's where filters come in. With filters you can easily select features that match specific criteria, such as all animal GPS locations from a certain day or all crabapple trees from a city tree inventory. Filters also let you limit features by spatial extent, so you could limit your crabapple trees to a specific neighborhood, or GPS locations to those within a kilometer of an animal feeding station. Filtering your data like this makes it easy to extract or process only the features you're interested in. I've used these techniques to extract features such as city boundaries for a single county from a larger dataset, or to extract highways and freeways from road datasets, while ignoring the smaller residential roads.

You can also use SQL queries to join attribute tables together from different layers. For example, if you had a layer containing all of the locations of your store franchises and each feature had an attribute denoting the city that the store was in, then you could join this layer with one containing cities. If the city layer contained demographic information for each city, then that data would be associated with the store data, and you could easily compare demographics between stores.

> **DEFINITION** SQL is short for Structured Query Language, although you'll rarely see it written out like that. If you've used a relational database, then you've probably used SQL, even if you didn't realize it. For example, if you build a graphical query in Microsoft Access, it still builds a SQL query behind the scenes, and you can see it if you switch to SQL View. SQL is featured more prominently in other database software such as PostgreSQL.

5.1 Attribute filters

If you need to limit the features by values contained in one or more attribute fields, then you want an attribute filter. To set one of these filters, you need to come up with a conditional statement that's much like the WHERE clause in a SQL statement. You compare the value of an attribute field to another value, and then all features where that comparison is true are returned. The standard logical operators, such as =, !=, <>, >, <, >=, and <=, allow you to use statements such as the following:

```
'Population < 50000'
'Population >= 25000'
'Type_code != 7'
'Name = "Cairo"'
"Name = 'Moscow'"
'Name != "Tokyo"'
```

Numeric comparisons don't require quotes around the number

String comparisons require either single or double quotes

You can probably guess what these comparisons do; they all test for equality or inequality. Notice that if you're comparing strings, you need to put quotes around the string values, but they can be either single or double. Make sure they're different from the quotes you use to surround the entire query string, or else you'll end your string prematurely and get a syntax error. Don't use quotes with numbers, because that turns them into string values, and you won't get the comparison you were expecting. Another thing you might have noticed is that you use a single equal sign to test for equality, which isn't the way programming languages typically work. But that's the way SQL does things, so who are we to argue? In addition, if you want to test if something doesn't equal another value, you can use either != or <>.

You can also combine statements using AND or OR:

```
'(Population > 25000) AND (Population < 50000)'
'(Population > 50000) OR (Place_type = "County Seat")'
```

The first of these selects features with a population value greater than 25,000 but less than 50,000. The second selects features that either have a population greater than 50,000 or are county seats (or both).

Conditions can be negated using NOT, and NULL is used to indicate a null or no data value in the attribute table:

```
'(Population < 50000) OR NOT (Place_type = "County Seat")'
'County NOT NULL'
```

That first example selects features that either have a population less than 50,000 or aren't county seats. Again, a feature will be selected if it meets one or both of those conditions. The second example selects features that have a value for the County attribute.

If you want to check if a value is between two other values, you can use BETWEEN instead of two different comparisons joined with AND. For example, the following two statements are equivalent, and both select features with a population between 25,000 and 50,000:

```
'Population BETWEEN 25000 AND 50000'
'(Population > 25000) AND (Population < 50000)'
```

You have an easy way to check if a value is equal to one of several different values. Once again, both of these select features where the Type_code value is 4, 3, or 7:

```
'Type_code IN (4, 3, 7)'
'(Type_code = 4) OR (Type_code = 3) OR (Type_code = 7)'
```

This also works for strings:

```
'Place_type IN ("Populated Place", "County Seat")'
```

Last, you can compare strings using the normal logical operators (*a* is less than *c*), or you can do fancier, case-insensitive, string matching using LIKE. This allows you to use wildcards to match any character in a string. An underscore matches any single character and a percent sign matches any number of characters. Table 5.1 shows examples, and this is how you'd use them:

```
'Name LIKE "%Seattle%"'
```

Table 5.1   Match examples using the LIKE operator

Pattern	Matches	Doesn't match
_eattle	Seattle	Seattle WA
Seattle%	Seattle, Seattle WA	North Seattle
%Seattle%	Seattle, Seattle WA, North Seattle	Tacoma
Sea%le	Seattle	Seattle WA
Sea_le	Seatle (note misspelling)	Seattle

If you want to read more about the SQL syntax available in OGR, check out the online documentation at http://www.gdal.org/ogr_sql.html and http://www.gdal.org/ogr_sql_sqlite.html. But for now, let's see how to put this newfound information to use. It will definitely be more fun if you fire up a Python interactive window for testing this out, because you can use the VectorPlotter class to interactively draw your

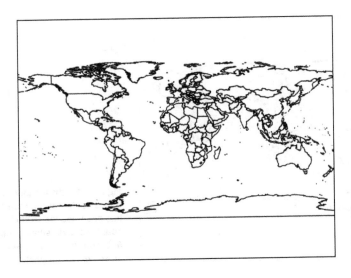

Figure 5.1
The ne_50m_admin_0_countries
shapefile layer in the global data
folder, with no filters applied

selections. After configuring an interactive vector plotter, open the global data folder and grab the low-resolution countries layer:

```
>>> ds = ogr.Open(r'D:\osgeopy-data\global')
>>> lyr = ds.GetLayer('ne_50m_admin_0_countries')
```

Then plot out the features, but be patient if it takes it a few seconds to draw the output shown in figure 5.1, since it has a fair amount of data to plot. Remember that setting fill=False tells it to draw only country outlines.

```
>>> vp.plot(lyr, fill=False)
```

Now inspect the layer attributes by printing out the names of the first few features:

```
>>> pb.print_attributes(lyr, 4, ['name'], geom=False)
FID     name
0       Aruba
1       Afghanistan
2       Angola
3       Anguilla
4 of 241 features
```

Notice that the feature IDs (FIDs) are in order and also the fact that there are 241 features in the layer. Now find out how many of those are in Asia by using an attribute filter. To do this, pass a conditional statement to SetAttributeFilter:

```
>>> lyr.SetAttributeFilter('continent = "Asia"')
0
>>> lyr.GetFeatureCount()
53
```

Now the layer thinks it has only 53 features. The zero that got spit out when you called SetAttributeFilter means that the query executed successfully. Now that you have

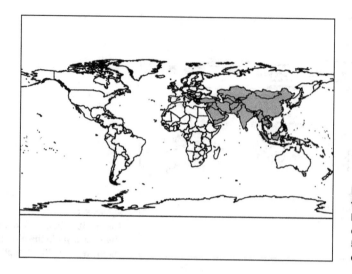

Figure 5.2 An attribute filter that selects countries in Asia has been applied to the countries layer shown in figure 5.1, and the results are plotted on top of the original.

selected the countries in Asia, try drawing them in yellow; your result should look like figure 5.2:

```
>>> vp.plot(lyr, 'y')
```

NOTE TO PRINT BOOK READERS: COLOR GRAPHICS   Many graphics in this book are best viewed in color. The eBook versions display the color graphics, so they should be referred to as you read. To get your free eBook in PDF, ePub, and Kindle formats, go to https://www.manning.com/books/geoprocessing-with-python to register your print book.

You can look a little more closely at what's happening with the filter by printing attributes for the first few features:

```
>>> pb.print_attributes(lyr, 4, ['name'], geom=False)
FID     name
1       Afghanistan
7       United Arab Emirates
9       Armenia
17      Azerbaijan
4 of 53 features
```

Huh. Now you're missing a bunch of FIDs. That's because those features aren't in Asia, so they're ignored while iterating through the layer. Getting features by specific FID doesn't honor the filter, however, because features aren't truly being deleted, and therefore the FID values don't change. You can prove it to yourself by getting a feature or two using FIDs:

```
>>> lyr.GetFeature(2).GetField('name')
'Angola'
```

You can see from this that even though Angola doesn't show up when you iterate through the filtered layer, it's still there. It should be obvious to you now that looping

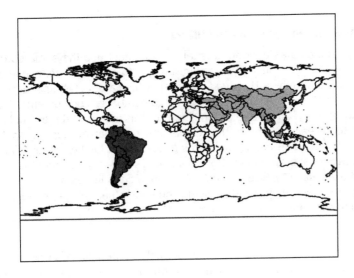

Figure 5.3 An attribute filter that selects countries in South America has been applied to the countries layer and the results plotted on top of the previous data from figure 5.2.

through a filtered layer using specific FIDs is a bad idea and you won't get the desired results. Instead, you need to iterate through the layer using a for loop.

If you set another attribute filter, it doesn't create a subset of the currently filtered features. Instead, the new filter is applied to the entire layer. To illustrate this, try applying a new filter that selects the countries in South America, and then draw them in blue, which results in the shading you see in figure 5.3.

```
>>> lyr.SetAttributeFilter('continent = "South America"')
>>> vp.plot(lyr, 'b')
```

You can, however, use both attribute and spatial filters together to refine your results, and you'll see an example of that in the next section. To clear out the attribute filter and get all 241 features back, simply pass None to SetAttributeFilter:

```
>>> lyr.SetAttributeFilter(None)
>>> lyr.GetFeatureCount()
241
```

Removing the filter also resets the current feature back to the beginning, as if you had just opened the layer.

5.2   *Spatial filters*

Spatial filters let you limit the features by spatial extent rather than attribute value. These filters can be used to select features within another geometry or inside a bounding box. For example, if you had a dataset of global cities with no attribute indicating the country that the cities are in, but you had another dataset with the same spatial reference system that contained the boundary of Germany, you could use a spatial filter to select the German cities.

Spatial reference systems and spatial filters

The geometries or coordinates used for spatial filtering must use the same spatial reference system as the layer you're trying to filter. Why is this? Pretend for a moment that you have a layer that uses a Universal Transverse Mercator (UTM) spatial reference system. Coordinates in that layer would be large numbers, much different than the latitude and longitude values we're all familiar with. This means that they wouldn't align if plotted on top of each other, and they'd appear to have non-overlapping spatial extents. For example, the UTM easting and northing coordinates for the capitol building in Salt Lake City, UT, are approximately 425045 and 4514422, but the corresponding longitude and latitude are -111.888 and 40.777. Those coordinates are awfully different from each other, and they wouldn't overlay on each other unless one of them was transformed to the same spatial reference system as the other.

Try selecting cities in Germany using the natural earth shapefiles. After setting up a vector plotter in an interactive window, open the folder data source and get the countries layer. Then use an attribute filter to limit the countries to Germany and grab the corresponding feature and geometry:

```
>>> ds = ogr.Open(r'D:\osgeopy-data\global')
>>> country_lyr = ds.GetLayer('ne_50m_admin_0_countries')
>>> vp.plot(country_lyr, fill=False)
>>> country_lyr.SetAttributeFilter('name = "Germany"')
>>> feat = country_lyr.GetNextFeature()
>>> germany = feat.geometry().Clone()
```

You can assume, in this case, that the attribute filter will return one and only one feature, so using GetNextFeature will get the first and only feature in the filtered results. Then you grab the geometry and clone it so that you can use the geometry even after

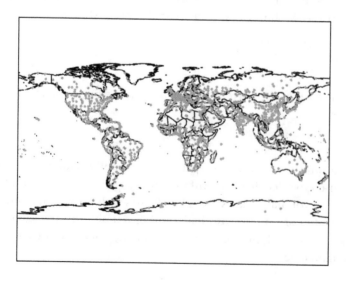

Figure 5.4 All of the cities in the populated_places layer in the natural earth dataset

the feature is removed from memory. Oh, and you also plot the world countries before applying the filter so that you have context for the cities later on. Now open the populated places layer and plot all cities (see figure 5.4) as yellow dots:

```
>>> city_lyr = ds.GetLayer('ne_50m_populated_places')
>>> city_lyr.GetFeatureCount()
1249
>>> vp.plot(city_lyr, 'y.')
```

The call to `GetFeatureCount` indicates there are 1,249 city features in the full layer. Now try applying a spatial filter by passing the `germany` geometry that you got earlier to `SetSpatialFilter`, and then plot the resulting cities as large dots:

```
>>> city_lyr.SetSpatialFilter(germany)
>>> city_lyr.GetFeatureCount()
5
>>> vp.plot(city_lyr, 'bo')
```

Now the layer claims to have only five features, so five cities fall within the German boundary polygon. You can also see from your plot that the circles fall in the correct geographical area. You can use the Zoom to rectangle tool on the bottom of the plot window to zoom in on Germany if you'd like (figure 5.5).

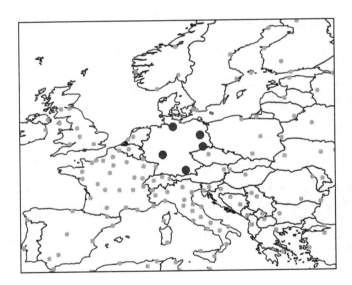

Figure 5.5 A spatial filter has been applied to the populated_places layer to limit the features to those within the boundaries of Germany. These filtered points are shown in as large dots.

To clone or not to clone?

Geometry objects have a `Clone` function, which makes a copy of the object. Why would you want to use this? When you get a geometry from a feature, that geometry is still associated with that feature. If that feature is then deleted (or the variable is populated

(continued)

with a different feature), then the geometry is no longer useable. In fact, if you try to use it, Python will crash instead of spit out an error. This problem is easy to solve, however, by cloning the geometry. Now you can store a copy of the feature or geometry that's no longer associated with other objects and will live on even if the parent objects disappear. Want to see this in action? Try this in an interactive window:

```
>>> ds = ogr.Open(r'D:\osgeopy-data\global\natural_earth_50m.sqlite')
>>> lyr = ds.GetLayer('countries')
>>> feat = lyr.GetNextFeature()
>>> geom = feat.geometry()
>>> geom_clone = feat.geometry().Clone()          Delete the feature that the
>>> feat = lyr.GetNextFeature()                    geometries came from
>>> print(geom_clone.GetArea())
0.014118879217099978
>>> print(geom.GetArea())          Python crashes
```

In this example, the geom variable holds a Geometry object that's still owned by the Feature object stored in the feat variable, but the geom_clone variable holds a geometry that has been disassociated from that feature. After you populate the feat variable with a different feature, you can still use the geom_clone geometry, but not the object stored in the geom variable, because you no longer have a handle to the feature that it came from.

Incidentally, this is related to why all of these examples would also cause Python to crash:

```
feat = ogr.Open(fn, 0).GetLayer(0).GetNextFeature()
# or
lyr = ogr.Open(fn, 0).GetLayer(0)
feat = lyr.GetNextFeature()
# or
ds = ogr.Open(fn, 0)
lyr = ds.GetLayer(0)
del ds
feat = lyr.GetNextFeature()
```

In each case, the data source has gone out of scope or been deleted before you try to use the layer. But the layer is associated with the data source and becomes unusable once the data source is gone, the same way a geometry becomes unusable if its parent feature disappears. You should never close your data source if you still need access to the layer.

As promised, you now get to combine a spatial and an attribute query. Further refine your selection by finding the cities with a population over 1,000,000, and draw them as the squares shown in figure 5.6:

```
>>> city_lyr.SetAttributeFilter('pop_min > 1000000')
>>> city_lyr.GetFeatureCount()
3
>>> vp.plot(city_lyr, 'rs')
```

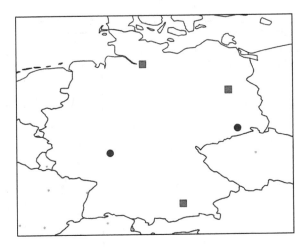

Figure 5.6 An attribute filter has been combined with a spatial filter to select the German cities with a population greater than 1,000,000 people. The selected features are shown as squares instead of circles.

Judging from these results, there are three German cities with populations of more than 1,000,000 people. Figure 5.6 shows the output plot zoomed in on Germany so you can see these features. But what if you decide that you want to know how many cities exist in the entire world with a population that large? All you have to do is remove the spatial filter by passing None to SetSpatialFilter. Note that the attribute filter will still be in effect. Go ahead and try it, drawing the results as triangles:

```
>>> city_lyr.SetSpatialFilter(None)
>>> city_lyr.GetFeatureCount()
246
>>> vp.plot(city_lyr, 'm^', markersize=8)
```

And now you know where the largest cities in the world are (figure 5.7).

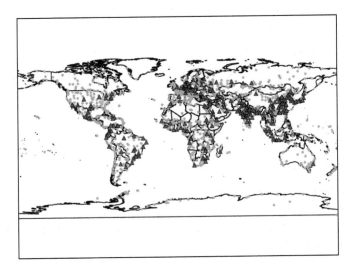

Figure 5.7 The spatial filter has been removed, but the attribute filter is still in effect, so now all of the cities in the world with a population of more than 1,000,000 are drawn as triangles over the top of the original dots for all cities.

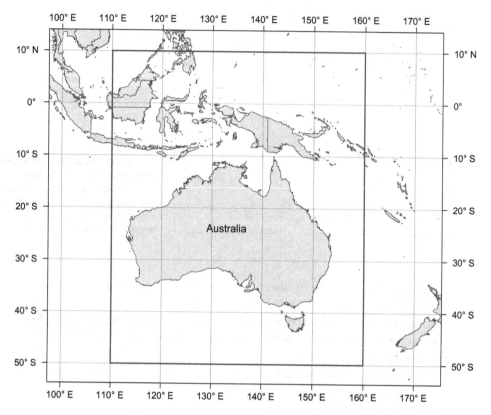

Figure 5.8 The minimum and maximum x and y values for the rectangle surrounding Australia can be used to set a spatial extent on the global countries layer.

You're not completely out of luck if you'd like to filter features spatially but don't have a geometry to use. You can also use a rectangular extent by providing the minimum and maximum x and y coordinates:

```
SetSpatialFilterRect(minx, miny, maxx, maxy)
```

You can use this to select the countries that fall within the box shown in figure 5.8. Again, start by plotting all of the countries:

```
>>> vp.clear()
>>> country_lyr.SetAttributeFilter(None)
>>> vp.plot(country_lyr, fill=False)
```

Now plug in the bounding coordinates shown in figure 5.8:

```
>>> country_lyr.SetSpatialFilterRect(110, -50, 160, 10)
>>> vp.plot(country_lyr, 'y')
```

Now you should have a plot that looks similar to figure 5.9, with Australia and a few surrounding countries shaded in.

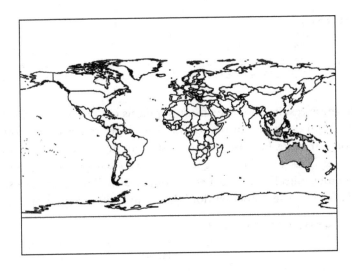

Figure 5.9 The shaded countries were selected using the rectangular extent shown in figure 5.8.

TIP To clear a spatial filter, whether it was created with a geometry or a bounding box, pass None to SetSpatialFilter. You can't clear the filter using SetSpatialFilterRect.

5.3 *Using SQL to create temporary layers*

If you're familiar with SQL, or are willing to learn, you can create more-complicated queries and do fun stuff using the ExecuteSQL function on a data source. This function applies to a data source instead of a layer because it allows you to use multiple layers if desired. It requires a SQL query and can optionally use a geometry as a spatial filter. In addition, you can also specify a different SQL dialect, but more on that later. Here's the signature:

```
ExecuteSQL(statement, [spatialFilter], [dialect])
```

- statement is the SQL statement to use.
- spatialFilter is an optional geometry to use as a spatial filter on the results. The default is no filter.
- dialect is a string specifying the SQL dialect to use. Available options are OGRSQL and SQLite. The default is to use the OGR dialect unless the data source has its own SQL engine (such as a SpatiaLite database).

This function is different from the filtering functions in that it returns a new layer containing the result set rather than only filtering features out of the existing layer. Let's look at a few examples using this technique, starting with a simple one that returns global countries sorted by population in descending order:

```
>>> ds = ogr.Open(r'D:\osgeopy-data\global')
>>> sql = '''SELECT ogr_geom_area as area, name, pop_est
...             FROM 'ne_50m_admin_0_countries' ORDER BY POP_EST DESC'''
>>> lyr = ds.ExecuteSQL(sql)
```

```
>>> pb.print_attributes(lyr, 3)
FID     Geometry        area                    name            pop_est
41      MULTIPOLYGON    950.9810937547769       China           1338612970.0
98      MULTIPOLYGON    278.3474038553223       India           1166079220.0
226     MULTIPOLYGON    1115.1781907153158      United States   313973000.0
3 of 241 features
```

As you can see from these results, the three most populous countries in the world are China, India, and the United States, in that order. The query returns each country's name and population attributes because you request them in the SQL statement. You also use the special ogr_geom_area field to get the area of each geometry (table 5.2), and the FID and geometry itself are returned automatically. This example uses the default OGR SQL dialect because shapefiles don't have any built-in SQL support.

Table 5.2 Special fields used in the OGR SQL dialect

Field	Returns
FID	The feature ID.
OGR_GEOMETRY	An OGR geometry type constant (see table 3.1). This is especially useful for data formats that support multiple geometry types in one layer.
OGR_GEOM_WKT	The well-known text (WKT) representation of the feature's geometry.
OGR_GEOM_AREA	The area of the feature's geometry. Returns zero for geometries with no area (for example, points or lines).
OGR_STYLE	The style string for the feature, if it exists. Very few applications use this.

If you're querying a data source that has its own SQL support, that native SQL version will be used. For example, if you have the SQLite driver, you could get the same information from the natural_earth_50m.sqlite database using the SQLite version of SQL. This dialect also allows you to limit the number of returned features, so you could limit the result set to the three countries with the highest populations:

```
>>> ds = ogr.Open(r'D:\osgeopy-data\global\natural_earth_50m.sqlite')
>>> sql = '''SELECT geometry, area(geometry) AS area, name, pop_est
...             FROM countries ORDER BY pop_est DESC LIMIT 3'''
>>> lyr = ds.ExecuteSQL(sql)
>>> pb.print_attributes(lyr)
FID     Geometry        area                    name            pop_est
0       MULTIPOLYGON    950.9810937547769       China           1338612970.0
1       MULTIPOLYGON    278.3474038553223       India           1166079220.0
2       MULTIPOLYGON    1115.1781907153158      United States   313973000.0
3 of 3 features
```

This time you could print attributes for the entire layer, because only three features are returned. You should also notice that now you use the area function instead of a special field name, and if you don't rename it with the AS area syntax, then it would be called area(geometry) instead. You also have to specifically request the geometry because the SpatiaLite engine doesn't return the geometry by default.

You can also use `ExecuteSQL` to join attributes from multiple layers. Take a look at this code and see if you can figure out what it's doing:

```
ds = ogr.Open(r'D:\osgeopy-data\global')
sql = '''SELECT pp.name AS city, pp.pop_min AS city_pop,
            c.name AS country, c.pop_est AS country_pop
        FROM ne_50m_populated_places pp
        LEFT JOIN ne_50m_admin_0_countries c
        ON pp.adm0_a3 = c.adm0_a3
        WHERE pp.adm0cap = 1'''
lyr = ds.ExecuteSQL(sql)
```

> **Rename the layers**

The first thing to notice is that you use the ne_50m_populated_places and ne_50m_admin_0_countries shapefiles and rename them to pp and c, respectively. You do this by putting the alias directly after the layer name. This isn't necessary, of course, but does make your SQL a lot shorter because those layer names are pretty long. You also link these two layers together by using a join, which allows you to link tables using a shared attribute. Here you use a LEFT JOIN to keep all records in the table on the left (populated places), and if a matching record exists in the table on the right (countries), then you'll also get data from that record. But how does it figure out what matches? That's where the ON clause comes in. For each feature in pp, it takes the adm0_a3 attribute value and tries to find a feature in the countries layer that has the same value for its adm0_a3 field. See figure 5.10 for an illustration.

ne_50m_populated_places

FID	NAME	ADM0CAP	ADM0_A3	POP_MIN
47	Douglas	0	IMN	26,218
48	San Marino	1	SMR	29,000
49	Willemstad	0	CUW	146,813
50	Oranjestad	0	ABW	33,000
51	Vaduz	1	LIE	5,342

ne_50m_admin_0_countries

FID	ADM0_A3	NAME	POP_EST	CONTINENT
126	LIE	Liechtenstein	34,761	Europe
127	LKA	Sri Lanka	21,324,791	Asia
...				
195	SLV	El Salvador	7,185,218	North America
196	SMR	San Marino	30,324	Europe

Figure 5.10 An illustration of a SQL query that selects records from the populated places table where adm0cap equals 1, and then gets related data from the countries table based on the adm0_a3 field in both tables.

Now that you know what tables the data are coming from, go back to the beginning of the SQL statement and look at what attribute fields are being requested. You ask for the NAME and POP_MIN fields from the populated places layer, as well as the NAME and POP_EST fields from the countries layer. Because the fields from the two layers have the same names, it makes sense to rename them so that you can tell what's what. Last, you use a WHERE clause to limit the results to features that represent capital cities (adm0cap = 1).

This technique is handy if you want to see related data from multiple layers at the same time. Without this, you could query city populations and country populations separately, but now you can see the country's population right beside the city's. To see this, look at the layer returned by this query:

```
pb.print_attributes(lyr, 3, geom=False)
FID    city              city_pop    country          country_pop
7      Vatican City      832         Vatican          832.0
48     San Marino        29000       San Marino       30324.0
51     Vaduz             5342        Liechtenstein    34761.0
3 of 200 features
```

I didn't print the geometry column because it wouldn't fit comfortably on the page, but because this uses the OGR SQL dialect, the geometry is returned automatically. But which one: the city or the country? It's the city, because that's the main table being used in the join, and corresponding country information is returned only if it existed for a city. You could plot the layer to prove it to yourself if you'd like.

Now check out a similar example using the SQLite dialect, but still shapefile data sources (you could use a SQLite database, of course, but I want to prove that the SQLite dialect will work with other data source types). See if you can spot the differences:

```
ds = ogr.Open(r'D:\osgeopy-data\global')
sql = '''SELECT pp.name AS city, pp.pop_min AS city_pop,
            c.name AS country, c.pop_est AS country_pop
         FROM ne_50m_populated_places pp
         LEFT JOIN ne_50m_admin_0_countries c
         ON pp.adm0_a3 = c.adm0_a3
         WHERE pp.adm0cap = 1 AND c.continent = "South America"'''
lyr = ds.ExecuteSQL(sql, dialect='SQLite')
pb.print_attributes(lyr, 3)
```

The most obvious difference is the inclusion of the dialect parameter to the ExecuteSQL function. But you also add one thing to the SQL that doesn't work with the OGR dialect. This time the results are limited to cities in South America by checking the value of the continent field in the countries layer. The OGR dialect doesn't support using fields from the joined table in the WHERE clause, so the only attributes allowed would be ones from the populated places layer. Also, because you need to specifically request geometries if you want them when using the SQLite dialect, no geometries are returned by this particular query. You could add them in by specifying pp.geometry along with the other fields.

If your version of OGR was built with SpatiaLite support (not only SQLite), you can also manipulate geometries within your SQL. Be warned that this could take a while, depending on what you try to do. As an example, if you have SpatiaLite support, try merging all of the counties in California into one big geometry. Start with drawing the individual counties so you have something to compare your results with:

```
>>> ds = ogr.Open(r'D:\osgeopy-data\US')
>>> sql = 'SELECT * FROM countyp010 WHERE state = "CA"'
>>> lyr = ds.ExecuteSQL(sql)
>>> vp.plot(lyr, fill=False)
```

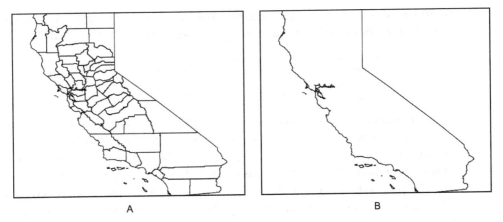

A   B

Figure 5.11   Part A, on the left, shows the counties in California drawn individually. Part B, on the other hand, shows the result of running the SpatiaLite st_union function on the counties. They're all joined together into one geometry.

This will draw a map of the counties in California, as shown in figure 5.11A. Now try using the SpatiaLite st_union function to merge all of the county polygons into one, as shown in figure 5.11B:

```
>>> sql = 'SELECT st_union(geometry) FROM countyp010 WHERE state = "CA"'
>>> lyr = ds.ExecuteSQL(sql, dialect='SQLite')
>>> vp.plot(lyr, 'w')
```

Geometry operations also work with data sources that have their own native SQL flavor and the ability to perform geometry manipulations. SpatiaLite and PostGIS are two obvious examples of this. For example, this is how you'd do the same thing with a Post-GIS data source:

```
conn_str = 'PG:host=localhost user=chrisg password=mypass dbname=geodata'
ds = ogr.Open(conn_str)
sql = "SELECT st_union(geom) FROM us.counties WHERE state = 'CA'"
lyr = ds.ExecuteSQL(sql)
vp.plot(lyr)
```

Don't worry if you want to perform operations like this but aren't using PostGIS or SpatiaLite, because you'll learn how to do it without databases in the next chapter.

5.4   *Taking advantage of filters*

Remember back in chapter 3 when you copied all of the capital cities in a global shapefile into a new shapefile? You looped through each feature in the shapefile, checked the appropriate attribute, and copied the feature if it was a capital city. This whole process can be made much easier if the features you want can be selected with filters. Do you remember the CopyLayer method that was introduced in section 4.2.4? As a reminder, it copies an existing layer into a new data source. How do you think you could use this to do something similar to the code back in listing 3.3, but much easier? Think about this problem for a minute and then look at the next example:

```
ds = ogr.Open(r'D:\osgeopy-data\global', 1)
in_lyr = ds.GetLayer('ne_50m_populated_places')
in_lyr.SetAttributeFilter("FEATURECLA = 'Admin-0 capital'")
out_lyr = ds.CopyLayer(in_lyr, 'capital_cities2')
```

Here the call to `CopyLayer` makes a copy of `in_lyr` in the `ds` data source. In this case, it happens to be the same data source as the original layer, but it could be any data source. Because you've already set an attribute filter on `in_lyr`, only the filtered features are copied. That's certainly easier than checking each one.

If you only want certain attributes, you could use a layer created using `ExecuteSQL`. Write a SQL query that pulls out the attributes you want and copy the results to a new layer:

```
sql = """SELECT NAME, ADM0NAME FROM ne_50m_populated_places
         WHERE FEATURECLA = 'Admin-0 capital'"""
in_lyr2 = ds.ExecuteSQL(sql)
out_lyr2 = ds.CopyLayer(in_lyr2, 'capital_cities3')
```

It should be obvious by now that you can simplify your life by taking advantage of filters and the `ExecuteSQL` function whenever possible.

5.5   *Summary*

- Attribute filters can be used to efficiently select specific features based on their attribute values.
- Spatial filters allow you to select features based on their location by using a bounding polygon or coordinates for a bounding box. The coordinates to set a spatial filter must use the same spatial reference system as the data to be filtered.
- Spatial and attribute filters can be combined.
- You can use SQL queries to create temporary layers made up of multiple layers joined on attribute values.
- You can't use objects once their owners go out of scope, so if you want to use a geometry after you've lost the handle to its feature, make sure you clone the geometry. Always keep your data source open if you want access to the layers. If you break one of these rules, Python will crash and burn.

Manipulating
geometries with OGR

This chapter covers

- Creating points, lines, and polygons from scratch
- Editing existing geometries

Thus far we've talked about using OGR to read and write vector datasets and how to edit attribute values, but you haven't manipulated the geometries in any way. If you want to create your own data and not use someone else's, you'll need to know how to work with the actual geometries. For example, if you have a time series of GPS coordinates from a hiking or bicycling trip, you can create a line that represents the route you took. You can even compare the timestamps from the GPS locations to the timestamps on the photos you took to create a point dataset showing where you stopped to take pictures.

You might even need to know how to manipulate geometries to better display existing data. For example, say you want to create a map using your photo points and link them to the actual photos. Certain locations will probably have multiple photos in the same spot. You can deal with this many ways, but one way is to offset each point a little in a different direction so that it looks like a cluster of points instead of only one. But to do this, you need to know how to manipulate the point geometries themselves.

You can also manipulate and combine geometries to create new ones. For example, if you want to create a simple map of riparian areas from a stream dataset, and you assume that the riparian zones stretch one meter on either side of a stream, you can create a polygon that surrounds each stream, with the edges of the polygon one meter out on each side of the water. If two streams join, then these polygons will overlap near their confluence, and you can combine the overlapping polygons into one using a union operation. You'll learn how to do all of this, and more, in the next two chapters.

Before you can do any of this, however, you need to be acquainted with the different types of geometries.

6.1 *Introduction to geometries*

You have several kinds of geometries with which you can work, points being the simplest. All other types are made up of points connected by straight line segments, and the points are what store the coordinate values, so points can be thought of as the building blocks of geometries. These points that are used to build other geometries are called *vertices*, and there can be thousands of vertices per geometry, if needed. For example, line geometries are an ordered collection of points that are connected by straight line segments, with a vertex at every location where the line needs to change direction. A line representing a short dead-end street wouldn't require many vertices, but one representing the Amazon River in much detail would need thousands. Polygons are somewhat similar to lines, but they're closed, meaning the first and last vertex are identical and they enclose a specific area. You'll start with creating and editing points and work your way up to the more complicated geometries as you go along.

> **DEFINITION** A vertex is a point where two line segments of a geometry meet. Vertices hold coordinates for the ends of each line segment.

Although many geometries live only in a two-dimensional (2D) Cartesian coordinate plane with x and y coordinates, it's also possible to have three-dimensional (3D) geometry objects with z values. These z values are typically used to represent elevation, but can also be used for other data, such as maximum annual temperature. Technically, these geometries are considered 2.5D instead of 3D in OGR, because OGR doesn't take the z values into account when performing spatial operations. One thing to be aware of is that although you can add z values to 2D geometries, they'll be ignored when writing the data to a file.

> **NOTE** Geometries with only x and y coordinates are considered 2D. Geometries with an additional z coordinate are considered 2.5D instead of 3D in OGR because the z values aren't taken into account when performing spatial operations.

It's probably easiest to work with simple geometries when you're beginning, so you'll learn to create different geometry types by re-creating the fictional yard shown in figure 6.1. If you use your imagination when looking at this figure, hopefully you can

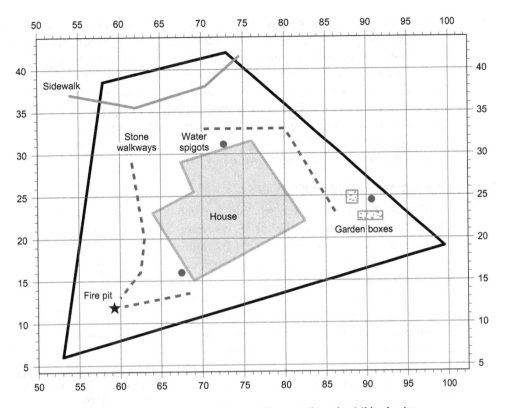

Figure 6.1 **The fictional yard whose geometries you'll create throughout this chapter.**
Coordinates are in meters.

envision a yard with a house in the middle, rectangular garden beds to the east, a sidewalk on the north side (solid line), stone pathways (dotted lines), a fire pit (star), and outdoor water spigots (circles). Although you'll create this scenario in two-dimensional space, the same concepts apply to 2.5D geometries.

Although the shapes shown in figure 6.1 are simple, the concepts are exactly the same as working with complex geometries. You can apply the material you learn here to real-world scenarios.

6.2 Working with points

Points consist of an east/west x coordinate, a north/south y coordinate, and sometimes a vertical z coordinate that's commonly used for elevation. You're probably familiar with the x coordinate being called *longitude* and the y coordinate called *latitude*. These terms are appropriate when a geographic coordinate system is being used, where latitude ranges from -90 to 90 and longitude is between -180 and 180. If the coordinates have been projected into a Cartesian coordinate system, such as UTM, then common terms are *easting* for the x coordinate and *northing* for the y.

Points are used to represent items that have only one set of coordinates. Points don't have a length, width, area, or any other measurement. Despite this, the features represented by points on a map vary depending on scale, and those features might have an area in real life. For example, a map of France would most likely represent Paris as a single point, while a map of the Île-de-France region would show more detail, with the Paris city boundaries represented as polygons and major landmarks such as the Eiffel Tower shown as points. As the scale changes, the areas of features represented by points will also change, similar to the way the area covered by the Eiffel Tower is much smaller than that covered by Paris.

6.2.1   Creating and editing single points

Looking at the yard diagram, you can see that the fire pit is a perfect candidate to be represented as a single point, so let's build it. Figure 6.2 shows a close-up of the area so that you can see the coordinates.

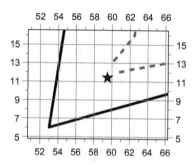

Figure 6.2   You can use a single point to hold the fire pit geometry, shown here as a star.

Unless you have a text representation of the geometry you want to build, the first step to building any type of geometry with OGR is to create an empty Geometry object using one of the constants from table 6.1. Go ahead and do this in an interactive window so that you can get immediate results:

```
>>> firepit = ogr.Geometry(ogr.wkbPoint)
```

Once you have the geometry, you can start adding vertices. Points only have one vertex, which you add with the AddPoint function. This function wants x, y, and an optional z.

```
>>> firepit.AddPoint(59.5, 11.5)
```

That's it! You now have a fully functional point object with a northing of 11.5 and an easting of 59.5. The coordinates can be retrieved if needed using GetX, GetY, and GetZ:

```
x, y = firepit.GetX(), firepit.GetY()
```

Remember that in Python you can set multiple variables at once, so x is assigned the results of GetX and y gets the results of GetY.

Table 6.1   OGR constants denoting geometry types

Geometry type	2D constant	2.5D constant
Point	wkbPoint	wkbPoint25D
Multipoint	wkbMultiPoint	wkbMultiPoint25D
Line	wkbLineString	wkbLineString25D

Table 6.1 OGR constants denoting geometry types *(continued)*

Geometry type	2D constant	2.5D constant
Multiline	wkbMultiLineString	wkbMultiLineString25D
Polygon ring	wkbLinearRing	n/a
Polygon	wkbPolygon	wkbPolygon25D
Multipolygon	wkbMultiPolygon	wkbMultiPolygon25D
Geometry collection	wkbGeometryCollection	wkbGeometryCollection25D

You can also print geometry objects in WKT format if you want to verify that things look okay, although this can get ugly pretty fast with geometry types other than points.

```
>>> print(firepit)
POINT (59.5 11.5 0)
```

Notice that the WKT shows a z value of 0, but you created a 2D point. That's not going to hurt anything, so there's no reason to worry about it. You could even set a z value yourself, although it would be ignored when it came time to write the geometry out to a file.

Unless you want to see the coordinate values, an easier way to visualize your geometries as you create them is to use the VectorPlotter class that we introduced in chapter 3, although this is boring with a single point:

```
>>> vp.plot(firepit, 'bo')
```

What if you realize later that your GPS was slightly off and the y coordinate is 13 instead of 11.5? The easiest way to solve the problem is to call AddPoint again, but with the correct coordinates. You can verify the results by plotting the new geometry with a different marker (figure 6.3) or by printing the WKT:

```
>>> firepit.AddPoint(59.5, 13)
>>> vp.plot(firepit, 'rs')
>>> print(firepit)
POINT (59.5 13.0 0)
```

Why doesn't this add a second set of coordinates to the geometry, as the name AddPoint implies? Points are a special case because they're only allowed one set of coordinates, so any existing ones are overwritten. You'll see later that AddPoint has different behavior when applied to other geometry types.

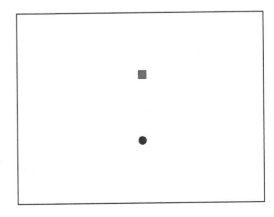

Figure 6.3 The original and edited fire pit geometries. The edited one has the square marker.

If you like to make life a little more complicated, or want to be more consistent with how vertices are edited with other geometry types, you can use SetPoint(point, x, y, [z]) instead, where point is the index of the vertex to edit. Because point geometries contain only one vertex, this parameter is always zero when dealing with points:

```
firepit.SetPoint(0, 59.5, 13)
```

To create a 2.5D point, specify the 2.5D type when you create it and then provide a z coordinate along with the x and y:

```
firepit = ogr.Geometry(ogr.wkbPoint25D)
firepit.AddPoint(59.5, 11.5, 2)
```

Other than the addition of a third coordinate value, working with 2.5D points is the same as working with 2D points.

6.2.2   Creating and editing multipoints: multiple points as one geometry

Multipoint geometries contain one or more points in a single object. This means that multiple points can be attached to a single feature rather than requiring a separate feature per point. For example, a dataset with multiple points might be the locations of all fire hydrants within city boundaries, where each hydrant is treated as a unique feature. Perhaps you also want to map outdoor water faucets in private yards. In this case, you might treat all spigots in an individual yard as one multipoint item so that you have only one feature per yard. In fact, that's what you'll do with the yard example. Three faucets are shown as circles in figure 6.4, and you'll build one multipoint object with three vertices to represent them.

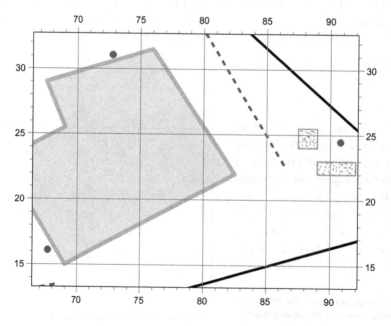

Figure 6.4   You can use a multipoint geometry to hold the water spigot geometries, shown here as dots.

To create a multipoint geometry, you need to create at least two geometries. You need at least one point object and also a multipoint object to hold the points. Referring back to table 6.1, you can see that the correct OGR constant for a multipoint object is wkbMultiPoint. You create the points exactly as before and then add them to your multipoint geometry. Here's one way to do this, using coordinates obtained from figure 6.4:

Create a 2D multipoint

```
>>> faucets = ogr.Geometry(ogr.wkbMultiPoint)
>>> faucet = ogr.Geometry(ogr.wkbPoint)
>>> faucet.AddPoint(67.5, 16)
>>> faucets.AddGeometry(faucet)
>>> faucet.AddPoint(73, 31)
>>> faucets.AddGeometry(faucet)
>>> faucet.AddPoint(91, 24.5)
>>> faucets.AddGeometry(faucet)
```

Create a point and add to multipoint

Reuse the point object

Notice that it's fine to reuse the same point geometry each time. A copy of the point object is added to the multipoint when AddGeometry is invoked, so the original point can be edited later without affecting the coordinates that have already been added to the multipoint. You could, of course, create a new point object for each vertex, but reusing the geometry saves a little overhead.

Once again, you can plot the geometry and print the WKT to see what it looks like:

```
>>> vp.clear()
>>> vp.plot(faucets, 'bo')
>>> vp.zoom(-5)
>>> print(faucets)
MULTIPOINT (67.5 16.0 0,73 31 0,91.0 24.5 0)
```

Zoom out so all points are visible

With a multipoint object, the WKT string separates each coordinate in the set with a comma. As with a regular point object, the x, y, and z coordinates for a single point are separated by spaces. Notice also that the vertices are listed in the same order you added them. That's extremely important if you need to access one later—you can always be sure of which point you're getting because their order doesn't change.

You can get a specific point from a multipoint geometry by passing the index of the desired point to GetGeometryRef. The first point added has index 0, the second has index 1, and so on. Once you have an individual point, you can edit it the same way as a single point. Because GetGeometryRef returns a reference to the point inside the multipoint instead of a copy, the multipoint is automatically updated when the point is changed. For example, this would get the second faucet and then edit its coordinates:

```
faucets.GetGeometryRef(1).AddPoint(75, 32)
```

You can also find out how many points are in a multipoint object, which is useful if you need to loop through them all. For example, to move all spigots two meters to the east, you'd need to loop through the points and add 2 to each x coordinate while leaving the y coordinates unchanged. The results are shown in figure 6.5.

```
>>> for i in range(faucets.GetGeometryCount()):
...         pt = faucets.GetGeometryRef(i)
...         pt.AddPoint(pt.GetX() + 2, pt.GetY())
...
>>> vp.plot(faucets, 'rs')
>>> vp.zoom(-5)
```

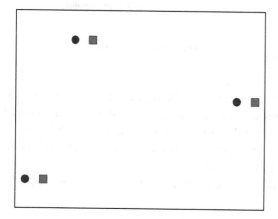

Figure 6.5   The original and edited water spigot multipoint geometries. The edited geometry uses square markers.

As you'll see in the rest of this chapter, working with other geometry types directly builds on these concepts that you've learned for points.

6.3   *Working with lines*

As mentioned earlier, lines are a sequence of vertices, or points, connected by straight line segments. Figure 6.6 shows a line with its vertices, although normally you don't see markers for the vertices when a line is drawn. A line can't change direction, no matter how slightly, without a vertex to end one segment and start another. Therefore, a line that looks like a smooth curve is a large number of short straight segments, all joined together by vertices.

Adding more vertices, and therefore a larger number of shorter segments, gives you more control over the shape of the line. Think about how you'd draw the coastline of

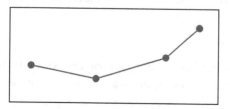

Figure 6.6   A line and the vertices connecting each segment

Great Britain using a series of straight lines. As you can see from figure 6.7, accuracy is greatly improved by using shorter lines. The same concept applies to any line geometry. The more detail required, the more vertices you need to add. Keep in mind that the more vertices you have, the more complicated the geometry object is and the more time it takes to process, so don't add unnecessary vertices. In fact, you might want to simplify geometries so that they use fewer vertices if you're going to serve data over the web. See the `Simplify` function in appendix C if you need to do this. (Appendixes C through E are available online at the Manning Publications website at https://www.manning.com/books/geoprocessing-with-python.)

Figure 6.7 The solid line follows the coast of Great Britain more closely because it has more vertices, and therefore more and shorter line segments, than the dotted line. More detail for lines and polygons can be achieved by using more vertices.

Lines can be used to represent linear features such as roads, streams, or pipelines. A line is a good choice if you want to show one coastline of an island, such as the example in figure 6.7, but a polygon is a better choice if you want to represent the entire island. You'll use a simple line object to model the sidewalk bordering the make-believe yard and the oddly shaped parking strip (figure 6.8).

The line you'll build for the sidewalk contains a small number of vertices, but the technique for working with longer and more-complex lines is exactly the same.

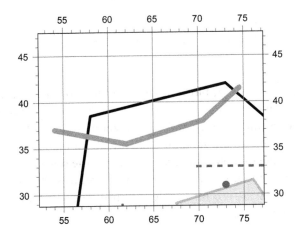

Figure 6.8 You can use a line to hold the sidewalk geometry, shown here as the thick green line.

6.3.1 *Creating and editing single lines*

As with points, the first step to creating a line geometry is to create an empty Geometry object and then add the vertices. Although the direction you traverse the line when adding coordinates isn't important, the vertices must be added in order. Try creating the line representing the sidewalk shown in figure 6.8, going from west to east:

```
>>> sidewalk = ogr.Geometry(ogr.wkbLineString)          ◄──── Create a 2D line
>>> sidewalk.AddPoint(54, 37)
>>> sidewalk.AddPoint(62, 35.5)          Add vertices from
>>> sidewalk.AddPoint(70.5, 38)          west to east
>>> sidewalk.AddPoint(74.5, 41.5)
```

Remember how AddPoint overwrote existing coordinates in point objects? That doesn't happen here because lines consist of many vertices instead of only one, so a new vertex is added to the end of the line instead of overwriting the only allowed point.

Again, you can verify that things are working as expected by plotting the geometry or printing the WKT:

```
>>> vp.plot(sidewalk, 'b-')
>>> print(sidewalk)
LINESTRING (54 37 0,62.0 35.5 0,70.5 38.0 0,74.5 41.5 0)
```

As before, the coordinates for a vertex are separated by spaces, and individual vertices are separated by commas. Because you know that the vertices are always in the same order that they were added to the line, you could use SetPoint to change the x coordinate for the last vertex (the one with index 3) in the sidewalk:

```
sidewalk.SetPoint(3, 76, 41.5)
```

You can find out how many vertices a line contains and then loop through all of them if necessary. For example, if you suddenly realize that the sidewalk is one meter too far south, you can nudge it north by looping through all of the vertices and adding one to each y coordinate (figure 6.9):

```
>>> for i in range(sidewalk.GetPointCount()):
...      sidewalk.SetPoint(i, sidewalk.GetX(i), sidewalk.GetY(i) + 1)
...
>>> vp.plot(sidewalk, 'r--')
```

Notice that you use GetX and GetY again, but now you need to provide the index of the vertex you want. You also might wonder why you use GetPointCount instead of GetGeometryCount as you did for multipoints, and that's a good question. The reason is because GetGeometryCount tells you how many individual geometry objects are combined to make up one multigeometry, and it returns zero if the object isn't a multigeometry. The GetPointCount function, on the other hand, returns the number of vertices in a geometry, and it returns zero for multigeometries because they're made of other geometries instead of vertices.

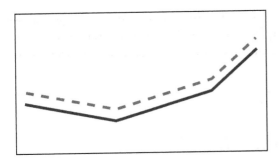

Figure 6.9 The original and edited sidewalk line geometries. The edited one is shown with the dotted line.

TIP Use `GetGeometryCount` to determine the number of geometries contained in a collection of geometries (such as a multigeometry or a polygon), and use `GetPointCount` to determine the number of vertices in a single geometry. The `GetGeometryCount` function will always return zero for a single geometry, and `GetPointCount` will always return zero for a collection of geometries.

What if you later realize that the shape of the sidewalk is all wrong because it's missing a vertex? If you still have the coordinates for each vertex, then the easiest thing is probably to create a new sidewalk geometry using all of the vertices. But if your landscaper only gave you the coordinates of the missing vertex and told you that it needed to be inserted between the second and third vertices, as shown in figure 6.10, you'd need to insert it in your line. You can solve this problem multiple ways, but I think the easiest is to get a list of all the vertices, insert a new set of coordinates into the list, and then use the list to create a new geometry. You can use `GetPoints` to get the list of vertices in a line, where each vertex is in the form of a tuple with x, y, and z coordinates. Here's what that list looks like for the original sidewalk:

```
>>> print(sidewalk.GetPoints())
[(54.0, 37.0, 0.0), (62.0, 35.5, 0.0), (70.5, 38.0, 0.0),
 (74.5, 41.5, 0.0)]
```

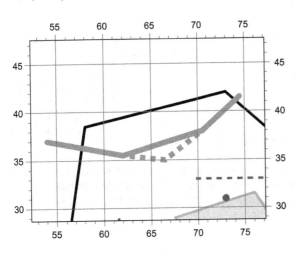

Figure 6.10 The dotted line shows changes to make to the sidewalk by inserting a new vertex between the existing second and third vertices.

A handy feature of lists is that you can easily insert items into the *i*th position using the
list[i:i] syntax. The following example gets the list of sidewalk vertices and then
inserts a tuple containing new x and y coordinates in between the second and third
vertices:

```
>>> vertices = sidewalk.GetPoints()
>>> vertices[2:2] = [(66.5, 35)]
>>> print(vertices)
[(54.0, 37.0, 0.0), (62.0, 35.5, 0.0), (66.5, 35),
➥ (70.5, 38.0, 0.0), (74.5, 41.5, 0.0)]
```

You can see that the original vertices are still there, but the new vertex has been inserted
at index 2. This vertex doesn't have a z coordinate because one wasn't provided in the
inserted tuple. The fact that the
original coordinates do have z values is unimportant, because this is a
2D geometry and they should all be
zero anyway.

Now you have a list of tuples, but
how do you turn it into a line geometry, like that in figure 6.11? The
easiest way is to take advantage of
the Python * operator to expand
the tuples to individual parameters
and pass them in turn to AddPoint:

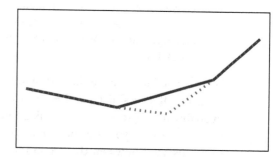

**Figure 6.11 The original sidewalk line geometry and
one with another vertex inserted in the middle**

```
>>> new_sidewalk = ogr.Geometry(ogr.wkbLineString)
>>> for vertex in vertices:
...     new_sidewalk.AddPoint(*vertex)
...
>>> vp.plot(new_sidewalk, 'g:')
```

The Python * operator

The * operator unpacks the contents of a tuple or list into separate items so they can
be passed as parameters to a function. Take this example:

```
>>> pt = ogr.Geometry(ogr.wkbPoint)              Resolves to
>>> vertex = (10, 20)                            pt.AddPoint(l0, 20),
>>> pt.AddPoint(*vertex)                    ◄    which works
>>> pt.AddPoint(vertex)
<snip Traceback>                                 Resolves to
TypeError: Required argument 'y' (pos 3) not found   pt.AddPoint((l0, 20)),
                                                 which fails
```

Using the * operator explodes vertex into two parameters that are successfully
passed to AddPoint. Forgetting the * operator only passes one parameter, a tuple.
But AddPoint expects at least two parameters, an x and a y, so it fails.

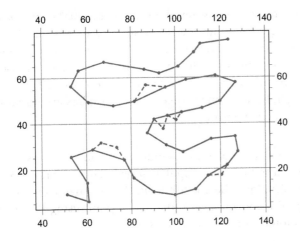

Figure 6.12 The results of inserting multiple vertices into a line geometry. The original is shown with a solid line, and the edited is drawn as a dotted line.

This is easy to expand to a larger number of edits if needed. For example, what if you want to add new points after the existing 5th, 11th, 19th, and 26th vertices, as shown by the dotted line in figure 6.12? As you look at the following code, see if you can figure out why you'd want to add the vertices at the end of the line first:

```
vertices = line.GetPoints()
vertices[26:26] = [(87, 57)]
vertices[19:19] = [(95, 38), (97, 43), (101, 42)]
vertices[11:11] = [(121, 18)]
vertices[5:5] = [(67, 32), (74, 30)]
new_line = ogr.Geometry(ogr.wkbLineString)
for vertex in vertices:
    new_line.AddPoint(*vertex)
```

Insert later vertices first so indexes don't change

Once you insert an item into a list, then the indices of the later items are changed. If you insert the points after the 5th vertex first, then the original 11th vertex will now have an index of 13, because two points were added earlier in the list. You'll have to keep track of how many items you inserted so that you can get the later indices right. That's certainly doable, but why bother if you can avoid the problem altogether by working backward?

> **TIP** If you need to insert or delete multiple items in a list (whether the list contains vertices or something else), you'll find that life is easier if you start from the end and work backward so you don't inadvertently change indexes that you still need to use.

If you don't want to create a new line geometry, you can modify the original instead. This adds one vertex to the sidewalk line without creating a copy:

```
vertices = sidewalk.GetPoints()
vertices[2:2] = [(66.5, 35)]
for i in range(len(vertices)):
    sidewalk.SetPoint(i, *vertices[i])
```

But this uses `SetPoint` to edit five vertices when the original `sidewalk` only has four. How can you possibly change a vertex that doesn't exist? It turns out that `SetPoint` will add a vertex at the requested index, along with any missing vertices in between. For example, if the line has ten vertices (so the highest index is 9) and you use `Set-Point` to create a vertex with index 15, it will also create vertices with indices 10 through 14. Be careful, though, because any vertices it adds as fillers are initialized to (0, 0), which is probably not what you want.

CREATING POINTS FROM LINES

Sometimes you need to get the vertices of a line as individual points. By this time, you know how to create points and also how to manipulate individual line vertices, so turning the vertices into points shouldn't be too difficult. All you need to do is loop through the line vertices, get the coordinates, and create a point using those coordinates. The following listing shows a function that does this.

Listing 6.1   Function to create a point layer from a line layer

```
def line_to_point_layer(ds, line_name, pt_name):
    """Creates a point layer from vertices in a line layer."""
    if ds.GetLayer(pt_name):
        ds.DeleteLayer(pt_name)
    line_lyr = ds.GetLayer(line_name)
    sr = line_lyr.GetSpatialRef()
    pt_lyr = ds.CreateLayer(pt_name, sr, ogr.wkbPoint)        Create the point layer
    pt_lyr.CreateFields(line_lyr.schema)
    pt_feat = ogr.Feature(pt_lyr.GetLayerDefn())
    pt_geom = ogr.Geometry(ogr.wkbPoint)
    for line_feat in line_lyr:
        atts = line_feat.items()
        for fld_name in atts.keys():
            pt_feat.SetField(fld_name, atts[fld_name])
        for coords in line_feat.geometry().GetPoints():     Loop through vertices
            pt_geom.AddPoint(*coords)                        and create points
            pt_feat.SetGeometry(pt_geom)
            pt_lyr.CreateFeature(pt_feat)
```

Copy attribute values

This function takes a data source, the name of an existing line layer, and the name of a new point layer. It creates the point layer and copies all of the attribute fields from the line to the point layer. Then it loops through all of the line features and creates a point feature for each vertex, which also contains the same attribute values as the line it came from.

6.3.2   *Creating and editing multilines: multiple lines as one geometry*

As with multipoints, multiline objects contain one or more lines that are treated as if they were one. The collection of channels in a braided river is a good candidate for this geometry type. As shown in figure 6.13, you can also treat the stone pathways running through the yard as a multiline object.

As with any multigeometry, you need to create each component separately and then add it to the main geometry, so you need at least one regular line object along

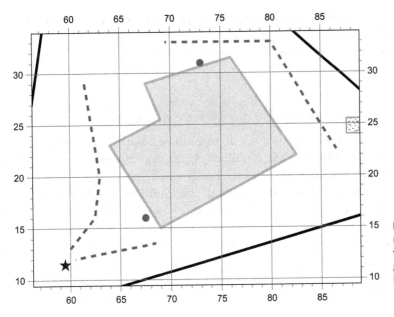

Figure 6.13 You can use a multiline to hold the garden path geometries, shown here as dotted lines.

with your multiline. The following code shows how to create the multiple paths from figure 6.13:

```
>>> path1 = ogr.Geometry(ogr.wkbLineString)
>>> path1.AddPoint(61.5, 29)
>>> path1.AddPoint(63, 20)
>>> path1.AddPoint(62.5, 16)
>>> path1.AddPoint(60, 13)
```
Create the first path

```
>>> path2 = ogr.Geometry(ogr.wkbLineString)
>>> path2.AddPoint(60.5, 12)
>>> path2.AddPoint(68.5, 13.5)
```
Create the second path

```
>>> path3 = ogr.Geometry(ogr.wkbLineString)
>>> path3.AddPoint(69.5, 33)
>>> path3.AddPoint(80, 33)
>>> path3.AddPoint(86.5, 22.5)
```

```
>>> paths = ogr.Geometry(ogr.wkbMultiLineString)
>>> paths.AddGeometry(path1)
>>> paths.AddGeometry(path2)
>>> paths.AddGeometry(path3)
```
Create the multiline geometry and add the paths

This is similar to creating a multipoint. The first thing you do is create the three separate line geometries that make up the pathways. After creating those, you create a multiline geometry and add the paths in order. It doesn't matter if you wait until all of the individual lines are created and then create the multiline, or if you create the multiline up front and add the individual lines as you go along. You can even reuse one line object for each path, but you'd need to add the path to the multiline immediately after adding its vertices, and call `Empty` on the path geometry to clear out the old vertices before starting on the next pathway.

Let's take a look at the guts of your new multiline:

```
>>> vp.plot(paths)
>>> print(paths)
MULTILINESTRING ((61.5 29.0 0,63 20 0,62.5 16.0 0,60 13 0),
 (60.5 12.0 0,68.5 13.5 0),(69.5 33.0 0,80 33 0,86.5 22.5 0))
```

Each inner line is inside its own set of parentheses, and they're listed in the same order as they were added to the multiline. Again, it's important that OGR preserves this order so that you always know which line is which.

To edit a vertex once it's been added to the multiline, you first need to grab the single line that you want to edit, the same way you did with multipoints. Once you have that, you can edit the vertices the same way you would a regular line. To edit the second vertex in the first path added to the multiline, you can do something like this:

```
paths.GetGeometryRef(0).SetPoint(1, 63, 22)
```

You can use the concepts you've already learned about getting inner geometries and editing lines to move the whole multiline two units to the east and three to the south, with the results shown in figure 6.14:

```
>>> for i in range(paths.GetGeometryCount()):         ◀──| Get inner geometry
...     path = paths.GetGeometryRef(i)
...     for j in range(path.GetPointCount()):
...         path.SetPoint(j, path.GetX(j) + 2, path.GetY(j) - 3)
...
>>> vp.plot(paths, 'r--')
```

Hopefully you feel comfortable constructing and editing lines by this point, and you'll see in the next section that working with polygons is only slightly more complicated.

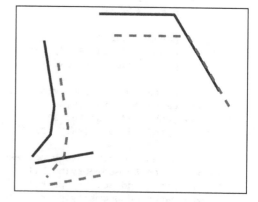

Figure 6.14   The original and edited pathway multiline geometries. The edited one is drawn as a dotted line.

6.4   *Working with polygons*

Polygons are used to represent things that have area, unlike a point or line. City boundaries and lakes are two examples of data that could be modeled as polygons. Instead of a polygon being made up of a list of vertices, like a line, they're made of rings. This is because polygons can have holes in them, like a donut, and a separate ring is required for the outer polygon and for each of the holes. A simple polygon with no holes in it is still made up of one ring. Like lines, rings are made up of a series of vertices connected by straight line segments, but the first and last vertices are the same so that they form a closed ring.

Like lines, ring vertices need to be added in order, but you have other considerations as well. The line segments making up a polygon's perimeter shouldn't touch or cross, as shown in figure 6.15. OGR will allow you to create a polygon like this, but calculations on it are apt to be wrong, even if they run without errors. You can check for problems like this by calling `IsValid` on a geometry, which you should make a habit of if you're building your own geometries. Note that polygons that don't have a width—so they look like a line—are also invalid.

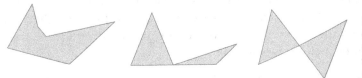

Figure 6.15 The polygon on the left is valid, but the other two are not because the line segments intersect and split the polygon.

Returning to the yard example, let's start by creating a polygon for the entire yard boundary (figure 6.16).

Once again, you'll work with simple polygons in the examples, but your new knowledge can be easily applied to more-complex geometries.

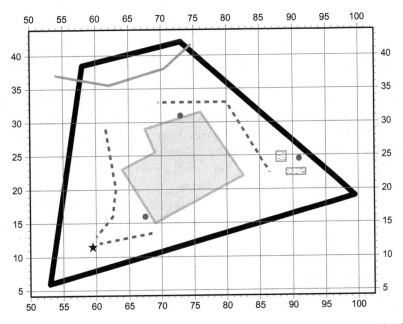

Figure 6.16 You can use a polygon to hold the yard boundary, shown here as the thick solid line.

6.4.1   Creating and editing single polygons

A polygon is like a multigeometry in the sense that it consists of a set of geometries. All polygons are made of rings, which in turn are made of vertices. A simple polygon only has one ring, but you still need to create a ring object and then add it to the polygon. As with lines, use `AddPoint` to add a vertex to a ring. Vertices need to be added in order, but the direction around the perimeter can vary depending on the format you want to use to store the data. For example, shapefiles specify that the outer rings are in clockwise order, but GeoJSON doesn't specify an order. Because of details like this, it's probably a good idea to read up about the format you intend to use. But no matter the direction, the first and last vertices must have the same coordinates so they close the ring. To do this, you can either make sure the last vertex added has identical coordinates to the first one, or you can call `CloseRings` on the ring or polygon after adding all vertices. The latter method is the one used here to create the yard outline shown in figure 6.16. The example starts with the upper left vertex and traverses the perimeter in counter-clockwise direction.

```
>>> ring = ogr.Geometry(ogr.wkbLinearRing)
>>> ring.AddPoint(58, 38.5)                     Create a ring and
>>> ring.AddPoint(53, 6)                         add vertices
>>> ring.AddPoint(99.5, 19)
>>> ring.AddPoint(73, 42)
>>> yard = ogr.Geometry(ogr.wkbPolygon)         Create a polygon
>>> yard.AddGeometry(ring)                        and add the ring
>>> yard.CloseRings()                            Close all rings
                                                   in the polygon
```

You can make sure things look okay by plotting the geometry and printing the WKT:

```
>>> vp.plot(yard, fill=False, edgecolor='blue')
>>> print(yard)
POLYGON ((58.0 38.5 0,53 6 0,99.5 19.0 0,73 42 0,58.0 38.5 0))
```

The WKT contains all of the coordinates for the ring, but notice that the coordinate list is inside a second set of parentheses. This is because the vertices make up a ring inside the polygon. You'll see later there can be more than one ring within a polygon, which is why the ring needs to be delineated with its own set of parentheses.

If you were to call `GetPointCount` on yard, the response would be zero because the vertices belong to the ring that's inside the polygon. This is similar to how multigeometries won't admit to having vertices, but they'll confess to containing other geometries. The yard variable would claim to have one geometry if you queried it with `GetGeometryCount`, and that one geometry is the ring. Because of this, to edit a polygon's vertices you need to get the ring first, and then edit the ring the same way you edit lines. This example grabs the ring and shifts it five map units to the west, which automatically moves the whole polygon, as seen in figure 6.17:

```
>>> ring = yard.GetGeometryRef(0)
>>> for i in range(ring.GetPointCount()):
...     ring.SetPoint(i, ring.GetX(i) - 5, ring.GetY(i))
...
>>> vp.plot(yard, fill=False, ec='red', linestyle='dashed')
```

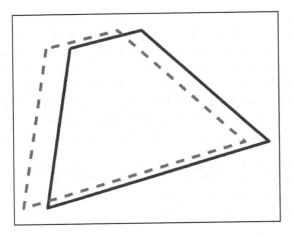

Figure 6.17 The original and edited yard polygon geometries. The original is drawn with a solid line.

You can insert vertices into polygon rings using the same method that you used for lines. For example, you can cut one of the sharp corners off of the yard by getting the ring and replacing the third vertex with two different vertices (figure 6.18):

```
>>> ring = yard.GetGeometryRef(0)
>>> vertices = ring.GetPoints()
>>> vertices[2:3] = ((90, 16), (90, 27))
>>> for i in range(len(vertices)):
...     ring.SetPoint(i, *vertices[i])
...
>>> vp.plot(yard, fill=False, ec='black', ls='dotted', linewidth=3)
```

> **WARNING** Creating a linestring with the same beginning and ending vertices won't create a ring that can be used to build a polygon. Instead, it will be a line that happens to stop at the same place it started. It still won't have an area, perimeter, or any other properties specific to polygons.

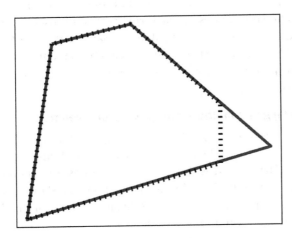

Figure 6.18 The original yard geometry and one with the third vertex replaced with two other vertices.

CREATING LINES FROM POLYGONS

Sometimes you need to convert polygons to lines. To do this, you need to create a line from each ring inside the polygon. I've had luck copying the rings into line features, but the following listing shows how to do it by creating a line from a ring instead. Similarly to the `line_to_point_layer` function in listing 6.1, this function requires a data source, the name of the existing polygon layer, and the name for a new line layer. It creates a new line layer with the same attributes as the polygon layer, and then for each polygon feature, copies each ring to a line and inserts a new feature in the line layer.

> **Listing 6.2   Function to create a line layer from a polygon layer**

```
def poly_to_line_layer(ds, poly_name, line_name):
    """Creates a line layer from a polygon layer."""
    if ds.GetLayer(line_name):
        ds.DeleteLayer(line_name)
    poly_lyr = ds.GetLayer(poly_name)
    sr = poly_lyr.GetSpatialRef()
    line_lyr = ds.CreateLayer(line_name, sr, ogr.wkbLineString)
    line_lyr.CreateFields(poly_lyr.schema)
    line_feat = ogr.Feature(line_lyr.GetLayerDefn())
    for poly_feat in poly_lyr:
        atts = poly_feat.items()
        for fld_name in atts.keys():
            line_feat.SetField(fld_name, atts[fld_name])
        poly_geom = poly_feat.geometry()
        for i in range(poly_geom.GetGeometryCount()):
            ring = poly_geom.GetGeometryRef(i)
            line_geom = ogr.Geometry(ogr.wkbLineString)      # Create a line from a ring
            for coords in ring.GetPoints():
                line_geom.AddPoint(*coords)
            line_feat.SetGeometry(line_geom)
            line_lyr.CreateFeature(line_feat)
```

After creating a new layer to hold the lines, the function starts iterating through the polygons in the original layer, and creates a new line for each ring in the polygon. To do this, each time it finds a ring, it creates an empty line object and then iterates through the ring's vertices. The coordinates for each ring vertex are used to create a new vertex in the line, so you get a line containing all of the same vertices as the ring. Lines, however, aren't closed even if the first and last vertices are the same. If you plot the line, it will look like a polygon, but it doesn't have the concept of an area or anything else specific to polygons.

6.4.2 *Creating and editing multipolygons: multiple polygons as one geometry*

If you've read this far, you can guess that a multipolygon is a geometry made of one or more individual polygons. A classic example of this is the Hawaiian Islands. This archipelago makes up the state of Hawaii and is usually represented as one geometry in datasets covering the United States, but it's obviously made of several islands. The collection of islands makes up one state, the same way a collection of polygons makes up one multipolygon. An example is shown in figure 6.19.

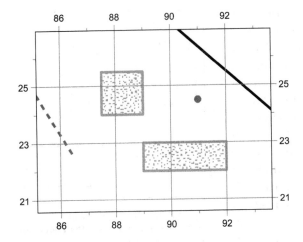

Figure 6.19 You can use a multipolygon to hold the garden boxes, shown here as the two rectangles.

You can probably also figure out how to create a multipolygon, because you don't need to know anything new. You create the individual polygons and add them to a multipolygon, and that's all there is to it. For example, the following listing shows how you can treat the garden boxes in figure 6.19 as a multipolygon made of two individual raised beds.

Listing 6.3 Creating a multipolygon

```
>>> box1 = ogr.Geometry(ogr.wkbLinearRing)
>>> box1.AddPoint(87.5, 25.5)
>>> box1.AddPoint(89, 25.5)
>>> box1.AddPoint(89, 24)
>>> box1.AddPoint(87.5, 24)
>>> garden1 = ogr.Geometry(ogr.wkbPolygon)
>>> garden1.AddGeometry(box1)
```
Create the polygon for the first garden box

```
>>> box2 = ogr.Geometry(ogr.wkbLinearRing)
>>> box2.AddPoint(89, 23)
>>> box2.AddPoint(92, 23)
>>> box2.AddPoint(92,22)
>>> box2.AddPoint(89,22)
>>> garden2 = ogr.Geometry(ogr.wkbPolygon)
>>> garden2.AddGeometry(box2)
```
Create the polygon for the second garden box

Close all the rings at once →
```
>>> gardens = ogr.Geometry(ogr.wkbMultiPolygon)
>>> gardens.AddGeometry(garden1)
>>> gardens.AddGeometry(garden2)
>>> gardens.CloseRings()
```
Create a multipolygon and add the two boxes

Let's look at the WKT for this multipolygon:

```
>>> vp.plot(gardens, fill=False, ec='blue')
>>> print(gardens)
MULTIPOLYGON (((87.5 25.5 0,89.0 25.5 0,89 24 0,87.5 24.0 0,87.5 25.5 0)),
➥ ((89 23 0,92 23 0,92 22 0,89 22 0,89 23 0)))
```

Here you have two polygons inside of a multipolygon, each inside its own set of parentheses, and each one of these contains one ring in another set of parentheses. Once again, everything is in the same order that you added it.

Editing a multipolygon is similar to what you've already seen, although it has one more step because you need to get each inner polygon and then get the ring from that before you can edit vertices. Figure 6.20 shows the result of moving the garden boxes one map unit to the east and half a unit to the north.

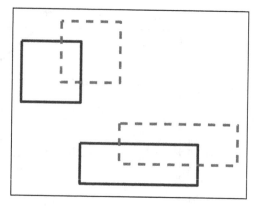

Figure 6.20 The original and edited garden box multipolygon geometries. The edited one is drawn with a dashed line.

```
>>> for i in range(gardens.GetGeometryCount()):
...     ring = gardens.GetGeometryRef(i).GetGeometryRef(0)
...     for j in range(ring.GetPointCount()):
...         ring.SetPoint(j, ring.GetX(j) + 1, ring.GetY(j) + 0.5)
...
>>> vp.plot(gardens, fill=False, ec='red', ls='dashed')
```

Now you know how to work with single and multi-geometries, but you still have the special case of polygons with holes. Keep reading to learn how these are different than multipolygons.

6.4.3 Creating and editing polygons with holes: donuts

What about polygons with holes in them, like donuts? These are different than multi-polygons, because the hole is the absence of a polygon, not a second polygon. This is why polygons need to be made of rings. One ring defines the outer edge of the donut, and another delineates the hole. You need to add the outer ring to the polygon first, and subsequent rings define holes in the geometry. To illustrate how to do this, try cutting the house out of the yard polygon (figure 6.21).

Listing 6.4 Creating a polygon with a hole

```
>>> lot = ogr.Geometry(ogr.wkbLinearRing)
>>> lot.AddPoint(58, 38.5)
>>> lot.AddPoint(53, 6)
>>> lot.AddPoint(99.5, 19)
>>> lot.AddPoint(73, 42)
```
Re-create the yard outline

```
>>> house = ogr.Geometry(ogr.wkbLinearRing)
>>> house.AddPoint(67.5, 29)
>>> house.AddPoint(69, 25.5)
>>> house.AddPoint(64, 23)
>>> house.AddPoint(69, 15)
>>> house.AddPoint(82.5, 22)
>>> house.AddPoint(76, 31.5)
```
Create a new ring for the hole

```
>>> yard = ogr.Geometry(ogr.wkbPolygon)
>>> yard.AddGeometry(lot)
>>> yard.AddGeometry(house)
>>> yard.CloseRings()
```

**Add the outer ring
before the inner one**

Likely, creating a donut was easier than you expected. Now if you take a look at the WKT, you'll see two rings shown inside the polygon:

```
>>> vp.plot(yard, 'yellow')
>>> print(yard)
POLYGON ((58.0 38.5 0,53 6 0,99.5 19.0 0,73 42 0,58.0 38.5 0),
    (67.5 29.0 0,69.0 25.5 0,64 23 0,69 15 0,82.5 22.0 0,76.0 31.5 0,
    67.5 29.0 0))
```

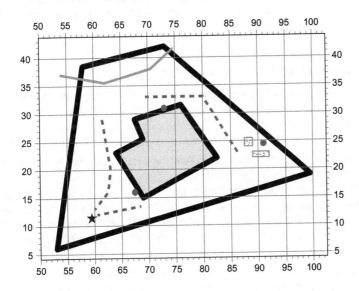

Figure 6.21 You can use a polygon with a hole for the yard boundary with the house cut out of the middle, shown here as the thick solid line.

The holes are taken into account when using the polygon. For example, the area of the yard polygon is equal to the area of the lot with the house subtracted out. The hole in the polygon isn't considered part of the geometry when using the spatial analysis tools shown in the next section, either.

The only difference when editing a polygon like this is that you need to loop through each of the rings instead of assuming only one exists (figure 6.22). In practice, you shouldn't ever assume only one ring exists, anyway, because that assumption could come back to haunt you later.

```
>>> for i in range(yard.GetGeometryCount()):
...     ring = yard.GetGeometryRef(i)
...     for j in range(ring.GetPointCount()):
...         ring.SetPoint(j, ring.GetX(j) - 5, ring.GetY(j))
...
>>> vp.plot(yard, fill=False, hatch='x', color='blue')
```

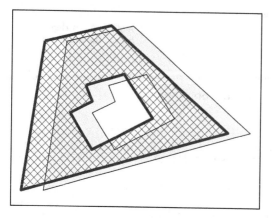

Figure 6.22 The original and edited yard polygon geometries with the house cut out. The original is filled and the edited one is hatched.

Using other modules to work with geometries

Now that you understand how to deal with geometries using OGR, other geometry libraries, such as Fiona, should be easy to understand. Lines are still created by an ordered collection of vertices, and polygons are still made up of rings. The underlying theory about geometries doesn't change, although the method of accessing them does.

Fiona, for example, is a library for reading and writing vector data that's built on top of OGR. Fiona doesn't use special geometry types, but instead uses Python lists to store vertices. The lists are filled with tuples that contain the vertex coordinates. For example, a ring is a list of tuples, and a polygon is a list of rings. A polygon with one ring is a list that contains another list that contains tuples for the vertices. The Fiona user manual is excellent and can be found online at http://toblerity.org/fiona/manual.html.

Shapely is another outstanding module that's designed for working with geometries, but not reading and writing data. Unlike Fiona, it does have special data types for geometries, but that's why it can do spatial analysis, unlike Fiona. Even though it has its own data types, the general ideas are still the same. The detailed user manual for Shapely is available online at http://toblerity.org/shapely/manual.html.

6.5   Summary

- A geometry consists of a collection of vertices. In the cases of lines and polygons, the vertices are connected by line segments to form shapes.
- Multigeometries consist of multiple geometries combined into one. This allows features such as Hawaii to be represented with a single geometry object.
- Geometries in OGR are either 2D or 2.5D. The 2.5D geometries have z values, but they're ignored during analyses, which is why they aren't considered 3D.
- All polygon geometries are made up of one or more rings.

Vector analysis with OGR

This chapter covers
- Determining if geometries share a spatial location
- Proximity relationships between geometries

Now you know how to access existing data and how to build your own geometries from scratch, but I see these as gateways to the more interesting task of spatial analysis. Without analysis capabilities, spatial data is only useful for making maps. Good cartography is essential for many things, but I imagine that even cartographers would get bored if new datasets weren't continually created from various types of analyses. Plus, spatial analyses can answer countless questions relating to pretty much every discipline. In fact, you're probably more likely to generate new data using the analysis functions described in this chapter than by creating geometries vertex by vertex as outlined previously in nauseating detail.

Spatial analysis with vector data comes down to looking at the spatial relationships between two or more geometries. Possible studies range from the extremely simple, such as the distance between two points, to much more complex algorithms such as network analyses. Have you ever wondered how certain mapping websites can provide you with various route options from point A to point B, and even provide travel times? That's network analysis. One easy exercise that I sometimes find entertaining is comparing the distance I hiked with the straight-line distance between the starting and ending points, because these two distances can be significantly different in mountainous terrain. There probably isn't much use for that particular example in my life, other than to satisfy my curiosity, but it's important information for search-and-rescue teams who need to know actual distances. There are plenty of other important questions out there waiting to be answered using spatial analysis techniques.

For example, biologists can use the information downloaded from GPS collars to study how animals use various habitat types or their reactions to roads or other man-made features. Businesses use spatial data to help determine the best location for new stores or factories. Utility companies can use this type of data to select the best routes to install pipelines or electrical transmission lines, and mining companies use geographic information to determine areas that are likely rich in resources. If you're reading this book, it's likely that you have a specific type of analysis in mind, and it's probably completely different from any of the examples mentioned. Spatial analyses are ubiquitous, and in fact, you use these sorts of analyses in your daily life when you choose where to live or what route to take to the office. OGR provides a good foundation for vector analysis, although it's left to you to implement more-complicated algorithms that you may be interested in. This section will introduce you to the basic tools that make up this foundation.

7.1   *Overlay tools: what's on top of what?*

One basic question in geographic analysis is what features occur at the same place. Certain entities, such as countries, don't occur in the same location, although they may share borders. Other types of areas, such as the home ranges of individual bears, can easily overlap, as can boundaries that aren't necessarily related, such as wetlands and land ownership. Many types of queries are concerned with this overlap idea. For example, insurance companies want to know if a parcel of land is on a floodplain before they set a premium, or even decide to insure it at all. A business looking for land to build a factory on would want to know which lots for sale are within an appropriate municipal land use zone. If you're making a map of Stockholm, you'll want to know which roads, train tracks, and parks, among other things, are within city limits.

What sorts of overlap tools exist? Several test certain conditions, such as Intersects, which tells you if two geometries share any space in common. For example, in figure 7.1, the line L2 intersects with the line L3 and the polygon L3. The polygons P2 and P4 also intersect. You can find out if two geometries touch edges, but don't actually share any area, with Touches. This is also true for lines L2 and L3, but not L2 and P3, because they do more than touch. How about discovering if one geometry is contained completely within another? You can test that with either Contains or Within. Polygon P5 is within polygon P1, and P1 contains P5. See table 7.1 for a list of the available operations, along with examples for each one from figure 7.1. Note that while these functions work with polygons, they don't work with linear rings. All functions return True or False. More information can be found in appendix C. (Appendixes C through E are available online on the Manning Publications website at https://www.manning.com/books/geoprocessing-with-python.)

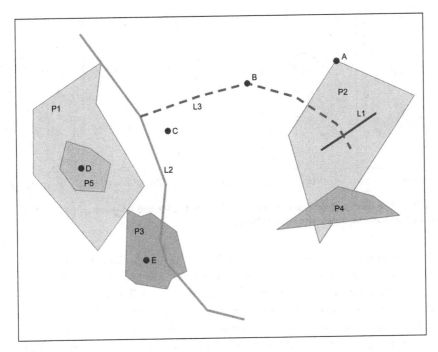

Figure 7.1 Geometries used to get the results of overlay operations that are shown in table 7.1 and figure 7.2.

Table 7.1 Functions to test relationships between geometries. These all return `True` or `False`.

Operation	Examples from figure 7.1
Intersects	Polygons P2 and P4 intersect. Line L3 and Point B intersect. Point A and Polygon P2 intersect. Lines L2 and L3 intersect. Line L2 and Polygon P3 intersect.
Touches	Polygon P2 and Point A touch. Polygon P5 and Point D do not touch. Lines L2 and L3 touch. Lines L1 and L3 do not touch.
Crosses	Lines L1 and L3 cross. Lines L2 and L3 do not cross.
Within	Line L1 is within Polygon P2. Line L3 is not within Polygon P2.
Contains	Polygon P1 contains Polygon P5. Polygon P2 does not contain Polygon P4.
Overlaps	Polygons P2 and P4 overlap. Polygons P1 and P5 do not overlap.
Disjoint	Polygon P1 and Line L1 are disjoint. Polygons P1 and P4 are disjoint.

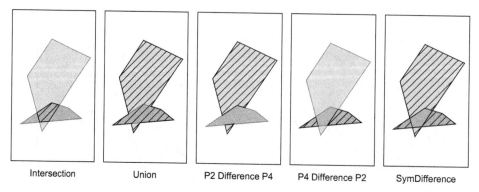

| Intersection | Union | P2 Difference P4 | P4 Difference P2 | SymDifference |

Figure 7.2   The results of several overlay operations on the P2 and P4 geometries from figure 7.1 are shown as hatched areas with dark outlines.

Several functions create new geometries based on the spatial relationships of existing geometries. For example, you can use `Intersection` to get a new geometry that represents only the area that two others have in common. In figure 7.1, the intersection of L1 and L3 is a single point; the intersection of L2 and P3 is a short segment from L2; and the intersection of P2 and P4 is shown in figure 7.2. You would probably use `Intersection` to create new datasets containing features only found within the Stockholm boundary when making the map mentioned earlier.

You can combine the areas of two existing geometries into one with `Union`, which may return a geometry collection if the input geometries are different types. You can treat a geometry collection kind of like a multigeometry, except that the parts don't all need to be the same kind of geometry. For example, the union of L2 and P3 is a geometry collection containing a polygon and two lines, as shown in figure 7.3. The section of L2 that intersects P3 no longer exists as a line, and instead the space it takes up is included in the polygon. The union of P2 and P4 is a single polygon, as shown in figure 7.2. You might use this function if you were given a roads dataset in which the roads were broken up into segments based on changes in speed limits, which would be required for an analysis looking at travel time, but you want each road to be a single feature so it's easier to use in a map.

Figure 7.3   The three parts of the geometry collection created by unioning L2 and P3 together.

It's also possible to clip an intersection out of a geometry so that you're left with the part of the geometry that doesn't intersect the second geometry. Unlike `Intersection` and `Union`, the results from `Difference` depend on which geometry the function is called on and which is passed to it. This is also illustrated in figure 7.2.

There's also `SymDifference`, which returns the union of two geometries with the intersection removed. If you were looking at the home ranges, or territories, of two different mountain lions, you might want to know the area that the first cat uses but

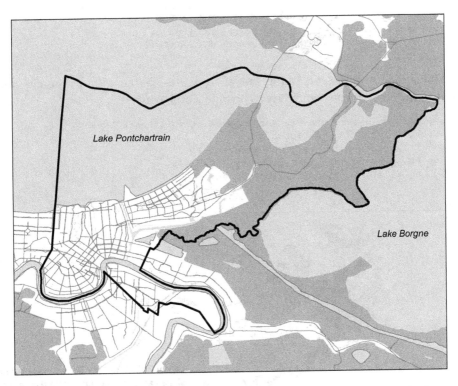

Figure 7.4   A simple map of New Orleans showing the city boundary, water, and wetlands

the second doesn't, or vice versa. You'd use Difference to get that information. You could use SymDifference to determine the area that was used by either lion, but not both. Intersection would give you the shared territory, and Union would provide the combined territories. Each type of information is likely useful to a cougar researcher, but each in a different way. In fact, it was a study similar to this, although on a threatened species of lizard and much more sophisticated than this simplified example, that got me hooked on GIS in the first place!

Let's look at a concrete example. Figure 7.4 might remind you of our discussion of wetlands within the boundaries of New Orleans back in chapter 3. You're about to look at two different ways of using intersections to determine the percentage of New Orleans made up by wetlands. But first, it will be helpful to do a little interactive exercise with the data to visualize what's happening. Open the water bodies shapefile for the United States, which contains features such as lakes, streams, canals, and marshes, and plot one specific feature that represents a marsh near New Orleans. This shapefile has approximately 27,000 features, so don't try to plot the entire file unless you want to wait all day.

```
>>> water_ds = ogr.Open(r'D:\osgeopy-data\US\wtrbdyp010.shp')
>>> water_lyr = water_ds.GetLayer(0)
```

```
>>> water_lyr.SetAttributeFilter('WaterbdyID = 1011327')
>>> marsh_feat = water_lyr.GetNextFeature()
>>> marsh_geom = marsh_feat.geometry().Clone()
>>> vp.plot(marsh_geom, 'b')
```

You should now see something similar to figure 7.5, but without the city boundary. Add the New Orleans boundary to provide a little context:

```
>>> nola_ds = ogr.Open(r'D:\osgeopy-data\Louisiana\NOLA.shp')
>>> nola_lyr = nola_ds.GetLayer(0)
>>> nola_feat = nola_lyr.GetNextFeature()
>>> nola_geom = nola_feat.geometry().Clone()
>>> vp.plot(nola_geom, fill=False, ec='red', ls='dashed', lw=3)
```

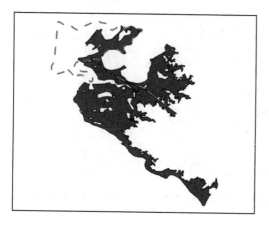

Figure 7.5   The New Orleans city boundary is shown as a dashed line overlaid on a single, but large, marsh polygon from the United States water bodies dataset.

Now you have two polygons, one for New Orleans and one for a marsh that's partly contained with the New Orleans boundary. Now intersect the two geometries:

```
>>> intersection = marsh_geom.Intersection(nola_geom)
>>> vp.plot(intersection, 'yellow', hatch='x')
```

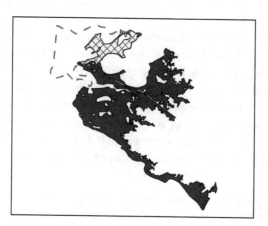

Figure 7.6   The result of intersecting the New Orleans boundary with the marsh is shown in the hatched area.

You can see from figure 7.6 that the intersection geometry consists of the area that's contained within both the city boundary and the marsh polygon. How can you use this to figure out how much of New Orleans is wetlands? Well, if you intersect the city boundary with all of the wetland polygons that it overlaps, then you'll end up with a bunch of polygons that represent wetlands within the boundary. All you need to do then is sum up their areas and divide by the area of the New Orleans geometry. Let's assume that anything in the water bodies dataset that's not a lake is a wetland, and try this:

```
>>> water_lyr.SetAttributeFilter("Feature != 'Lake'")        Limit features to anything not
>>> water_lyr.SetSpatialFilter(nola_geom)                    a lake within New Orleans
>>> wetlands_area = 0
>>> for feat in water_lyr:
...     intersect = feat.geometry().Intersection(nola_geom)  Sum up the intersecting area
...     wetlands_area += intersect.GetArea()                 of each wetland feature
...
>>> pcnt = wetlands_area / nola_geom.GetArea()
>>> print('{:.1%} of New Orleans is wetland'.format(pcnt))
28.7% of New Orleans is wetland
```

The first thing you do is change the attribute filter on the water bodies so that lakes, specifically Lake Pontchartrain, are ignored. Then you use a spatial filter to toss out all of the features not in the vicinity of New Orleans, which gets rid of almost everything in the shapefile. This step isn't technically necessary, but it speeds up processing time considerably because you get to ignore most of the dataset. Then you loop through the remaining water bodies, intersect each one with the New Orleans geometry, and add the intersection area to a running total. When done with the loop, all you needed to do was divide by the area of New Orleans to get your answer.

TIP Filtering out unneeded features, either with spatial or attribute filters, can significantly decrease your processing time.

You have an easier way to do this, however, if you want to work with layers instead of individual geometries. In this case, OGR takes care of looping through the geometries in the layers for you. Let's intersect the New Orleans boundary with the water layer to get the area in common between the two:

```
Keep         >>> water_lyr.SetAttributeFilter("Feature != 'Lake'")
excluding
lakes        >>> memory_driver = ogr.GetDriverByName('Memory')        Create a temporary
             >>> temp_ds = memory_driver.CreateDataSource('temp')     layer in memory
             >>> temp_lyr = temp_ds.CreateLayer('temp')
             >>> nola_lyr.Intersection(water_lyr, temp_lyr)   Intersect the layers and
                                                              store result in temp_lyr
```

As before, you limit the water bodies to the non-lakes, but you don't perform a spatial filter because the layer intersection handles that. An empty layer is required for a layer intersection, however, so you do need the extra step of creating that. Because there's no reason to save the layer, you use the memory driver to create the data source and layer. This driver doesn't write anything out to disk, so it's a good choice for temporary data. Once you have the empty layer, you pass it to the layer `Intersection` function, which populates it with the intersection of `nola_lyr` and `water_lyr`.

Once you have the intersected area, you can use a SQL statement to sum up the areas of all geometries in `temp_lyr`. Remember that `ExecuteSQL` returns a new layer object, so you need to get the first feature from it in order to access the results of the `SUM` function:

```
>>> sql = 'SELECT SUM(OGR_GEOM_AREA) AS area FROM temp'
>>> lyr = temp_ds.ExecuteSQL (sql)
>>> pcnt = lyr.GetFeature(0).GetField('area') / nola_geom.GetArea()
>>> print('{:.1%} of New Orleans is wetland'.format(pcnt))
28.7% of New Orleans is wetland
```

One more important detail is that functions that operate on entire layers instead of individual geometries preserve the attribute values from the input layers. This is handy if you still need the information about each feature. In this case you don't need it, but think about the mountain lion home range example, but with even more cats. The researcher would almost definitely want to know which two cougars were sharing the same habitat, and a layer intersection would keep this information in the results, assuming it was in the original attribute tables.

7.2   *Proximity tools: how far apart are things?*

Another common problem when analyzing geographic features is determining how far apart they are from one another. For example, many municipalities have regulations concerning the types of businesses allowed within a certain distance of a church or school, and proximity to a large customer base is another important factor when considering business locations. Or how about an ornithologist trying to determine how roads affect the nesting sites chosen by various species of birds? He would need to measure the distance between each nest and the closest roads as part of his study.

Two proximity tools are included with OGR, one to measure distance between geometries and one to create buffer polygons. A buffer is a polygon that extends out a certain distance from the original geometry. Figure 7.7 shows the yard geometries from chapter 6 with buffers around them, although they're not in their true yard configuration so that you can see the buffers better. You could use a buffer to visualize which businesses were within walking distance of your location, or to make sure that you didn't build a pizza joint within a certain distance of an existing one. You could also buffer a stream geometry to get an idea of the riparian area surrounding it, or to show where cattle aren't allowed to graze and risk damaging the ecosystem.

TIP   Unprojected datasets (those using latitude and longitude) are fine for displaying data in many cases, but can be a poor choice when it comes to analysis. Think about how the longitudinal lines on a globe converge on the poles. One longitudinal degree at 40° latitude is shorter than one degree at the equator, which makes comparing distances at different latitudes extremely problematic. You're much better off converting your data to a different coordinate system with a constant unit of measure.

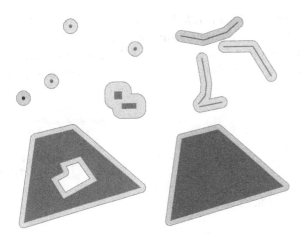

Figure 7.7 The geometries from the make-believe yard shown along with buffer geometries. Notice how the buffer for multigeometries becomes a single polygon if the individual buffers overlap.

As a buffering example, let's figure out how many cities in the United States are within 10 miles of a volcano. We'll use datasets that have an Albers projection so that the map units are meters instead of decimal degrees. We'll also use this example to highlight a potential source of error when doing analyses like this. The first step in your analysis will be to buffer a volcano dataset by 16,000 meters, which is roughly equivalent to 10 miles. Because there isn't a buffer function on an entire layer, you'll buffer each volcano point individually and add it to a temporary layer. Once that's done, you can intersect the buffer layer with the cities layer to get the number of cities that fall within that 10-mile radius. All of this is shown in the following listing.

Listing 7.1 A flawed method for determining the number of cities near volcanoes

```
>>> shp_ds = ogr.Open(r'D:\osgeopy-data\US')
>>> volcano_lyr = shp_ds.GetLayer('us_volcanos_albers')
>>> cities_lyr = shp_ds.GetLayer('cities_albers')

>>> memory_driver = ogr.GetDriverByName('memory')
>>> memory_ds = memory_driver.CreateDataSource('temp')      Make a temporary
>>> buff_lyr = memory_ds.CreateLayer('buffer')              layer to store buffers
>>> buff_feat = ogr.Feature(buff_lyr.GetLayerDefn())

>>> for volcano_feat in volcano_lyr:
...     buff_geom = volcano_feat.geometry().Buffer(16000)   Buffer each volcano and
...     tmp = buff_feat.SetGeometry(buff_geom)              add it to the buffer layer
...     tmp = buff_lyr.CreateFeature(buff_feat)
...
>>> result_lyr = memory_ds.CreateLayer('result')            Intersect the cities
>>> buff_lyr.Intersection(cities_lyr, result_lyr)           and volcano buffers
0
>>> print('Cities: {}'.format(result_lyr.GetFeatureCount()))
Cities: 83
```

From this you could conclude that that are 83 cities in the United States that are within 10 miles of a volcano. But for good measure, try doing the same thing with the slightly different method shown in listing 7.2. This time you'll add the buffers to a multipolygon instead of a temporary layer. A function called UnionCascaded efficiently unions all of the polygons in a multipolygon together; you'll use this to create one polygon from all of the volcano buffers and then use the result as a spatial filter on the cities layer.

Listing 7.2   A better method for determining the number of cities near volcanoes

```
>>> volcano_lyr.ResetReading()
>>> multipoly = ogr.Geometry(ogr.wkbMultiPolygon)          ◄── Go back to the start
>>> for volcano_feat in volcano_lyr:                            of the volcanoes layer
...       buff_geom = volcano_feat.geometry().Buffer(16000)
...       multipoly.AddGeometry(buff_geom)
...
>>> cities_lyr.SetSpatialFilter(multipoly.UnionCascaded())
>>> print('Cities: {}'.format(cities_lyr.GetFeatureCount()))
Cities: 78
```

Huh, somehow you lost five cities in the last few minutes, which is a little disconcerting. What happened? In the first example, a copy of a city is included in the output every time it falls within a volcano buffer. This means a city will be included more than once if it's within 16,000 meters of multiple volcanoes. This happened with a few cities, which is why the count from the intersection method was wrong, and higher than from the spatial filter method. This is a good example of why you should always think through your methodology carefully, because the "obvious" solution might be incorrect and provide the wrong results.

TIP   Use UnionCascaded when you need to union many geometries together. It will be much faster than joining them one by one.

We'll look at one last example. Perhaps you want to know how far a particular city is from a certain volcano. The first thing you need to do is get the geometries for the city and volcano of interest:

```
>>> volcano_lyr.SetAttributeFilter("NAME = 'Rainier'")
>>> feat = volcano_lyr.GetNextFeature()
>>> rainier = feat.geometry().Clone()

>>> cities_lyr.SetAttributeFilter("NAME = 'Seattle'")
>>> feat = cities_lyr.GetNextFeature()
>>> seattle = feat.geometry().Clone()
```

Once you have the geometries, you can use the Distance function to ask them how far apart they are from each other:

```
>>> meters = round(rainier.Distance(seattle))
>>> miles = meters / 1600
>>> print('{} meters ({} miles)'.format(meters, miles))
92656 meters (57.91 miles)
```

The city of Seattle is approximately 58 miles from Mount Rainier, which is considered an active volcano. Of course, you'd get a different answer if you used actual city boundaries instead of a point, but I doubt that the fine people of Seattle would appreciate the distinction if the mountain did erupt.

2.5D geometries

You may remember from the last chapter that geometries with z values are considered 2.5D in OGR because the z values aren't used when performing spatial operations. To illustrate this, let's look at the distance between two points:

```
>>> pt1_2d = ogr.Geometry(ogr.wkbPoint)
>>> pt1_2d.AddPoint(15, 15)
>>> pt2_2d = ogr.Geometry(ogr.wkbPoint)
>>> pt2_2d.AddPoint(15, 19)
>>> print(pt1_2d.Distance(pt2_2d))
4.0
```

That returns a distance of 4 units, as expected. Now try the same thing but with 2.5D points:

```
>>> pt1_25d = ogr.Geometry(ogr.wkbPoint25D)
>>> pt1_25d.AddPoint(15, 15, 0)
>>> pt2_25d = ogr.Geometry(ogr.wkbPoint25D)
>>> pt2_25d.AddPoint(15, 19, 3)
>>> print(pt1_25d.Distance(pt2_25d))
4.0
```

That also returned a distance of 4, but taking the elevation values into account, the real distance is 5. Clearly, the z values weren't used in the calculation. How about an area example? This polygon is 10 units long on each side, so it should have an area of 100:

```
>>> ring = ogr.Geometry(ogr.wkbLinearRing)
>>> ring.AddPoint(10, 10)
>>> ring.AddPoint(10, 20)
>>> ring.AddPoint(20, 20)
>>> ring.AddPoint(20, 10)
>>> poly_2d = ogr.Geometry(ogr.wkbPolygon)
>>> poly_2d.AddGeometry(ring)
>>> poly_2d.CloseRings()
>>> print(poly_2d.GetArea())
100.0
```

You got the expected result there, but try moving the right-most edge to a higher elevation so that the rectangle is in the 3D plane:

```
>>> ring = ogr.Geometry(ogr.wkbLinearRing)
>>> ring.AddPoint(10, 10, 0)
>>> ring.AddPoint(10, 20, 0)
>>> ring.AddPoint(20, 20, 10)
>>> ring.AddPoint(20, 10, 10)
>>> poly_25d = ogr.Geometry(ogr.wkbPolygon25D)
>>> poly_25d.AddGeometry(ring)
>>> poly_25d.CloseRings()
>>> print(poly_25d.GetArea())
100.0
```

(continued)

This new rectangle also claims to have an area of 100 but in reality, the area is closer to 141.

Overlay operations also ignore the elevation values. For example, if elevation were accounted for, pt1_2d would be contained in the 2D polygon but not in the 2.5D one, which isn't what we see:

```
>>> print(poly_2d.Contains(pt1_2d))
True
>>> print(poly_25d.Contains(pt1_2d))
True
```

Now you know the basics of spatial analysis with vector data. You might not need to do anything more complicated than what you've seen here, but if you do, these tools are the building blocks with which to start.

7.3 *Example: locating areas suitable for wind farms*

Let's do a simple analysis to look for suitable wind farm locations in Imperial County, California. The United States National Renewal Energy Laboratory provides a wind dataset that shows areas in the United States that are suitable for wind farms based on wind speed and abundance, and geographical factors such as terrain (figure 7.8).

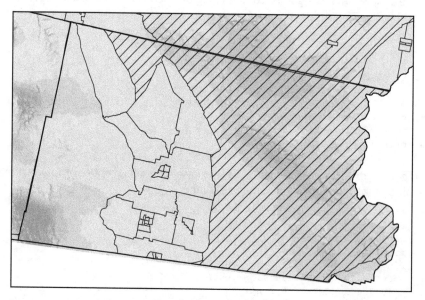

Figure 7.8 Census and wind data for Imperial County, CA. The darker the shading, the better the wind conditions for a wind farm. The hatched area shows census tracts with a population density less than 0.5/km2.

Areas are rated on a scale of 1 to 7, where anything 3 and above is generally considered suitable. We'll combine this with census data to locate areas with an appropriate wind rating and a population less than 0.5 per square kilometer.

The census dataset contains population per census tract, but doesn't have a population density attribute. You can calculate that given the tract area and the population, however, so the first thing to do is add a field containing that information:

```
census_fn = r'D:\osgeopy-data\California\ca_census_albers.shp'
census_ds = ogr.Open(census_fn, True)
census_lyr = census_ds.GetLayer()
density_field = ogr.FieldDefn('popsqkm', ogr.OFTReal)
census_lyr.CreateField(density_field)
for row in census_lyr:
    pop = row.GetField('HD01_S001')
    sqkm = row.geometry().GetArea() / 1000000
    row.SetField('popsqkm', pop / sqkm)
    census_lyr.SetFeature(row)
```

You open the census shapefile for editing and add a floating-point field. Then you loop through each row and calculate the population density. The map units for this dataset are meters, so the geometry's area is square meters, but you convert that to square kilometers by dividing by 1,000,000. You grab the tract population from the HD01_S001 field and divide by the calculated area to get population per km^2.

Now get the geometry for Imperial County so that you can use it to spatially limit your analysis. You don't need to keep the county data source open after cloning the geometry.

```
county_fn = r'D:\osgeopy-data\US\countyp010.shp'
county_ds = ogr.Open(county_fn)
county_lyr = county_ds.GetLayer()
county_lyr.SetAttributeFilter("COUNTY ='Imperial County'")
county_row = next(county_lyr)
county_geom = county_row.geometry().Clone()
del county_ds
```

But one problem exists, though. The county data uses coordinates that are latitude and longitude values, but the census and wind datasets use meters. You'll learn how to work with different spatial reference systems like these in the next chapter, but for now please trust me that this bit of code will convert the county geometry to the correct coordinate system:

```
county_geom.TransformTo(census_lyr.GetSpatialRef())
census_lyr.SetSpatialFilter(county_geom)
census_lyr.SetAttributeFilter('popsqkm < 0.5')
```

Once the geometry is converted, you use it to set a spatial filter on the census tract data so you'll only be considering tracts in the correct part of the state. You also set an attribute filter to further limit the tracts to those with a low population density.

Now open the wind dataset and use an attribute filter to limit it to the areas with a rating of 3 or better:

```
wind_fn = r'D:\osgeopy-data\California\california_50m_wind_albers.shp'
wind_ds = ogr.Open(wind_fn)
wind_lyr = wind_ds.GetLayer()
wind_lyr.SetAttributeFilter('WPC >= 3')
```

It makes sense to create a data source to put the results in before starting any analysis, so let's do that now. Create a new shapefile that uses the same spatial reference system as the wind data, and then add fields for the wind rating and the population density. You might as well use the layer's definition to create an empty feature for inserting data later, too.

```
out_fn = r'D:\osgeopy-data\California\wind_farm.shp'
out_ds = ogr.GetDriverByName('ESRI Shapefile').CreateDataSource(out_fn)
out_lyr = out_ds.CreateLayer(
    'wind_farm', wind_lyr.GetSpatialRef(), ogr.wkbPolygon)
out_lyr.CreateField(ogr.FieldDefn('wind', ogr.OFTInteger))
out_lyr.CreateField(ogr.FieldDefn('popsqkm', ogr.OFTReal))
out_row = ogr.Feature(out_lyr.GetLayerDefn())
```

You're finally ready to look for possible wind farm locations. In the next listing, you'll loop through the census tracts, intersect them with the suitable wind polygons, and put the results in your new shapefile.

Listing 7.3   Intersecting census and wind data

```
for census_row in census_lyr:
    census_geom = census_row.geometry()                  │ Intersect census
    census_geom = census_geom.Intersection(county_geom)  │ tract with county
    wind_lyr.SetSpatialFilter(census_geom)

    print('Intersecting census tract with {0} wind polygons'.format(
        wind_lyr.GetFeatureCount()))

    if wind_lyr.GetFeatureCount() > 0:
        out_row.SetField('popsqkm', census_row.GetField('popsqkm'))
        for wind_row in wind_lyr:
            wind_geom = wind_row.geometry()
            if census_geom.Intersect(wind_geom):
                new_geom = census_geom.Intersection(wind_geom)
                out_row.SetField('wind', wind_row.GetField('WPC'))
                out_row.SetGeometry(new_geom)
                out_lyr.CreateFeature(out_row)
del out_ds
```

Check if there are wind polygons ➡ (points to `if wind_lyr.GetFeatureCount() > 0:`)

Check if the tract and wind polygons intersect ➡ (points to `if census_geom.Intersect(wind_geom):`)

You have an extra step to get the results you want, however. Unfortunately, the census and county boundaries don't line up exactly (figure 7.9), which means that a census tract that barely overlaps the county because of this data error will be used to select wind polygons even though you don't need it. One way to deal with this is to intersect the census and county polygons so that you only use the part of the census polygon

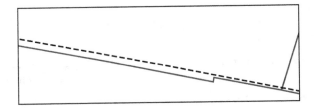

Figure 7.9 The solid census tract boundary doesn't line up perfectly with the dotted county boundary.

that falls within the county polygon (for example, the tiny sliver in figure 7.9). Once you've found this intersection, then you can use a spatial filter to select the wind polygons that it contains or overlaps.

After setting the spatial filter, you iterate through the selected wind polygons and intersect each of them with the census polygon. This throws out parts of a census tract that don't get enough wind or suitable wind areas with too high of a population density. The attribute filter remains in effect, even with the spatial filter changes, so this is always limited to the suitable wind polygons. You add each of these intersection polygons to the new dataset, along with attributes for wind class and population density.

Figure 7.10 is zoomed in on part of the results. You're close, but it would be nice to have large polygons instead of many small ones. This will lose the information about wind suitability class and population density, but at this point you know that all of your polygons are appropriate, anyway.

The fastest way to combine the little polygons into one large one is to use the UnionCascaded function, which requires that the polygons to be joined are all contained in a single multipolygon. It works correctly only if you add individual polygons to the multipolygon, however. If you add a multipolygon, then you'll get incorrect results

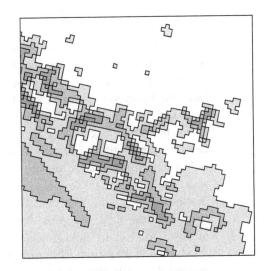

Figure 7.10 Suitable wind farm locations according to our analysis. The darker the shading, the higher the wind rating.

later, so you need to break up any multipolygons created by your earlier intersections and add each one individually. The following listing shows this process.

Listing 7.4 Combining small polygons into large ones

```
folder = r'D:\osgeopy-data\California'
ds = ogr.Open(folder, True)
in_lyr = ds.GetLayerByName('wind_farm')
```

```
    out_lyr = ds.CreateLayer(
        'wind_farm2', in_lyr.GetSpatialRef(), ogr.wkbPolygon)
    out_row = ogr.Feature(out_lyr.GetLayerDefn())

    multipoly = ogr.Geometry(ogr.wkbMultiPolygon)

    for in_row in in_lyr:
        in_geom = in_row.geometry().Clone()
        in_geom_type = in_geom.GetGeometryType()
        if in_geom_type == ogr.wkbPolygon:
            multipoly.AddGeometry(in_geom)
        elif in_geom_type == ogr.wkbMultiPolygon:
            for i in range(in_geom.GetGeometryCount()):
                multipoly.AddGeometry(
                    in_geom.GetGeometryRef(i))

    multipoly = multipoly.UnionCascaded()

    for i in range(multipoly.GetGeometryCount()):
        poly = multipoly.GetGeometryRef(i)
        if poly.GetArea() > 1000000:
            out_row.SetGeometry(poly)
            out_lyr.CreateFeature(out_row)
del ds
```

◀ **The multipolygon to hold everything**

Add single polygons

◀ **Break up multipolygons**

◀ **Union everything together**

◀ **Only keep large polygons**

After you union all of the polygons together into one large multipolygon, you go through it and break it up into individual polygons that you add to the new shapefile. Small islands of land that aren't big enough to hold a wind farm can be thrown out, so you only keep the polygons with an area of at least a square kilometer. The results are shown in figure 7.11, and you can see that some of these little polygons that were off by themselves are now gone.

A dataset like this, with only large polygons, is probably easier to work with than one with many small polygons, as long as you don't need the information that's lost by joining them all together.

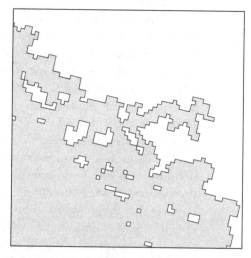

Figure 7.11   The results from unioning the small polygons in figure 7.10 together and throwing out the small island polygons

7.4   *Example: animal tracking data*

The website https://www.movebank.org/ has a database of animal tracking data for studies all over the world. I downloaded GPS location data for Galapagos Albatrosses as a CSV file, but let's convert it into a shapefile and then play with the data a bit. You can use the x and y coordinates from the location-long and location-lat columns to create a point and copy that and the individual-local-identifier and timestamp columns as attributes. The shapefile format doesn't support true date/time fields, so you'll keep the timestamp information as a string. The code for this is shown in the following listing.

> **Listing 7.5   Create a shapefile from a .csv file**

```
from osgeo import ogr, osr

csv_fn = r"D:\osgeopy-data\Galapagos\Galapagos Albatrosses.csv"
shp_fn = r"D:\osgeopy-data\Galapagos\albatross_dd.shp"
sr = osr.SpatialReference(osr.SRS_WKT_WGS84)

shp_ds = ogr.GetDriverByName('ESRI Shapefile').CreateDataSource(shp_fn)
shp_lyr = shp_ds.CreateLayer('albatross_dd', sr, ogr.wkbPoint)
shp_lyr.CreateField(ogr.FieldDefn('tag_id', ogr.OFTString))
shp_lyr.CreateField(ogr.FieldDefn('timestamp', ogr.OFTString))
shp_row = ogr.Feature(shp_lyr.GetLayerDefn())

csv_ds = ogr.Open(csv_fn)
csv_lyr = csv_ds.GetLayer()
for csv_row in csv_lyr:
    x = csv_row.GetFieldAsDouble('location-long')
    y = csv_row.GetFieldAsDouble('location-lat')
    shp_pt = ogr.Geometry(ogr.wkbPoint)              # Create the point
    shp_pt.AddPoint(x, y)
    tag_id = csv_row.GetField('individual-local-identifier')
    timestamp = csv_row.GetField('timestamp')
    shp_row.SetGeometry(shp_pt)
    shp_row.SetField('tag_id', tag_id)
    shp_row.SetField('timestamp', timestamp)
    shp_lyr.CreateFeature(shp_row)

del csv_ds, shp_ds
```

Unfortunately, if you plot this new shapefile or open it up in a GIS, you'll see bad points over by Africa (figure 7.12). There must have been an error with the data collection for these points, so their latitude and longitude values are set to 0. Let's get rid of them.

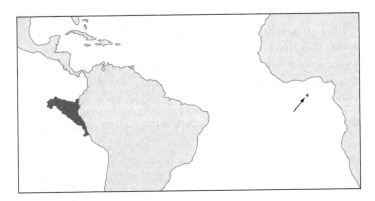

Figure 7.12   A few bad GPS locations by Africa instead of South America

Because you know that the bad points have coordinates of (0, 0), you can set a spatial filter to select those points and then delete them one by one:

```
shp_ds = ogr.Open(shp_fn, True)
shp_lyr = shp_ds.GetLayer()
shp_lyr.SetSpatialFilterRect(-1, -1, 1, 1)
for shp_row in shp_lyr:
    shp_lyr.DeleteFeature(shp_row.GetFID())
shp_lyr.SetSpatialFilter(None)
shp_ds.ExecuteSQL('REPACK ' + shp_lyr.GetName())
shp_ds.ExecuteSQL('RECOMPUTE EXTENT ON ' + shp_lyr.GetName())
del shp_ds
```

Don't forget to use `REPACK` to permanently delete the points and `RECOMPUTE EXTENT` to recalculate the shapefile's spatial extent. Now all of the points are between the Galapagos Islands and South America, as shown in figure 7.13.

Now that the bad points are gone, you can think about doing some analysis. The first things I think of doing with GPS tracking data from animals are to see how far they move and to look at the area they use. Unfortunately, latitude/longitude data in degrees isn't ideal for this, but that's the coordinate system used by these points. Because you won't learn how to work with spatial references and coordinate systems until the next chapter, let's see how to convert between

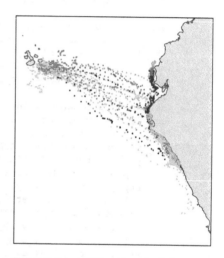

Figure 7.13   GPS locations for Galapagos Albatrosses

coordinate systems using the ogr2ogr command-line utility. Remember that you need to run this from a terminal window or command prompt, not from Python. You'll also need to make sure that you're in the same folder as the albatross_dd shapefile.

You'll convert the coordinates to a system that uses meters rather than degrees as units of measure. Not only are meters easier to understand (most people probably can't visualize half a degree very well), but they're constant, unlike degrees that change with latitude. The system you'll use is called Lambert Conformal Conic, and you'll use a variation of it that's specific for South America. The parts of this command after -t_srs and up to +no_defs are what define the coordinate system. The output will be a shapefile called albatross_lambert.shp.

```
ogr2ogr -f "ESRI Shapefile" -t_srs "+proj=lcc +lat_1=-5 +lat_2=-42
➥ +lat_0=-32 +lon_0=-60 +x_0=0 +y_0=0 +ellps=aust_SA +units=m +no_defs"
➥ albatross_lambert.shp albatross_dd.shp
```

Now you've got a shapefile that uses meters, so let's calculate the distance between each location. To do this, you need to select the points for an individual bird, so let's write a function that will get unique values from an attribute column. You can use the function from the following listing to get `tag_id` values in later listings.

Listing 7.6 Function to get unique values from an attribute column

```
def get_unique(datasource, layer_name, field_name):
    sql = 'SELECT DISTINCT {0} FROM {1}'.format(field_name, layer_name)
    lyr = datasource.ExecuteSQL(sql)
    values = []
    for row in lyr:
        values.append(row.GetField(field_name))
    datasource.ReleaseResultSet(lyr)
    return values
```

To calculate distances, you'll iterate through the points for each bird in order and then calculate the distance between each location and the previous one, so you'll need to keep track of the previous point as you loop. The points should be in the correct order in the original .csv file, which means they're also in order in the shapefile you created, but you'll add code to check, just in case. If it does find something out of order, it will bail so that you can correct the problem. The following listing shows the process.

Listing 7.7 Calculating distance between adjacent points

```
ds = ogr.Open(r'D:\osgeopy-data\Galapagos', True)
lyr = ds.GetLayerByName('albatross_lambert')
lyr.CreateField(ogr.FieldDefn('distance', ogr.OFTReal))    ← Add a distance field

tag_ids = get_unique(ds, 'albatross_lambert', 'tag_id')    ⎫ Loop through
for tag_id in tag_ids:                                     ⎭ individual ids
    print('Processing ' + tag_id)
    lyr.SetAttributeFilter(                                ⎫ Limit points to
        "tag_id ='{}'".format(tag_id))                     ⎭ current id
    row = next(lyr)
    previous_pt = row.geometry().Clone()                   ⎫ Save first time
    previous_time = row.GetField('timestamp')              ⎭ and point
    for row in lyr:
        current_time = row.GetField('timestamp')
        if current_time < previous_time:                   ⎫ Make sure points
            raise Exception('Timestamps out of order')     ⎭ are ordered
        current_pt = row.geometry().Clone()
        distance = current_pt.Distance(previous_pt)        ⎫ Calculate distance
        row.SetField('distance', distance)                 ⎭ to previous point
        lyr.SetFeature(row)
        previous_pt = current_pt                           ⎫ Save current
        previous_time = current_time                       ⎭ time and point
del ds
```

Before starting the loop, you save the timestamp and point geometry for the first location so that you can use it to calculate distance between it and the second feature. This increments the current feature, so the loop starts with the second feature instead of the first, and you calculate a distance between the first and second points in the first iteration. After saving the distance, you store the timestamp and geometry for the current feature in your "previous" variables, so that the next time through the loop you'll have this information. If you hadn't stored the current values, you'd always calculate the distance to the first point, because that's the one originally stored in previous_pt.

Now it's an easy matter to get information about the distances. For example, you could use SQL to find out which bird had the longest distance between GPS fixes:

```
ds = ogr.Open(r'D:\osgeopy-data\Galapagos')
for tag_id in get_unique(ds, 'albatross_lambert', 'tag_id'):
    sql = """SELECT MAX(distance) FROM albatross_lambert
            WHERE tag_id = '{0}'""".format(tag_id)
    lyr = ds.ExecuteSQL(sql)
    for row in lyr:
        print '{0}: {1}'.format(tag_id, row.GetField(0))
```

The first few lines of output look like this:

```
4264-84830852: 106053.530233
4266-84831108: 167097.198703
1103-1103: 69342.7642097
```

What if later you want to know the maximum travel speed from one point to the next? You've got the distances, but you need to know the amount of time in between GPS readings to calculate speed. The fact that the timestamp field is a string, not a date/time, presents a small but easily surmountable problem. Fortunately, you can create Python datetime objects from a string as long as you can tell it how the string is formatted. The timestamps in your dataset look like this:

```
timestamp = '2008-05-31 13:30:02.001'
```

You can create a format string that matches this using the information at https://docs.python.org/2/library/datetime.html#strftime-strptime-behavior, and then use the strptime function to convert the string to a datetime:

```
date_format = '%Y-%m-%d %H:%M:%S.%f'
my_date = datetime.strptime(timestamp, date_format)
```

The following listing shows how to use this information to find the maximum travel speed between each location. This won't be completely accurate because it's probably rare that a bird has been flying the entire time between readings, but at least it's a start.

Listing 7.8 Find maximum speed from locations and elapsed time

```
from datetime import datetime
from osgeo import ogr

date_format = '%Y-%m-%d %H:%M:%S.%f'
ds = ogr.Open(r'D:\osgeopy-data\Galapagos')
lyr = ds.GetLayerByName('albatross_lambert')
for tag_id in get_unique(ds, 'albatross_lambert', 'tag_id'):
    max_speed = 0
    lyr.SetAttributeFilter("tag_id ='{}'".format(tag_id))
```

```
row = next(lyr)
ts = row.GetField('timestamp')
previous_time = datetime.strptime(ts, date_format)
for row in lyr:
    ts = row.GetField('timestamp')
    current_time = datetime.strptime(ts, date_format)
    elapsed_time = current_time - previous_time
    hours = elapsed_time.total_seconds() / 3600
    distance = row.GetField('distance')
    speed = distance / hours
    max_speed = max(max_speed, speed)
print 'Max speed for {0}: {1}'.format(tag_id, max_speed)
```

Convert string to datetime

Calculate speed

As with finding distance, you need to keep track of the previous point so that you can find the length of time between GPS fixes. After getting that information, you divide the distance by the number of hours between readings to get speed in meters per hour.

Now let's take a look at the areas used by each bird. Sophisticated methods are available for determining an animal's home range, but we'll use convex hull polygons because they're simple and OGR has them built in. To do this, you need to put the points for each bird into a multipoint geometry that can then be used to create the convex hull polygons, as shown in the following listing.

Listing 7.9 Create convex hull polygons for each bird

```
ds = ogr.Open(r'D:\osgeopy-data\Galapagos', True)
pt_lyr = ds.GetLayerByName('albatross_lambert')
poly_lyr = ds.CreateLayer(
    'albatross_ranges', pt_lyr.GetSpatialRef(), ogr.wkbPolygon)
id_field = ogr.FieldDefn('tag_id', ogr.OFTString)
area_field = ogr.FieldDefn('area', ogr.OFTReal)
area_field.SetWidth(30)
area_field.SetPrecision(4)
poly_lyr.CreateFields([id_field, area_field])
poly_row = ogr.Feature(poly_lyr.GetLayerDefn())

for tag_id in get_unique(ds, 'albatross_lambert', 'tag_id'):
    print('Processing ' + tag_id)
    pt_lyr.SetAttributeFilter("tag_id = '{}'".format(tag_id))
    locations = ogr.Geometry(ogr.wkbMultiPoint)
    for pt_row in pt_lyr:
        locations.AddGeometry(pt_row.geometry().Clone())

    homerange = locations.ConvexHull()
    poly_row.SetGeometry(homerange)
    poly_row.SetField('tag_id', tag_id)
    poly_row.SetField('area', homerange.GetArea())
    poly_lyr.CreateFeature(poly_row)

del ds
```

Make area field big enough to hold data

Multipoint to hold locations

◀ **Create convex hull**

The results are shown in figure 7.14, where the polygon for one bird is filled in and the rest are hollow. Maybe it's me, but those big polygons don't tell me a whole lot. I'd like to see the area each bird used around the islands and around the mainland, but without the middle of the ocean. In fact, comparing the area used between different visits to the archipelago or mainland might be interesting.

I can think of a few different ways to separate the points into different visits to the continent or islands, but we'll only look at one of them. Listing 7.10 does it by ignoring all locations that are more than 100 kilometers from land, and every time the birds cross an imaginary vertical line in the middle of the ocean, a new set of points is created so the two geographical areas are separated. In the interest of space, the code

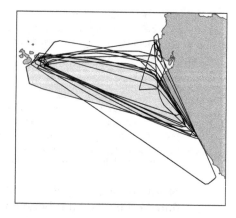

Figure 7.14   The ranges for each bird. The polygon for the bird with ID 4264-84830852 is filled in, but the rest are hollow.

to create the new polygon shapefile itself is omitted and only the code to create the polygons is shown.

Listing 7.10   Create convex hull polygons separated by geographic area

```
land_lyr = ds.GetLayerByName('land_lambert')            Buffer the
land_row = next(land_lyr)                                land polygon
land_poly = land_row.geometry().Buffer(100000)

for tag_id in get_unique(ds, 'albatross_lambert', 'tag_id'):
    print('Processing ' + tag_id)
    pt_lyr.SetAttributeFilter("tag_id = '{}'".format(tag_id))
    pt_locations = ogr.Geometry(ogr.wkbMultiPoint)
    last_location = None
    for pt_row in pt_lyr:
        pt = pt_row.geometry().Clone()
        if not land_poly.Contains(pt):
            continue
        if pt.GetX() < -2800000:
            location = 'island'               Figure out side
        else:                                 of the ocean
            location = 'mainland'
        if location != last_location:                      ◄─  Save points if area
            if pt_locations.GetGeometryCount() > 2:            has changed
                homerange = pt_locations.ConvexHull()
                poly_row.SetGeometry(homerange)
                poly_row.SetField('tag_id', tag_id)
                poly_row.SetField('area', homerange.GetArea())
                poly_row.SetField('location', last_location)
                poly_lyr.CreateFeature(poly_row)
            pt_locations = ogr.Geometry(ogr.wkbMultiPoint)
            last_location = location
        pt_locations.AddGeometry(pt)
```

Skip points not inside the land buffer (annotation pointing to `if not land_poly.Contains(pt):`)

```
if pt_locations.GetGeometryCount() > 2:
    homerange = pt_locations.ConvexHull()
    poly_row.SetGeometry(homerange)
    poly_row.SetField('tag_id', tag_id)
    poly_row.SetField('area', homerange.GetArea())
    poly_row.SetField('location', last_location)
    poly_lyr.CreateFeature(poly_row)
```

◀── **Save the last set of points**

In this example you need a land dataset so you can tell which points are within 100 kilometers of land. After getting the land polygon, you buffer it by 100,000 meters, which is the same as 100 km. When iterating through the points, the first thing you do is check to see if the point falls within the land buffer. If it doesn't, then you move on to the next point without doing anything more. If a point is within the buffer, and therefore within 100 kilometers of land, you check to see which side of the imaginary line the point is on and set a location variable to keep track of whether the point is on the island or mainland. If the location has changed since the previous point you looked at, and you encounter at least three locations (the minimum required for a polygon), then you create a new convex hull polygon with the collected points. After checking the number of points and possibly creating a polygon, you create a new multipoint object to store the next set of points. If you hadn't created a new multipoint, then your next convex hull would use all locations you'd saved so far, but you want to start over now that you're in a different geographic area. When you finish iterating through the points for a specific animal, the points since the last location change still need to be turned into a polygon, so you've got another bit of code to take care of those last locations.

The result for one bird is shown in figure 7.15. This is the same animal whose range was shaded in figure 7.14, so you can see the difference in the calculated ranges.

I don't know about you, but I'm curious how much of the same area is used by an individual on separate visits to the islands or the mainland—do they haunt the same locations or do they switch it up a bit? A simple way to get at this, which ignores the fact that some polygons might be created from a day's worth of data as opposed to others with a week or two, might be to look at the ratio of common area to total area (if you're an albatross biologist, please don't cringe too much at

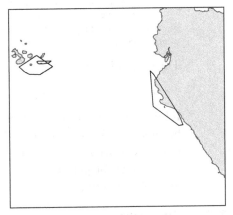

Figure 7.15 The area-specific ranges for bird 4264-84830852. Compare this with the large polygon for the same animal shown in figure 7.14.

my idea). Figure 7.16 shows the difference between these two for one of the bird's visits to the islands.

All area (union)   Common area (intersection)

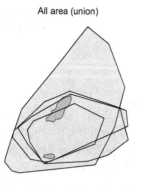 

Figure 7.16   The outlines show the areas used on four different visits to the islands by bird 1163-1163, and the shaded areas show the results of union and intersection operations on these polygons.

Let's look at how you'd calculate this ratio, and then we'll leave the albatrosses alone. The following listing does it for the bird shown in figure 7.16, but you could easily adapt the code to calculate this value for all of the birds.

Listing 7.11   Calculate percentage of total area used in all island visits

```
ds = ogr.Open(r'D:\osgeopy-data\Galapagos')
lyr = ds.GetLayerByName('albatross_ranges2')
lyr.SetAttributeFilter("tag_id = '1163-1163' and location = 'island'")
row = next(lyr)
all_areas = row.geometry().Clone()
common_areas = row.geometry().Clone()
for row in lyr:
    all_areas = all_areas.Union(row.geometry())
    common_areas = common_areas.Intersection(row.geometry())
percent = common_areas.GetArea() / all_areas.GetArea() * 100
print('Percent of all area used in every visit: {0}'.format(percent))
```

The output looks like this:

```
Percent of all area used in every visit: 25.1565197202
```

It looks like a quarter of this bird's total range is used during each visit to the islands, but it's clear from figure 7.16 that this area is in the middle of the range. A next step might be to see how many points were used to create each polygon—maybe the larger polygons are bigger because they have more points and not because the bird changed its habits each time.

7.5   Summary

- Overlap tools tell you the spatial relationship of geometries to each other, such as whether or not they intersect in space.
- Proximity tools are used to determine distances between geometries or create buffers around them.
- As with any type of analysis, it's important to carefully consider your methodologies and assumptions when creating your workflow.

Using spatial reference systems

This chapter covers

- Understanding spatial reference systems
- Transforming data using OSR
- Transforming data using pyproj
- Great-circle calculations using pyproj

Most people are familiar with the concept of using latitude and longitude to specify a location on the earth's surface. Would you be surprised to learn that many other coordinate systems are also used, and that these different spatial reference systems are used for different purposes? To make things even more complicated, the earth isn't a perfect sphere, and multiple models, called *datums*, are used to represent the planet's shape. Given this, coordinates from any system, including latitude and longitude, aren't absolute—a set of coordinates can specify a slightly different location depending on the datum used.

Because so many coordinate systems exist, it's unlikely that all of your data will use the one you need, so the ability to convert data between them is critical. Not only that, but it's impossible to transform data from one spatial reference system to another if you don't know which system they currently use, so you must ensure that this information is documented or risk rendering your data unusable. To effectively work with coordinate systems, you need to understand why so many of them exist in the first place and how to select an appropriate one for your purposes, so we'll start with background information and then move on to transforming data.

8.1   Introduction to spatial reference systems

A spatial reference system is made of three components—a coordinate system, a datum, and a projection—all of which affect where on the earth a set of coordinates refers to. Briefly, datums are used to represent the curvature of the earth, and projections transform coordinates from a three-dimensional globe to a two-dimensional map. Different projections are appropriate for different purposes, such as web mapping, accurately measuring distances, or calculating areas. There's more to it than that, however, and it's important to understand the role that both datums and projections play. Let's back up and review how coordinates are repre-

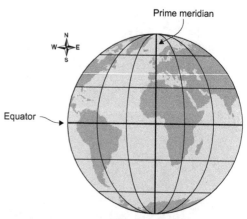

Figure 8.1   Latitude and longitude lines at 30° intervals. Positive latitude values are north of the equator, and positive longitudes are east of the prime meridian.

sented on a globe. Latitude and longitude are the distance, in degrees, from the equator and the prime meridian, respectively. Latitude values range from -90 to 90, with positive values north of the equator. Longitudes range from -180 to 180, with positive values east of the Greenwich prime meridian (shown in figure 8.1). Using degrees makes perfect sense on a spherical surface, and although the earth isn't a perfect sphere, it's close enough for this to be a convenient way to specify a precise location on the planet.

> **DEFINITION**   The prime meridian is the line of longitude that passes through the Royal Observatory, Greenwich, in London. This has been recognized by much of the world as the reference meridian since 1884. The equator is the line of latitude that is equal distances from both the north and south poles.

Methods of specifying latitude and longitude

Multiple methods exist for specifying latitude and longitude coordinates. For example, these are all equivalent:

- *Decimal degrees (DD)*—37.8197° N, 122.4786° W
- *Degrees decimal minutes (DM)*—37° 49.182' N, 122° 28.716' W
- *Degrees minutes seconds (DMS)*—37° 49' 11" N, 122° 28' 43" W

These different notations are based on the fact that angles are divided up into minutes, where one degree in an angle is made up of 60 minutes, and each minute is made up of 60 seconds. Because latitude and longitude are degree measurements, they're also divided up into minutes and seconds. To get decimal minutes from decimal degrees, multiply the fractional part of the DD value by 60, so for example, 60×0.8197 = 49.182. Therefore, 37.8197 degrees equals 37 degrees and 49.182 minutes. Similarly, you can multiply the fractional part of the minutes value by 60 to get seconds. Because $60 \times 0.182 = 10.92$, now you have 37 degrees, 49 minutes, and about 11 seconds.

To go the opposite direction and convert DMS to DD, divide the minutes by 60 and the seconds by 3600 and add those results to the hours value, like this (notice the rounding error):

$$DD = 37 + \frac{49}{60} + \frac{11}{360} = 37 + 0.8167 + 0.0031 = 37.8198$$

Additionally, south and west values are represented as negative numbers if the directions aren't specified. For example, -122.4786° is the same as 122.4786° W.

To use latitude and longitude values in your Python code, you'll need to make sure that they use the decimal degrees format and specify directions using positive and negative values instead of N, S, E, or W.

However, a complication arises from the fact that the earth isn't a perfect sphere or even a perfect ellipsoid. As you probably learned in geometry class, but then promptly forgot, simple equations can model the shape of ellipsoids, including spheres. But these equations assume a perfect geometry with a nice smooth surface and no protrusions and dips. It would be quite something if a planet were to form that perfectly, and ours certainly didn't. Have you ever seen a worn-out ball, like a volleyball, that has developed a weak spot and has a bulge that wasn't there when the ball was new? Not only does the earth have mountains and valleys, but it's a little lopsided like that volleyball, which definitely makes describing its surface with a simple set of equations more complicated.

Because of these anomalies in the planet's surface, and also because measurement accuracies vary, the earth's ellipsoid has multiple models. These models are called datums, and every spatial reference system is based on one of them. One widely used global datum, the *World Geodetic System,* was last revised in 1984. This datum, called WGS84 for short, is the one used for data with a global coverage, including the Global Positioning System (GPS). Most datums are designed to model the curvature of the earth in a more localized area, such as a continent or even a smaller area. A datum designed for one area will not work well elsewhere. For example, the North American Datum of 1983 (NAD83) shouldn't be used in Europe.

Depending on which datum is being used, the same set of latitude and longitude coordinates can refer to slightly different locations, because the underlying ellipsoids are different shapes. Sometimes the difference between coordinates using two different datums is negligible, but other times it can be hundreds of meters. Because of this, you always need to know which datum your geographic data is based on.

Until now we've only talked about three-dimensional ellipsoids, but what you really want in most cases is a two-dimensional map because they tend to be more convenient for most purposes. After all, it's hard to fold up a globe and put it in your pocket or embed one inside of a book! How do mapmakers go from three to two dimensions? One way to solve the problem is with what's called an *interrupted map*, like that shown in figure 8.2. You've probably seen these before, and perhaps you've even cut one out and bent the paper to make a globe. That's kind of cool, but in its two-dimensional form, the map would be much easier to use if land masses weren't split up into chunks and separated by wasted space. This is where projections come in. As their name implies, they're used to project, or transform, locational data into a different coordinate system. These map projections use Cartesian coordinate systems, so locations are specified with x,y coordinate pairs based on two perpendicular axes, like scatterplots or line graphs. The tricky part is converting coordinates on a sphere to a two-dimensional plane.

In fact, many ways to accomplish this exist, and they all have their own strengths and weaknesses. Think about stretching the different parts of the interrupted map shown in figure 8.2 so that the map was a single rectangle with no cutouts. Geographic features would obviously get warped, especially near the poles where you had to stretch farther. No matter how you project geographic data to two dimensions, you'll get distortion, but the type of distortion depends on how you do the conversion. Depending on what you plan to use the data for, some types of distortion may be acceptable while others won't. Figure 8.3 shows a couple of ways a piece of paper could be wrapped around a globe and used to convert the geographic data to 2D. Even with those methods shown here, the angle of the paper could be changed to get a different effect.

Certain projections, called *conformal*, preserve local shapes. For example, the shape of Lake Titicaca on the border of Bolivia and Peru wouldn't change between the globe and the 2D map. No mathematical trickery can preserve the shape of a large area, such as all of Eurasia, however. Mercator projections, including the Universal Transverse Mercator (UTM), are examples of conformal projections. Others, called *equal-area projections*, keep the amount of area the same, so the measured area of

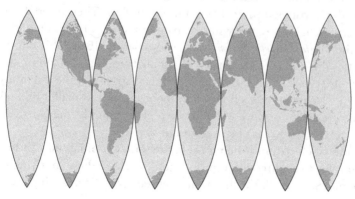

Figure 8.2
An interrupted map

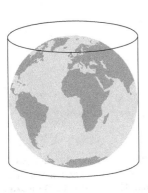 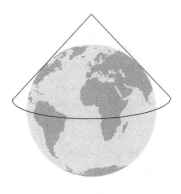

Figure 8.3 Two different ways that a piece of paper could be wrapped around a globe and used to project geographic data onto a two-dimensional surface. The example on the left is cylindrical, and the one on the right is conical.

Greenland wouldn't change, although the shape might. The Lambert equal-area and Gall-Peters projections are two examples of this. Equidistant projections, such as the Azimuthal equidistant, keep distances and scales the same, but only for a certain part of the map, such as the equator. The farther you get from this true line, the greater the distortion. Figure 8.4 shows examples of different projections.

TIP Several terms exist for data that use latitude and longitude coordinate values. You might see them referred to as having a geographic projection or see them called unprojected or geographic.

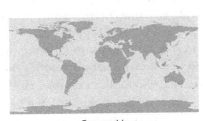

Geographic

Web Mercator (conformal)

Figure 8.4 Examples of different types of projections

Gall-Peters (equal-area)

Azimuthal equidistant

Why should you care about all of these differences? Depending on your purposes, maybe you don't. I doubt I'd be worried about it if I was making a map of the small town I live in. But if I was making a map of the state I live in, I might care if it looked short and fat or a little taller and skinnier, as shown in figure 8.5. What if you cared more about measurements and less about appearances? Let's consider a dramatic example and think about what would happen if you needed to compare the amount of forested area in Columbia and Chile. Sticking with lati-

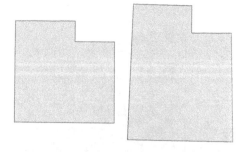

Figure 8.5   The state of Utah shown using geographic (lat/lon) coordinates on the left and UTM Zone 12N on the right. Both examples use the NAD83 datum.

tude and longitude wouldn't work, because the lines of longitude converge at the poles, so one degree of longitude doesn't represent a constant distance. In fact, one degree of longitude is equal to approximately 111 kilometers at the equator, but only about 79 km at a latitude of 45 degrees. Although latitude distance can vary slightly because the earth isn't a perfect sphere, it's generally around 111 kilometers per degree. Therefore, a square 100 km long on each side would measure about 0.8 square degrees in Columbia, but closer to 0.5 square degrees at the southern tip of Chile. Using latitude and longitude to compare the amount of forested area in the two countries would obviously give inaccurate results. Instead, you'd want to choose an appropriate equal-area projection for this purpose.

Projections aren't tied to specific datums, so knowing the projection of your data isn't enough. You also need to know the datum, and it's the combination of the two that defines the spatial reference system. For example, most of the data I get for Utah uses a UTM projection and the NAD83 datum, but I can't safely assume that all UTM data I receive uses NAD83. It could easily be NAD27 or WGS84 instead, so I don't have a complete spatial reference system unless I know both the projection and the datum. If you don't know both components, you might map your data in the wrong location. I've known people who unknowingly set their GPS to display coordinates in an unusual spatial reference system and then collected data by writing down the coordinates shown on their screen. Unfortunately, their data were then unusable because they didn't know what spatial reference system the GPS had been set to display at the time. On the other hand, I also know people who lacked spatial reference information for their data, but fortunately the data were in a common system and they figured it out. If you're collecting data, please simplify your life by paying attention to this crucial information at the beginning of the process, no matter how boring it might seem.

TIP   If you collect geographic data, it's crucial to know from the beginning what projection and datum your coordinates use. If you don't pay attention to this, then your data may end up useless, and nobody wants that.

8.2   *Using spatial references with OSR*

Because spatial reference systems (SRSs) are so important, most vector data formats provide a way to store this information with the data, and you need to know how to work with it. When using spatial data, one common task is to convert the dataset from one spatial reference system to another so that it can be used with other datasets or for a particular analysis. The analysis techniques discussed in the last chapter, for example, only work if the geometries all use the same SRS. Another reason you might need to convert between SRSs is if you're using an online mapping solution that requires a Web Mercator projection to display data.

> **WARNING**   Many GIS software packages will project on-the-fly, which means that they'll automatically convert data to a different SRS when displaying it. For example, if you load in a dataset that uses an Albers equal-area projection, the map will be drawn using that projection. But if you then load a second file that uses UTM, it will be converted to Albers so it can be displayed correctly with the first. Of course, this only happens if both of the datasets have SRS information stored with them, because without that the software doesn't know what to do. Also, this process only changes what's in memory and doesn't alter what's stored on disk in any way. Although this behavior can be helpful when you're using a GIS, sometimes it leads people to assume that datasets use the same SRS when in reality they don't.

The osgeo package contains a module called OSR (short for OGR Spatial Reference) that's used to work with SRSs. This section will show you how to use OSR to assign SRS information to your data so that GIS software, including OSR, knows how to work with it. You'll also learn how to convert data between different SRSs so that you can transform data to whichever SRS you need for a particular project.

8.2.1   *Spatial reference objects*

To work with a spatial reference system, you need a SpatialReference object that represents it. If you already have a layer that uses the SRS you want, then you can get the SRS from it using the GetSpatialRef function. A similar function, GetSpatial-Reference, will get an SRS from a geometry. Both of these functions will return None if the layer or geometry doesn't have an SRS stored with it.

Let's look at the information contained in one of these SRS objects. Perhaps the easiest way is to print it out, which will display a nicely formatted description of the SRS in WKT format and doesn't require the OSR module to be imported. The states_48 shapefile uses a geographic, or unprojected, coordinate system along with the North American Datum of 1983 (NAD83):

```
>>> ds = ogr.Open(r'D:\osgeopy-data\US\states_48.shp')
>>> print(ds.GetLayer().GetSpatialRef())
GEOGCS["GCS_North_American_1983",
    DATUM["North_American_Datum_1983",
        SPHEROID["GRS_1980",6378137.0,298.257222101]],
    PRIMEM["Greenwich",0.0],
    UNIT["Degree",0.0174532925199433]]
```

You can tell that this isn't a projected SRS because it doesn't have a PROJCS entry, only a GEOGCS one. If you were looking at a projected SRS, there would be more information describing the parameters of the coordinate system, such as the UTM example show in figure 8.6.

WKT isn't the only string representation of an SRS. I like PROJ.4 strings because they're especially concise. For example, this is the PROJ.4 string for the UTM SRS shown in figure 8.6:

```
'+proj=utm +zone=12 +ellps=GRS80 +towgs84=0,0,0,0,0,0,0 +units=m +no_defs '
```

The PROJ.4 Cartographic Projections Library is a popular open source library for converting data between projections, and you can read about the details of PROJ.4 definitions at http://trac.osgeo.org/proj/. See appendix D for other functions you can use to get text representations of spatial reference systems. (Appendixes C through E are available online on the Manning Publications website at https://www.manning.com/books/geoprocessing-with-python.) Several of the results are wordy; try out Export-ToXML to see what I mean.

DEFINITION   Spatial Reference System Identifiers (SRIDs) are used to uniquely identify each spatial reference system, datum, and several other related items within a GIS. The software can use its own set of IDs, or it can use a common set such as EPSG (short for European Petroleum Survey Group) codes. These are the AUTHORITY entries in the WKT examples.

```
PROJCS["NAD83 / UTM zone 12N",
    GEOGCS["NAD83",
        DATUM["North_American_Datum_1983",
            SPHEROID["GRS 1980",6378137,298.257222101,
                AUTHORITY["EPSG","7019"]],
            TOWGS84[0,0,0,0,0,0,0],
            AUTHORITY["EPSG","6269"]],
        PRIMEM["Greenwich",0,
            AUTHORITY["EPSG","8901"]],
        UNIT["degree",0.0174532925199433,
            AUTHORITY["EPSG","9122"]],
        AUTHORITY["EPSG","4269"]],
    PROJECTION["Transverse_Mercator"],
    PARAMETER["latitude_of_origin",0],
    PARAMETER["central_meridian",-111],
    PARAMETER["scale_factor",0.9996],
    PARAMETER["false_easting",500000],
    PARAMETER["false_northing",0],
    UNIT["metre",1,
        AUTHORITY["EPSG","9001"]],
    AXIS["Easting",EAST],
    AXIS["Northing",NORTH],
    AUTHORITY["EPSG","26912"]]
```

Figure 8.6   Well-known text for the NAD83 UTM Zone 12N spatial reference system

Fortunately, you don't have to print anything out to discover if an SRS is geographic or projected, because handy functions called `IsGeographic` and `IsProjected` can provide that information. You can also get other information about the SRS, although you do need to know the structure of an SRS to do so. Go back and look at the WKT in figure 8.6. You can use the `GetAttrValue` function to get the text corresponding to the first occurrence of each keyword such as `PROJCS` or `DATUM`, where the keywords aren't case-sensitive. Assuming that the utm_sr variable holds the SRS from figure 8.6, you could get the projection name like this:

```
>>> utm_sr.GetAttrValue('PROJCS')
'NAD83 / UTM zone 12N'
```

Several `AUTHORITY` entries are in the UTM SRS. Which one do you think will be returned by `GetAttrValue`? Let's try it and see:

```
>>> utm_sr.GetAttrValue('AUTHORITY')
'EPSG'
```

That didn't tell you much, because the first value of each one happens to be `'EPSG'`. An optional parameter to `GetAttrValue` lets you specify which child you want returned using its index. The string `'EPSG'` is the first child, but the second is a number, so try getting it:

```
>>> utm_sr.GetAttrValue('AUTHORITY', 1)
'26912'
```

Why did it get the last one shown in figure 8.6 instead of the first? This is because items are nested inside each other in the SRS, and this one is the least nested, so it's the first one returned by the function.

If `GetAttrValue` only grabs the first item that appears with a given keyword, how do you get the others? To get authority codes, or SRIDs, pass the key for the SRID that you're interested in to `GetAuthorityCode`:

```
>>> utm_sr.GetAuthorityCode('DATUM')
'6269'
```

You can get values with a `PARAMETER` key using `GetProjParm`, which takes one of the `SRS_PP` constants from appendix D as an argument:

```
>>> utm_sr.GetProjParm(osr.SRS_PP_FALSE_EASTING)
500000.0
```

Many other functions are available for getting information from an SRS, several of which only apply to certain types of SRSs. See appendix D for a full list.

8.2.2 Creating spatial reference objects

Because you can't always get an appropriate spatial reference object from a layer or geometry, you may need to create your own. Because I like short representations, my two favorite ways to do this are to use a standard EPSG code if it exists for the SRS I want to use, or a PROJ.4 string. As you saw with the UTM example, the EPSG code for

NAD83 UTM 12N is 26912, which you can pass to the `ImportFromEPSG` function after importing OGR and creating a blank `SpatialReference` object:

```
>>> from osgeo import osr
>>> sr = osr.SpatialReference()
>>> sr.ImportFromEPSG(26912)
0
>>> sr.GetAttrValue('PROJCS')
'NAD83 / UTM zone 12N'
```

The SRS you create is equivalent to the UTM example from earlier. The zero returned by `ImportFromEPSG` means that the SRS was imported successfully. Interestingly, watch what happens if you use the PROJ.4 string you saw earlier:

```
>>> sr = osr.SpatialReference()
>>> sr.ImportFromProj4('''+proj=utm +zone=12 +ellps=GRS80
...                       +towgs84=0,0,0,0,0,0,0 +units=m +no_defs ''')
0
>>>
>>> sr.GetAttrValue('PROJCS')
'UTM Zone 12, Northern Hemisphere'
```

The datum name is no longer part of the SRS name because the datum wasn't specified in the PROJ.4 string. However, the GRS80 ellipsoid used by the NAD83 datum was part of the string, so the required information is still there (if you want to prove it to yourself, print the WKT and compare the SPHEROID values to the ones from figure 8.6). To include the datum, add +datum=NAD83 to the PROJ.4 representation.

TIP You can look up EPSG codes, WKT, PROJ.4 strings, and several other representations of SRSs, at www.spatialreference.org.

Several different functions for exporting SRS information exist, and so do multiple methods for importing this information into a spatial reference object, including one to import information from a URL such as a definition on www.spatialreference.org (again, see appendix D). You can also create a spatial reference object from a WKT string without having to use one of the importer functions:

```
>>> wkt = '''GEOGCS["GCS_North_American_1983",
...             DATUM["North_American_Datum_1983",
...               SPHEROID["GRS_1980",6378137.0,298.257222101]],
...             PRIMEM["Greenwich",0.0],
...             UNIT["Degree",0.0174532925199433]]'''
>>>
>>> sr = osr.SpatialReference(wkt)
```

You can also build a spatial reference object yourself, and several projection-specific functions can help you with this. Let's branch out from UTM and build the Albers Conic Equal Area SRS that the United States Geological Survey uses for the lower 48 states (figure 8.7). The projection-specific function for Albers looks like this:

```
SetACEA(stdp1, stdp2, clat, clong, fe, fn)
```

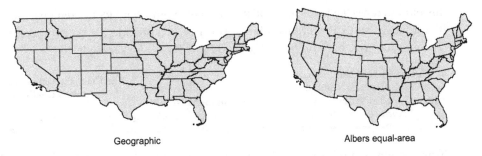

Figure 8.7   **The lower 48 states are shown using geographic coordinates and an Albers equal-area projection.**

The parameters are standard parallel 1, standard parallel 2, latitude of center, longitude of center, false easting, and false northing, in that order. You could use this to build the USGS Albers spatial reference:

```
>>> sr = osr.SpatialReference()
>>> sr.SetProjCS('USGS Albers')
>>> sr.SetWellKnownGeogCS('NAD83')
>>> sr.SetACEA(29.5, 45.5, 23, -96, 0, 0)
>>> sr.Fixup()
>>> sr.Validate()
0
```

The first thing you do after creating an empty SRS is set a name for it, then specify a datum, and last you provide the required parameters for the Albers projection. The call to Fixup adds default values for missing parameters and reorders items so that they match the standard. The last thing you need to do is call Validate to make sure that you didn't forget anything. In this case, Validate returns a zero, which means everything is fine (in fact, many of the other functions in this example also returned zero, but I cut the returned values out of the examples in the interest of space). Try leaving out the datum and see what happens when you call Validate. In that case, it should return 5, which means that the SRS is corrupt. This is because an SRS needs either a datum or a spheroid, neither of which is specified if you leave out the call to SetWellKnownGeogCS. If you turn on exception handling with osr.UseExceptions (True) then Validate will throw an exception instead of return a number.

8.2.3   *Assigning an SRS to data*

It's a good idea to attach SRS information to your dataset whenever possible, so that you always know what coordinate system it uses. You can assign an SRS to layers and individual geometries, although all geometries in a layer share the same SRS. A data source can't be assigned an SRS because individual layers are allowed to have different spatial reference systems.

Do you remember when you created new layers in chapter 3? One of the parameters for the CreateLayer function was a spatial reference object. The default value for this parameter is None, because OGR can't figure out what SRS the data use on its own. If you have a spatial reference object, you need to provide it when you create a new layer because you have no function to add an SRS to an existing layer.

```
sr = osr.SpatialReference()
sr.ImportFromEPSG(26912)
driver = ogr.GetDriverByName('ESRI Shapefile')
ds = driver.CreateDataSource(r'D:\osgeopy-data\output\testdata.shp')
lyr = ds.CreateLayer('counties', sr, ogr.wkbPolygon)
```

The SRS is the second parameter when creating the layer.

Now the new counties shapefile and all of the geometries contained in it will know that they use a UTM SRS (EPSG 26912). Of course, you must create the geometries using UTM coordinates. Assigning an SRS to a layer doesn't magically convert all of the data to that coordinate system. All it does is provide information, so if you assign a different SRS than you're using, you're basically lying and causing confusion because nothing will know how to work with the data.

If you're working with individual geometries instead of layers, you might want to assign an SRS to a geometry. You can do this with the `AssignSpatialReference` method:

```
geom.AssignSpatialReference(sr)
```

Again, this doesn't force the geometry to use the assigned spatial reference, but instead provides information about the SRS, whether right or wrong.

8.2.4 Reprojecting geometries

If you need your data to use a different SRS than the one they're already using, you'll need to reproject them to the new SRS. I have to do this most commonly when I get a new dataset from somewhere but it doesn't use the SRS that I usually use. If I want to use the new data with my existing files, I need to project it so that the SRSs match. I've also needed to do this recently when using software that required data to use Web Mercator but my originals used UTM.

You have two different ways of projecting a geometry. One assumes that the geometry already has an SRS assigned to it, and the other doesn't. We'll look at them both, but first let's get data to work with. This book's data has a shapefile called ne_110m_land_1p.shp that contains the world's landmasses as one multipolygon, and the ospybook module has a function called `get_shp_geom` that pulls the first geometry out of a shapefile. You can use these to get the global multipolygon, and for good measure, why don't you also create a point containing the latitude and longitude of the Eiffel Tower?

```
>>> world = pb.get_shp_geom(r"D:\GeoData\Global\ne_110m_land_1p.shp")
>>> tower = ogr.Geometry(wkt='POINT (2.294694 48.858093)')
>>> tower.AssignSpatialReference(
...     osr.SpatialReference(osr.SRS_WKT_WGS84))
```

Latitude and longitude coordinates

Because WGS84 is so common, the OSR module has a constant that contains the WKT for that geographic coordinate system, which you use here to add an SRS to the tower

geometry. The world geometry also
has a WGS84 SRS associated with it
because the shapefile it came from
does. If you plot the multipolygon,
you should see something similar to
figure 8.8.

Because both geometries know
their SRS, you can reproject them
using their `TransformTo` function,
where you only need to provide the
target SRS. We'll use this to transform them both to a Web Mercator
projection. Certain points, such as
the North and South poles, can't
always be successfully reprojected,

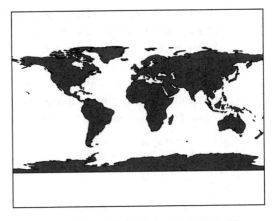

**Figure 8.8 The world's landmasses plotted with
geographic coordinates**

however. This is the case when transforming the world geometry to Web Mercator, so
you also need to use the built-in module to set an environment variable telling it that
it's okay to skip those points. Without that, the world geometry won't be successfully
transformed, and you'll get an error message that says "ERROR 1: Full reprojection
failed, but partial is possible if you define OGR_ENABLE_PARTIAL_REPROJECTION
configuration option to TRUE." You can fix this by importing the gdal module and
using its `SetConfigOption` method:

```
>>> from osgeo import gdal
>>> gdal.SetConfigOption('OGR_ENABLE_PARTIAL_REPROJECTION', 'TRUE')
>>> web_mercator_sr = osr.SpatialReference()
>>> web_mercator_sr.ImportFromEPSG(3857)
>>> world.TransformTo(web_mercator_sr)
>>> tower.TransformTo(web_mercator_sr)
>>> print(tower)
POINT (255444.16760638013 6250816.9576802524)
>>> vp.plot(world)
```

← **Web Mercator
coordinates**

As you can see, the coordinates for
the Eiffel Tower no longer fall in the
range for latitude and longitude values, and the world geometry should
look like figure 8.9 when plotted.
Notice also that the world and tower
geometries themselves were changed
instead of returning new geometries,
which is different behavior than many
other functions we've looked at.

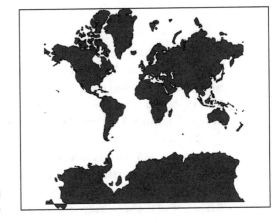

**Figure 8.9 The world's landmasses plotted
with Web Mercator coordinates**

If you use `TransformTo` on a geometry that doesn't have an SRS assigned to it, it won't change and you'll get an error code of 6. You can still transform it provided that you know what its SRS is, however. To do this, you need a `CoordinateTransformation` object, which you can create using a source and a target spatial reference. For example, let's pretend that the `world` geometry doesn't have spatial reference data and use this technique to convert it from Web Mercator to Gall-Peters. This time you'll use the `Transform` function, which requires a `CoordinateTransformation` object:

```
>>> peters_sr = osr.SpatialReference()
>>> peters_sr.ImportFromProj4("""+proj=cea +lon_0=0 +x_0=0 +y_0=0
...                              +lat_ts=45 +ellps=WGS84 +datum=WGS84
...                              +units=m +no_defs""")
>>>
>>> ct = osr.CoordinateTransformation(web_mercator_sr, peters_sr)
>>> world.Transform(ct)
>>> vp.plot(world)
```

Now your plot should look like figure 8.10. If you wanted to reverse this and go from Gall-Peters to Web Mercator, you'd switch the order of the arguments when creating the coordinate transformation.

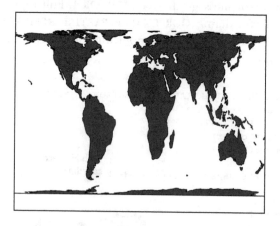

Figure 8.10   The world's landmasses plotted with Gall-Peters coordinates

CHANGING DATUMS

Sometimes you'll also need to change the datum that your dataset uses. For example, sometimes I'm given data that uses the NAD27 datum, which I then need to convert to NAD83 so that it matches the rest of my data. The `TransformTo` and `Transform` functions will convert between datums if the necessary information is present.

Because mathematical equations to convert between datums don't always exist, many times GIS uses data files called *grid shift files* to help with the conversion. These contain the information needed to accurately transform coordinates from one datum to another. The OSR module will use the appropriate files for datum transformations if it can find them on your system, although you must make sure that both your source

SRS with datum and ellipsoid

```
PROJCS["UTM Zone 12, Northern Hemisphere",
    GEOGCS["NAD83",
        DATUM["North_American_Datum_1983",
            SPHEROID["GRS 1980",
                6378137,298.257222101,
                AUTHORITY["EPSG","7019"]],
            TOWGS84[0,0,0,0,0,0,0],
            AUTHORITY["EPSG","6269"]],
        PRIMEM["Greenwich",0,
            AUTHORITY["EPSG","8901"]],
        UNIT["degree",0.0174532925199433,
            AUTHORITY["EPSG","9108"]],
        AUTHORITY["EPSG","4269"]],
    PROJECTION["Transverse_Mercator"],
    PARAMETER["latitude_of_origin",0],
    PARAMETER["central_meridian",-111],
    PARAMETER["scale_factor",0.9996],
    PARAMETER["false_easting",500000],
    PARAMETER["false_northing",0],
    UNIT["Meter",1]]
```

SRS with ellipsoid but no datum

```
PROJCS["UTM Zone 12, Northern Hemisphere",
    GEOGCS["GRS 1980(IUGG, 1980)",
        DATUM["unknown",
            SPHEROID["GRS80",
                6378137,298.257222101]],
        PRIMEM["Greenwich",0],
        UNIT["degree",0.0174532925199433]],
    PROJECTION["Transverse_Mercator"],
    PARAMETER["latitude_of_origin",0],
    PARAMETER["central_meridian",-111],
    PARAMETER["scale_factor",0.9996],
    PARAMETER["false_easting",500000],
    PARAMETER["false_northing",0],
    UNIT["Meter",1]]
```

Figure 8.11 Examples of two spatial references that use the same spheroid, but only one has the datum specified

and target spatial references contain datum information. Figure 8.11 shows an example of two spatial references that use the same projection and ellipsoid, but one has a datum included and the other doesn't. Although both are valid spatial references, only the one with the datum will work for datum transformations. See appendix D for more information about making grid shift files available to OSR.

If you don't have the appropriate grid shift files for your datum transformation, you can set the `towgs84` parameters for your source and target spatial references. These parameters describe an approximate transformation from a particular datum to the WGS84 datum. If you don't know the parameters that you need, you can look them up at www.epsg-registry.org. Make sure you set the search type to Coordinate Transformation, select a geographic area, and enter the name of the datum you're interested in. I searched for *nad27* in the United States, and then selected the *NAD27 to WGS 84 (4)* result because it was described as being appropriate for the lower 48 states. This gave me x, y, and z translation values equal to -8, 160, and 176, respectively. Once you have the appropriate parameters, you can use `SetTOWGS84` to add them to your SRS:

```
sr = osr.SpatialReference()
sr.SetWellKnownGeogCS('NAD27')
sr.SetTOWGS84(-8, 160, 176)
```

It's easier to rely on grid shift files if you can get them, however.

8.2.5   *Reprojecting an entire layer*

No function exists for projecting an entire layer at once, but it's not hard to do. After creating the new layer, you need to loop through each of the features in the original layer, get the geometry and transform it, and then insert a feature containing the transformed geometry into the new layer. Chances are, you'll also want to keep all of the attribute fields, so you'll need to copy the field definitions from the original layer when creating the new one. The following listing shows a simple example of this that assumes the new layer doesn't already exist and that the layer being reprojected contains point geometries.

Listing 8.1   Projecting a point layer

```
from osgeo import ogr, osr

sr = osr.SpatialReference()
sr.ImportFromProj4('''+proj=aea +lat_1=29.5 +lat_2=45.5 +lat_0=23
                      +lon_0=-96 +x_0=0 +y_0=0 +ellps=GRS80
                      +datum=NAD83 +units=m +no_defs''')

ds = ogr.Open(r'D:\osgeopy-data\US', 1)
in_lyr = ds.GetLayer('us_volcanos')

out_lyr = ds.CreateLayer('us_volcanos_aea', sr,
                         ogr.wkbPoint)
out_lyr.CreateFields(in_lyr.schema)            ◀── Copy field definitions

out_feat = ogr.Feature(out_lyr.GetLayerDefn())
for in_feat in in_lyr:
    geom = in_feat.geometry().Clone()
    geom.TransformTo(sr)
    out_feat.SetGeometry(geom)                     Transform geometry
    for i in range(in_feat.GetFieldCount()):       and copy attributes
        out_feat.SetField(i, in_feat.GetField(i))
    out_lyr.CreateFeature(out_feat)
```

The first thing the code in this listing does is create an output spatial reference. Then it opens a data source for writing and gets the existing layer to reproject. Next, a new layer is created and the field definitions are copied from the input layer to the output layer. If you don't do this, then you can't copy attribute values into the new layer. After the new layer is ready to go, then you loop through each of the features in the original layer, and for each one you get its geometry and transform it using the spatial reference created at the beginning of the listing. Notice that you don't provide an input spatial reference and instead are assume that the input layer has an SRS associated with it. After transforming the geometry, you add it to a new feature, copy all of the attribute values to this feature, and then use it to insert the data into the new layer.

8.3   *Using spatial references with pyproj*

As briefly mentioned earlier, the PROJ.4 Cartographic Projections Library is a C library for converting data between SRSs. It's used by a variety of open source projects, including OSR. You don't need to install all of GDAL and OGR in order to take advantage of PROJ.4 with Python, however, because the pyproj module provides a Python wrapper for PROJ.4. Instead of working with geometries, like OSR does, this module works with lists of coordinate values, which can be provided as Python lists, tuples, arrays, NumPy arrays, or scalars (NumPy is a Python module designed to work with large arrays, and you'll learn more about it in chapter 11). If you had a collection of coordinates in a text file, the functions contained in the pyproj module would be an ideal way to convert them to other coordinate systems.

> **TIP** You can find online documentation and downloads for the pyproj module at https://code.google.com/p/pyproj/.

8.3.1 *Transforming coordinates between spatial reference systems*

There are a couple of different ways to convert coordinates between spatial reference systems using pyproj. You can use the `Proj` class to convert between geographic and projected coordinates or the module-level `transform` function to convert between two spatial reference systems. Let's start with converting the Eiffel Tower coordinates from latitude and longitude to UTM Zone 31N. The first thing to do is initialize a `Proj` object with the UTM spatial reference system using a PROJ.4 string, and then use that to transform the coordinates. The syntax might look a bit odd to you, because you don't need to call a specific function on the `Proj` object to do the conversion:

```
>>> import pyproj
>>> utm_proj = pyproj.Proj('+proj=utm +zone=31 +ellps=WGS84')
>>> x, y = utm_proj(2.294694, 48.858093)          ◀── Do the conversion
>>> print(x, y)
448265.9146797105 5411920.651462567
```

Here you pass a single x and single y coordinate to `utm_proj`, and in return it gives you one x and one y. You could also pass in a list of x values and a list of y values (where `x[i]` and `y[i]` are a coordinate pair), and then you'd get two lists in return.

To go the other direction, from projected to geographic coordinates, set the optional `inverse` parameter to `True` and pass in the UTM coordinates. If you use the UTM coordinates just calculated, you'll get the original latitude and longitude values, except with a slight bit of rounding error:

```
>>> x1, y1 = utm_proj(x, y, inverse=True)
>>> print(x1, y1)
2.294693999999985 48.85809299999999
```

Initializing Proj objects

You can initialize `Proj` objects using PROJ.4 strings, arguments corresponding to the parameters in the PROJ.4 string, or with an EPSG code. For example, these are all equivalent:

```
p = pyproj.Proj('+proj=utm +zone=31 +ellps=WGS84')
p = pyproj.Proj(proj='utm', zone=31, ellps='WGS84')
p = pyproj.Proj(init='epsg:32631')
```

If you need to convert between two projected coordinate systems, then it's easiest to use the pyproj `transform` function instead. In addition, you're required to use `transform` if you want to convert between datums. Let's use UTM coordinates for the Statue of Liberty in New York City to compare the difference between the WGS84 and NAD27 datums. The `transform` function takes four required parameters: source SRS, target SRS, x, and y, where the spatial reference information is contained in `Proj` objects. This example converts coordinates from the WGS84 datum to NAD27. Both sets of coordinates use the UTM Zone 18N projection.

```
>>> wgs84 = pyproj.Proj('+proj=utm +zone=18 +datum=WGS84')
>>> nad27 = pyproj.Proj('+proj=utm +zone=18 +datum=NAD27')
>>> x, y = pyproj.transform(wgs84, nad27, 580744.32, 4504695.26)
>>> print(x, y)
580711.5381565462 4504472.13698683
```

WGS84 coordinates →

← **NAD27 coordinates**

Comparing the input and output numbers, it looks like the two datums differ by 30 meters or so in the east/west direction, but the north/south difference is over 200 meters, at least in New York City. As shown in figure 8.12, using the NAD27 coordinates as if they were WGS84 puts the Statue of Liberty in the water rather than on Liberty Island.

This example is a good illustration of why you should always know your datum.

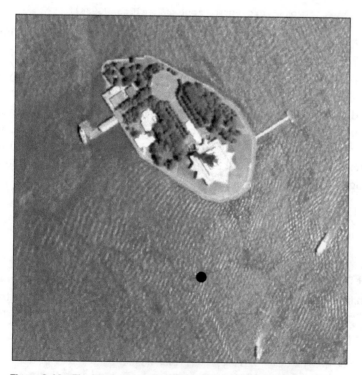

Figure 8.12   The black dot shows where NAD27 coordinates would place the Statue of Liberty if they were treated as if they were WGS84.

8.3.2 *Great-circle calculations*

The shortest distance between two points on the globe is called the *great-circle distance*. Because travelers don't like to travel farther than necessary, these have been important for navigation for centuries. You can use pyproj to get this distance between two sets of latitude and longitude coordinates, along with the starting and ending bearings of the great-circle line between them. To illustrate how this is done, let's look at the distance between Los Angeles and Berlin (figure 8.13). The first thing you need to do is instantiate an object of the Geod class with the ellipsoid you want to use. A list of options is available on the pyproj website mentioned earlier. Once you have the Geod, pass the starting and ending coordinates in decimal degrees to its inv function in order to get the forward bearing, backward bearing, and distance:

```
>>> la_lat, la_lon = 34.0500, -118.2500
>>> berlin_lat, berlin_lon = 52.5167, 13.3833
>>> geod = pyproj.Geod(ellps='WGS84')
>>> forward, back, dist = geod.inv(la_lon, la_lat, berlin_lon, berlin_lat)
>>> print('forward: {}\nback: {}\ndist: {}'.format(forward, back, dist))
forward: 27.23284045673668
back: -38.49148498662066
dist: 9331934.878166698
```

What exactly do these results mean? If you were to head out from Los Angeles (the first set of coordinates passed to inv) at a bearing of 27.2328 degrees and travel 9,331,935 meters, you'd find yourself in Berlin. Or if you wanted to travel the other way, leave Berlin at a bearing of -38.4915° and travel the same distance to arrive in Los Angeles.

You can also find out where you'd end up if you followed a bearing for a certain distance. To do this, pass the starting coordinates, bearing, and distance in meters to the fwd function. This will return the ending longitude, latitude, and bearing back to where you came from. For example, if you plug in

Figure 8.13 The great-circle path between Los Angeles and Berlin

the Berlin coordinates, backward bearing, and distance you got a minute ago, it should spit out the coordinates for Los Angeles along with the bearing from LA to Berlin:

```
>>> x, y, bearing = geod.fwd(berlin_lon, berlin_lat, back, dist)
>>> print('{}, {}\n{}'.format(x, y, bearing))
-118.25000000000001, 34.05000000000002
27.23284045673668
```

You can also get a list of equally spaced coordinates along the great-circle line by passing starting and ending coordinates and the number of desired points to the npts function:

```
>>> coords = geod.npts(la_lon, la_lat, berlin_lon, berlin_lat, 100)
>>> for i in range(3):
...     print(coords[i])
```

```
. . .
(-117.78803196383676, 34.78972514500416)
(-117.31774994946879, 35.52757560403803)
(-116.83878951054419, 36.2634683783333)
```

I used the npts function to generate the points used to draw the great circle path between Los Angeles and Berlin in figure 8.13.

8.4   Summary

- Several main types of map projections exist, and each is used to preserve a specific property of the data. Make sure you choose a projection appropriate for your use.
- Always make sure you know both the projection and datum of your datasets.
- You can't transform data to another spatial reference system if you don't know the one it currently uses.
- You can use either OSR or pyproj to transform data between spatial reference systems.
- Use pyproj for great-circle calculations.

Reading and writing raster data

If you have a geographic dataset that's made of continuous data such as elevation or temperature, it's probably a raster dataset. Spectral data such as aerial photographs and satellite imagery are also stored this way. These types of datasets don't assume strict boundaries exist between objects in the way that vector datasets do. Think of a digital photograph and how each pixel can be a slightly different color than the pixels next to it. The fact that pixel values can vary continuously like this makes for a much better-looking photo than if there were only a few colors to choose from. This trait also makes rasters appropriate for continuously varying data such as elevation.

Working with raster datasets is different from working with vectors. Instead of having individual geometries, you have a collection of pixels which is essentially a large two- or three-dimensional array of numbers. A raster dataset is made of bands

173

instead of layers, and each of these bands is a two-dimensional array. The collection of bands becomes a 3D array. It's a different way of thinking about spatial data, and if you have a math phobia, this description might make it sound scary, too. But you'll soon see that it's not.

In this chapter you'll learn basic theory of raster data, including tips for keeping them to manageable sizes. Then you'll see how to use Python and GDAL to read these datasets into memory and how to write them back out to disk. The easiest case is to read and write an entire dataset at once, but sometimes you don't need the entire spatial extent, and other times the amount of memory is a limiting factor, so you'll also learn how to deal with a spatial subset of the data. It's also possible to change the pixel size while reading or writing, and you'll see how to do that as well.

9.1   *Introduction to raster data*

As mentioned, raster datasets can hold pretty much any type of data you'd like. That doesn't mean that it's always a good idea to use rasters, however. Objects that can be thought of as points, lines, or polygons are usually better left as vectors. For example, country boundaries lend themselves perfectly to a polygon vector dataset. This same data could be stored in a raster, but it would take up more disk space and the boundaries wouldn't be nice, smooth lines. You also couldn't use vector data analysis functions such as buffering and intersecting. These would still be possible using raster techniques, but you'd be better off sticking to vector in this case.

Raster is a perfect choice when values change continuously instead of at sharply defined boundaries. This includes common datasets such as elevation, slope, aspect, precipitation, temperature, and satellite data, but it can include many other things, too. It could be evapotranspiration, distance from roads, soil moisture, or anything else you might need to model as a continuous variable. Sometimes you need what would normally be vector data to be represented as a raster. For example, rivers and streams are good candidates for vector data, but rasters are required for modeling flow accumulation or groundwater flow, such as what would be needed to track the flow of a contaminant in the water supply.

Also, raster datasets don't have to contain continuous data. In fact, I see many rasters made of categorical data such as land cover type. One reason for this is that rasters are used in the models used to produce these datasets in the first place. For example, land cover models typically use visible and nonvisible wavelengths of light from satellite imagery, along with ancillary data such as elevation. The model output is a raster because the inputs are rasters, and it makes sense to leave it that way. It also makes the data easy to use as an input to other raster-based models.

Other examples include viewshed analysis, which takes topology into account when determining what's visible from a certain location. Maybe a ski resort would use this to decide where to locate a restaurant for the best views, and I know of a case where this type of analysis is being used to determine if ground bird nesting sites are visible to hawks perched on power lines (unpublished, but see Hovick et al.[1] for related research). Speaking of wildlife, rasters can also be used for habitat modeling, which might be done purely for the sake of knowledge, or to help select conservation areas. You might think of elevation as a fairly static dataset, but I know researchers who use ground-based lasers (in the form of LIDAR systems) to create elevation models of riverbeds, and then they do the same thing after a flood event so they can compare the before and after elevation profiles (Schaffrath et al.[2]). The possibilities are endless, really.

Now let's talk a bit about the details of raster datasets. You can probably envision a digital photograph as a two-dimensional array of pixels. In fact, that's what we talk about when discussing the dimensions of a photo—the numbers of rows and columns in that 2D array. This is what raster datasets are, except that they aren't limited to two dimensions. They can have a third dimension in the form of bands. Digital photos have multiple bands, too, although you don't usually think of them that way. But they have one each for red, green, and blue wavelengths of light. Your computer (or printer) combines these together to produce the colors you see on your monitor screen. You're familiar with this concept if you've ever created a webpage and specified a color using HEX notation, where the first two numbers correspond to red, the second two to green, and the last two to blue, or RGB notation where you provide a separate number for each of these colors.

Also, just as with a photograph, each band in a dataset has the same numbers of rows and columns, so the pixels from one fall in the same spatial location as the pixels in another. If the pixels for each individual color in a photograph didn't line up correctly, I imagine the results would look fairly blurry.

Obviously not all raster datasets are photographs, so pixel values don't have to correspond to colors. Pixel values in a digital elevation model (DEM), for example, correspond to elevation values. Generally DEM datasets only contain one band, because elevation is the only value required to create a useful dataset. Figure 9.1 shows a single-band raster landcover map for the state of Utah, where each unique pixel value represents a different landcover classification. This dataset contains discrete, rather than continuous, data.

Figure 9.1 A landcover classification map, where each unique pixel value corresponds to a specific landcover classification

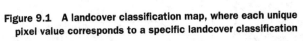

[1] Hovick, T. J., Elmore, R. D., Dahlgren, D. K., Fuhlendorf, S. D., Engle, D. M. 2014. REVIEW: Evidence of negative effects of anthropogenic structures on wildlife: a review of grouse survival and behavior. Journal of Applied Ecology, 51: 1680–1689. DOI: 10.1111/1365-2664.12331.

[2] Schaffrath, K. R., P. Belmont, and J.M Wheaton. 2015. Landscape-scale geomorphic change detection: Quantifying spatially variable uncertainty and circumventing legacy data issues. Geomorphology, 250: 334-348. DOI: 10.1016/j.geomorph.2015.09.020.

NOTE TO PRINT BOOK READERS: COLOR GRAPHICS   Many graphics in this book are best viewed in color. The eBook versions display the color graphics, so they should be referred to as you read. To get your free eBook in PDF, ePub, and Kindle formats, go to https://www.manning.com/books/geoprocessing-with-python to register your print book.

Satellite imagery, on the other hand, contains measurements of various wavelengths, many of which aren't visible to our eyes. While the image might contain bands corresponding to visible red, green, and blue wavelengths, there might also be bands for infrared or thermal radiation. False color images such as those shown in figure 9.2 are created by displaying an infrared band along with visible light. This figure illustrates another use case for satellite imagery in the form of raster data. The image on the left was created using visible light wavelengths, like a traditional photograph. The camera also captured a near infrared wavelength as another band at the same time, and it was used along with the visible red and green bands to create the false color image on the right. The near infrared band is brighter for growing vegetation, and it's displayed as red, so red areas are vegetation. This band combination is useful for monitoring vegetation. The field in the stadium and the practice fields outside are both green and look like grass in the natural color image, but the field inside the stadium is dark gray in the false color image, so it must be artificial. The practice fields outside the stadium are bright red, however, so they must be grass.

Figure 9.2   Two images of Gillette Stadium, home of the New England Patriots, in Foxborough, Massachusetts. The field inside the stadium looks like grass in the natural color image on the left, and it looks gray in the false color image on the right, which makes it clear that it's actually artificial. The practice fields, on the other hand, also look like grass in the natural color image on the left, but are red in the image on the right, signifying that they are indeed grass.

Let's make another comparison to photographs. Although pictures you take with your phone might be geotagged, meaning that metadata in the image specifies where you were standing when you took the photo, each pixel doesn't correspond to a specific location on the ground. That wouldn't even make sense for most of the photos you take, but what if you took one from an airplane, looking straight down? A photo like that could be overlaid on a map if you had the appropriate spatial information. Having only the geotagged coordinates isn't enough, even if you knew exactly what part of the photo the coordinates corresponded to, such as a corner or the center. For example, think how different your photo would look if you were in a Cessna flying relatively close to the ground as opposed to a 747 much higher up. Your photo would cover a much larger area in the second case, even if the photo was taken with the same camera and had the same number of rows and columns. The difference is the pixel size or the area on the ground that a single pixel covers. The pixels in the photo taken from the Cessna would each cover a smaller area than a pixel in the 747 photo. If you knew how much area they each covered, along with the coordinates of one of them, you could figure out how to stretch the photo so that it overlaid on a map. This is assuming that your camera was pointed exactly straight down so the pixels aren't skewed to one side. It also assumes that you had things aligned perfectly so that the top of the photo was exactly north, although you could rotate the image to compensate for that.

Knowing the pixel size is important if you want to overlay your photo on a map, but you obviously need coordinates to go with it. With vector data it's enough to know the spatial reference system because the coordinates for each feature are stored in the vertices. Given the SRS, each vertex can be placed in the correct location with lines drawn between them, and you have your geometry. Raster datasets don't use vertices, and instead commonly use one set of coordinates, the pixel size, and the amount the dataset is rotated to determine coordinates for the rest of the image. This is called an *affine transformation*, and is a common way to georeference a raster dataset, although it isn't the only way. The set of coordinates is generally for the upper-left corner of the image and is called the *origin*. For the simple (and common) case of a raster that has the top of the dataset facing north, you only need these coordinates and the pixel size to find the coordinates of any pixel in the image. All you need to do is figure out the offsets from the origin, multiply those by the pixel size, and add that to the origin coordinates. Figure 9.3 shows how to get the upper-left coordinates for the pixel in the fifth column and fourth row. The first row and column have offsets 0, so

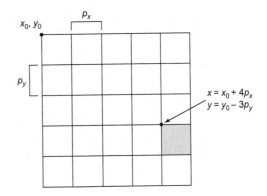

Figure 9.3 An example showing how to get the upper-left coordinates for the pixel in the fifth column and fourth row

your offsets in this case are 4 and 3. To get the easting coordinate, multiply the pixel width by 4 to get the distance across those four columns. Then add that to the origin's easting coordinate, and you're done. You can get northings the same way, but by using the row offset instead of the column. As you can see, it's extremely important that the origin coordinates are correct and using the right SRS, or else you can't calculate coordinates for any part of the dataset.

As you've no doubt noticed, the more pixels contained in a photo, the more disk space that photo requires. Similarly, raster datasets can take up a lot of space on disk and in RAM, so you should ensure that you're not using smaller pixels or larger data types than are necessary. For example, if your data are only good down to 10 meters, it doesn't make sense to have 1-meter pixels because the smaller pixels inside your 10-meter block will all have the same value. As a comparison, you may have seen compact digital cameras with more megapixels than high-end digital SLRs. The SLR still takes better photos, though, because higher quality data is collected. More pixels aren't a substitute for quality data or effective resolution. Not only would they fail to improve your data, all of these extra pixels would greatly increase the size of your file. Doubling the number of rows and columns doesn't double the size of the image. Instead, it quadruples it! For example, an image with 250 rows and 250 columns would have 250 x 250 = 62,500 pixels, while an image with 500 rows and columns has 250,000 pixels.

The data type you choose for your data is also important when it comes to storage space. For example, if your pixel values all fall in the range of 0 to 254, then you should use a byte data type (254 is the largest value a byte can hold). In this case, each pixel will take up a byte, or 8 bits, of disk space, no matter the value, unless you're using compression. If you were to store this same data as 32-bit integer, each pixel would take up four times as much space as before. You'd be taking up four times as much memory with absolutely no benefit. With a small dataset, this might not matter so much, but it certainly does with large datasets.

If rasters can be so large and take a while to process, how can they be drawn on your screen in a reasonable amount of time? This is where *overview layers* come in. You might have also heard them called *pyramid layers* or *reduced resolution datasets* (hence the .rrd extension that several types of overviews have). Overview layers are reduced resolution layers—they're rasters that cover the same area as the original, but are resampled to larger pixel sizes. A raster dataset can have many different overviews, each with a different resolution. When you're zoomed out and looking at the whole image, the coarse resolution layer is drawn. Because the pixels are so large, that layer doesn't take much memory and can draw quickly, but you can't tell the difference at that zoom level. As you zoom in, a higher resolution layer is drawn, but only the part you're viewing needs to be loaded and shown. If you zoom in enough, you'll see the original pixels, but because you're only looking at a small subset of the image, it still draws quickly.

Figure 9.4 shows how this works. Each successive overview layer has a pixel size twice as large as the previous one. All resolutions look the same when viewing the entire image, but you can see the difference when you're zoomed into a smaller area. The

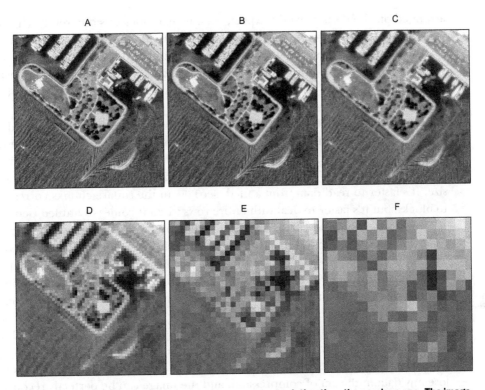

Figure 9.4 Example of how each overview is a coarser resolution than the previous one. The image in the upper left (A) is full resolution, and each successive image uses a pixel size twice as large as the previous one. The last image (F) only looks good when zoomed way out, but it draws much faster than the full image at that scale.

upper-left image in the figure is a full resolution (1 x 1-meter pixel) image of San Diego Harbor. You can see individual cars, boats, and trees. The middle-top image is the first set of overviews, with pixels 2 meters on a side. The pixels in the upper-right image are 4 x 4 meters. Along the bottom row, the pixels are 8, 16, and 32 meters on a side. If you're zoomed out so you can see all of San Diego, you can't tell the difference between the first and the last image, but the last one is a small fraction of the size of the full-resolution one and draws much faster. A good rule of thumb is to create overview layers of decreasing resolution until the coarsest one has 256 pixels or less in one dimension.

Incidentally, the ubiquitous web-mapping services use this same technique to display aerial photography, except that the reduced resolution layers are stored as collections of individual tiles. Your browser downloads whichever tiles are required to cover the area you're looking at, and as you zoom in, you get tiles that have a higher resolution and cover a smaller area, until eventually the resolution is so good that you can see your house or car.

Another aspect of raster datasets that influences access speed is how they're stored on disk. Rasters are made up of blocks, which have to do with how the data are arranged on the disk. As you'd expect, each format does this differently. (If not, they wouldn't really be different formats, would they?) Blocks of pixels are all physically stored next to each other on disk, so they can be accessed together efficiently. It's possible that other blocks belonging to the same image are stored on another part of the disk; this is the sort of problem you solve by defragmenting your drive. It's faster to grab data that are close to each other physically, just as it's faster for you to pull two books off the same shelf instead of from two different bookshelves. If you need to read or write data, it's most efficient to use blocks. For example, GeoTIFFs come in tiled and untiled formats. Untiled GeoTIFFs store each row of pixels as a block, but tiled ones use square sets of pixels instead, with 256 x 256 being a common block (or tile) size. It's faster to read data from a tiled GeoTIFF in the square chunks corresponding to blocks, but it's faster to deal with entire rows when it comes to untiled GeoTIFFs.

You're probably also wondering about data compression, because this is regularly used with digital photos in .jpeg format and can significantly reduce the size of a file. This is certainly possible, and multiple types of compression are available, depending on the data format. You might have heard of *lossy* versus *lossless* compression. Lossy compression loses information in the act of compressing the data. When saving .jpegs, you've probably noticed the compression quality option. The higher the quality, the less data you lose and the better the resulting image looks. The .png format, however, is lossless, which is why you aren't asked for a compression quality when saving one of those. That doesn't mean that the data can't be compressed. It means that you won't lose any data in the act of compression, and the image can be perfectly reconstructed into the original uncompressed dataset. If you plan to compress your data but you also need to perform analyses on it, make sure you select a lossless algorithm. Otherwise, your analysis won't be operating on the actual pixel values because several will have been lost. GeoTIFF is a popular lossless format.

I have one more important concept you need to understand if you're going to use raster data, and that's the difference between resampling methods. You don't have a one-to-one mapping between pixels when a raster is resampled to a different cell size or reprojected to another spatial reference system, so new pixel values have to be calculated. The simplest and fastest method, called *nearest-neighbor*, is to use the value from the old pixel that's closest to the new one. Another possible algorithm, shown in figure 9.5, is to take the average of the four closest pixels. Several other methods use

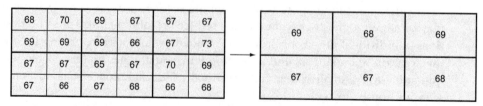

Figure 9.5   Simple resampling example, where the average of four pixels is used to calculate the value for a new pixel that covers the same extent as the four smaller pixels

multiple input pixels, such as bilinear, which uses a weighted average of the four closest input pixels.

You should always use nearest-neighbor resampling when dealing with datasets containing discrete values, such as the landcover classifications from figure 9.1. Otherwise, you might end up with a value that doesn't correspond to a classification, or with a number that denotes a completely unrelated classification. Continuous data, on the other hand, are well-suited to the other resampling methods. For example, taking an average elevation value makes perfect sense, and you'll get smoother output that way than if you used nearest-neighbor.

9.2   *Introduction to GDAL*

Now that the theory is out of the way, let's learn how to work with these datasets using GDAL. Numerous different file formats exist for raster data, and GDAL is an extremely popular and robust library for reading and writing many of them. The GDAL library is open source, but has a permissive license, so even many commercial software packages use it. Unfortunately, they don't necessarily use it to read as many formats as possible, so I have students and colleagues ask me on a regular basis if I can convert their data into a format their software can read. Every time I get asked this, I point the person to GDAL and its command-line utilities. They're usually amazed with what they can do with free software, and if they knew how to write their own code, they could do even more.

The GDAL library is well known for its ability to read and write so many different formats, but it also contains a few data processing functions such as proximity analysis. You'll still have to write your own processing code in many cases, but this is relatively easy for many types of analyses. There's a Python module called NumPy that's designed for processing large arrays of data, and you can use GDAL to read data directly into NumPy arrays. After manipulating the data however you need, using NumPy or another module that works with these arrays, you can write the array back out to disk as a raster dataset. It's a pretty painless process. You'll learn how to work directly with NumPy arrays in chapter 11.

I only use a small handful of raster formats on a regular basis, and I imagine that most readers of this book do the same, but nonetheless there are well over 100 different format drivers available for GDAL, listed online at http://www.gdal.org/formats_list.html. Each one of these drivers handles reading and writing a specific data format. You probably won't have all of them available with your version of GDAL, but they do exist. If you need a particular driver and can't find a precompiled GDAL binary with that specific one, you can always compile your own customized version of GDAL (although this might be tricky, especially if you have no experience with that sort of thing). Not all drivers support the same operations, however. While many support reading and writing, some are read-only, and others won't let you modify existing datasets, although you can create new ones. Assuming the driver supports the operation you have in mind, you use all of the drivers the same way. Most of my examples will use GeoTIFFs, but they'd work fine with other formats as well.

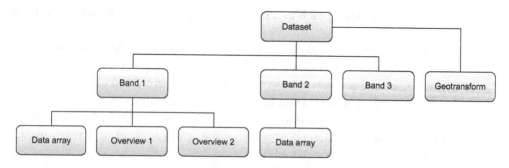

Figure 9.6   The basic structure of a GDAL dataset. Each dataset contains one or more bands, which in turn contain the pixel data.

The basic structure of a GDAL dataset is shown in figure 9.6 and matches what you've learned about raster datasets in general. Each dataset contains one or more bands, which in turn contain the pixel data and possibly overviews. The georeferencing information is contained in the dataset because all of the bands use the same information for this.

To illustrate how to use GDAL to read and write raster data, let's start with an example that combines three individual Landsat bands into one stacked image, as shown in figure 9.7. The Landsat program is a joint initiative between the United

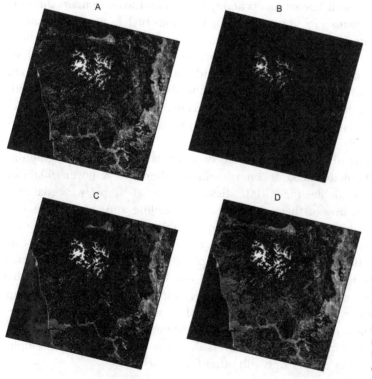

Figure 9.7   Red (A), green (B), and blue (C) Landsat bands are shown in black and white, and you can see that they look a bit different from each other. Part D shows these three bands stacked into an RGB image like that created by listing 9.1.

States Geological Survey (USGS) and the National Aeronautics and Space Administration (NASA), and has been collecting moderate-resolution satellite imagery worldwide since 1972. Landsat images are distributed by the USGS as a collection of GeoTIFFs, one for each collected band. With the exception of bands 6 (thermal) and 8 (panchromatic), each of these has a 30-meter resolution and because they're from the same Landsat scene, the same dimensions. This makes things easy, because the bands go directly on top of one another with no fiddling required. Listing 9.1 shows how to create a three-band dataset with these same dimensions, and then copy bands 3, 2, and 1 into it. These three bands correspond to red, green, and blue wavelengths of visible light, respectively, so putting them in this order will result in an RGB (red, green, blue) image that will look much like it would to your own eyes. Figure 9.7 shows the individual red, blue, and green bands in black and white, along with the resulting three-band natural color image. If you preview your images in a GIS, they'll probably look similar to this because the GIS will most likely stretch them. If you view them somewhere else, they'll look washed out compared to this. I chose to include the stretched images because you can't see any detail at all in the others on the printed page. Let's take a look at the following code.

Listing 9.1   Stacking individual raster bands into one image

```
import os
from osgeo import gdal                                    ← Import GDAL

os.chdir(r'D:\osgeopy-data\Landsat\Washington')
band1_fn = 'p047r027_7t20000730_z10_nn10.tif'
band2_fn = 'p047r027_7t20000730_z10_nn20.tif'
band3_fn = 'p047r027_7t20000730_z10_nn30.tif'

in_ds = gdal.Open(band1_fn)                               Open the band 1
in_band = in_ds.GetRasterBand(1)                          GeoTIFF

gtiff_driver = gdal.GetDriverByName('GTiff')              Create a three-band
out_ds = gtiff_driver.Create('nat_color.tif',            GeoTIFF with same
    in_band.XSize, in_band.YSize, 3, in_band.DataType)    properties as band 1
out_ds.SetProjection(in_ds.GetProjection())
out_ds.SetGeoTransform(in_ds.GetGeoTransform())

in_data = in_band.ReadAsArray()                           Copy pixel data from input
out_band = out_ds.GetRasterBand(3)                        band to band 3 of output
out_band.WriteArray(in_data)

in_ds = gdal.Open(band2_fn)                               Copy pixel data from a
out_band = out_ds.GetRasterBand(2)                        dataset instead of a band
out_band.WriteArray(in_ds.ReadAsArray())

out_ds.GetRasterBand(1).WriteArray(
    gdal.Open(band3_fn).ReadAsArray())

out_ds.FlushCache()                                       Compute statistics on
for i in range(1, 4):                                     each output band
    out_ds.GetRasterBand(i).ComputeStatistics(False)
```

```
out_ds.BuildOverviews('average', [2, 4, 8, 16, 32])
```
← **Build over
views/pyramid layers**
```
del out_ds
```

What happens here? Well, the first thing you do is import the gdal module. Then you set your current directory and specify which file corresponds to which Landsat band. Then you open the GeoTIFF containing the first band by passing the filename to gdal.Open. You also grab a handle to the first and only band inside the dataset, although you haven't read any data in yet. Notice that you use an index of 1 instead of 0 to get the first band. Band numbers always start with 1 when you use GetRasterBand, although I frequently forget and use 0 and then have to fix my error. Anyway, you need this band object before creating the output image because it has information you need.

   TIP   Remember that band indices start at 1 instead of 0.

Next you create a new dataset to copy the pixel data into. You have to use a driver object to create a new dataset, so you find the GeoTIFF driver and then use its Create function. Here's the full signature for that function:

```
driver.Create(filename, xsize, ysize, [bands], [data_type], [options])
```

- filename is the path to the dataset to create.
- xsize is the number of columns in the new dataset.
- ysize is the number of rows in the new dataset.
- bands is the number of bands in the new dataset. The default value is 1.
- data_type is the type of data that will be stored in the new dataset. The default value is GDT_Byte.
- options is a list of creation option strings. The possible values depend on the type of dataset being created.

Because you use the GeoTIFF driver, the output file will be a GeoTIFF no matter what file extension you give it. The extension isn't added automatically, however, so you do need to provide it. In this case, you call it nat_color.tif and save it in the D:\osgeopy-data\Landsat\Washington folder, because that's the current folder set with os.chdir. You're also required to provide the numbers of columns and rows when creating a new dataset, so you use the XSize and YSize properties, respectively, to get that information from the input band. The next argument to Open is the number of bands, and you want this new raster to have three of them. The next optional parameter is the data type, which has to be one of the values from table 9.1. You obtain this information from the input band, although you could have ignored it in this case because these images use the default type of GDT_Byte. You can also provide format-specific creation options, but you don't do that here. Because every format has its own options, you need to consult www.gdal.org/formats_list.html for your format of interest.

Table 9.1 GDAL data type constants

Constant	Data type
GDT_Unknown	Unknown
GDT_Byte	Unsigned 8-bit integer (byte)
GDT_UInt16	Unsigned 16-bit integer
GDT_Int16	Signed 16-bit integer
GDT_UInt32	Unsigned 32-bit integer
GDT_Int32	Signed 32-bit integer
GDT_Float32	32-bit floating point
GDT_Float64	64-bit floating point
GDT_CInt16	16-bit complex integer
GDT_CInt32	32-bit complex integer
GDT_CFloat32	32-bit complex floating point
GDT_CFloat64	64-bit complex floating point
GDT_TypeCount	Number of available data types

At this point you have an empty three-band dataset, but you probably want it to know what SRS it uses and where it's located on the planet. The next two lines take care of those details, and they're repeated here:

```
out_ds.SetProjection(in_ds.GetProjection())
out_ds.SetGeoTransform(in_ds.GetGeoTransform())
```

You get the projection (SRS) from the input dataset and copy it to the new dataset, and then you do the same for the *geotransform*. The geotransform is important because it provides the origin coordinates and pixel sizes, along with rotation values if the image isn't situated so the top faces north. As you learned earlier, the origin and pixel size are extremely important when it comes to placing the dataset in the correct spatial location. Although you don't have to add the projection and geotransform information before adding pixel values, I prefer to get this out of the way as soon as I create the new dataset.

After setting up your dataset, it's time to add pixel values. Because you already have the band object from the GeoTIFF for Landsat band 1, you can read the pixel values from it into a NumPy array. If you don't provide any parameters to ReadAsArray, then all pixel values are returned in a two-dimensional array with the same dimensions as the raster itself. At this point your in_data variable holds a two-dimensional array of pixel values:

```
in_data = in_band.ReadAsArray()
```

Now, because band 1 of a Landsat image is the blue band, you need to put that into the third band of your output image to get the bands in RGB order. The next thing you do is get the third band from out_ds and then use WriteArray to copy the values in the in_data array into the third band of your new dataset:

```
out_band = out_ds.GetRasterBand(3)
out_band.WriteArray(in_data)
```

You still need to add the green and red Landsat bands to your dataset, so then you open the second band's GeoTIFF. Notice that you don't get the band object from the dataset, though, because you're going to read pixel data directly from the dataset itself this time. Because the second Landsat band is the green one, you then get a handle to the second (green) band in your stacked dataset, and copy the data from the Landsat file to your stacked dataset:

```
in_ds = gdal.Open(band2_fn)
out_band = out_ds.GetRasterBand(2)
out_band.WriteArray(in_ds.ReadAsArray())
```

When you call ReadAsArray on a dataset, you get a three-dimensional array if the dataset you're reading from has multiple bands. Because the Landsat file only has one band, ReadAsArray on the dataset returns the same two-dimensional array that you'd get from the band object. Instead of saving the data into an intermediate variable, this time you immediately send it to the output band. Then you do the same thing for the red pixel values, but compress it into even less code. The result is the same, however:

```
out_ds.GetRasterBand(1).WriteArray(gdal.Open(band3_fn).ReadAsArray())
```

In the next bit of code, you compute statistics on each band in your dataset. This isn't strictly necessary, but it makes it easier for some software to display it nicely. The statistics include mean, minimum, maximum, and standard deviation. A GIS can use this information to stretch the data on the screen and make it look better. You'll see an example of how to stretch data manually in a later chapter. Before computing statistics, you have to ensure that the data have been written to disk instead of only cached in memory, so that's what the call to FlushCache does. Then you loop through the bands and compute the statistics for each one. Passing False to this function tells it that you want actual statistics instead of estimates, which it might get from overview layers (which don't exist yet) or from sampling a subset of the pixels. If an estimate is acceptable, then you can pass True instead; this will also make the calculation go faster because not every pixel needs to be inspected:

```
out_ds.FlushCache()
for i in range(1, 4):
    out_ds.GetRasterBand(i).ComputeStatistics(False)
```

The last thing you do is build overview layers for the dataset. Because these pixel values are continuous data, you use average interpolation instead of the default of nearest-neighbor. You also specify five levels of overviews to build. It happens that five levels are what you'd need to get tiles of size 256 for this particular image:

```
out_ds.BuildOverviews('average', [2, 4, 8, 16, 32])
```

Oh, and don't forget to delete the output dataset. This will happen automatically when the variable goes out of scope, but this may not be when your script finishes running if you're using an interactive Python environment. This is a regular occurrence when my students are working on their homework. They don't flush the cache or delete the variable, and their IDE doesn't release the dataset object when the script finishes, so they end up with an empty image and don't know why.

Other modules for working with raster data

If you'd like to play with a module that uses more "Pythonic" syntax but still harnesses the power of GDAL, check out rasterio at https://github.com/mapbox/rasterio. This module depends on GDAL and uses it internally to read and write data, but it tries to make the process of working with raster data a little easier.

Another module that might be of interest is imageio. This one is written in pure Python and doesn't rely on GDAL. It doesn't focus on geospatial data, but it can read and write many different raster formats, including video formats. You can read more about it at http://imageio.github.io/.

9.3 Reading partial datasets

In listing 9.1 you read and wrote entire bands of data at a time. You can break it up into chunks if you need to, however. This might be because you only need a spatial subset of the data to begin with, or maybe you don't have enough RAM to hold it all at once. Let's a take a look at how you can access subsets instead of the entire images.

The `ReadAsArray` function has several optional parameters, although they differ depending on whether you're using a dataset or a band.

Here's the signature for the band version:

```
band.ReadAsArray([xoff], [yoff], [win_xsize], [win_ysize], [buf_xsize],
                 [buf_ysize], [buf_obj])
```

- `xoff` is the column to start reading at. The default value is 0.
- `yoff` is the row to start reading at. The default value is 0.
- `win_xsize` is the number of columns to read. The default is to read them all.
- `win_ysize` is the number of rows to read. The default is to read them all.
- `buf_xsize` is the number of columns in the output array. The default is to use the `win_xsize` value. Data will be resampled if this value is different than `win_xsize`.

- buf_ysize is the number of rows in the output array. The default is to use the win_ysize value. Data will be resampled if this value is different than win_ysize.
- buf_obj is a NumPy array to put the data into instead of creating a new array. Data will be resampled, if needed, to fit into this array. Values will also be converted to the data type of this array.

The xoff and yoff parameters specify the column and row offsets, respectively, to start reading at. The default is to start reading at the first row and column. The win_xsize and win_ysize parameters indicate how many rows and columns to read, and the default is to read them all. The buf_xsize and buf_ysize parameters allow you to specify the size of the output array. If these values are different than the win_xsize and win_ysize values, then the data will be resampled as it's read to match the output array size. The buf_obj parameter is a NumPy array that the data will be stored in instead of a new array being created. The pixel data type will be changed to match the data type of this array. You'll get an error if you provide buf_xsize and buf_ysize values that don't match the dimensions of this array, but there's no reason to provide sizes in that case anyway, because they can be determined from the array itself.

For example, to read the three rows and six columns starting at row 6000 and column 1400 shown in figure 9.8, you could do something like the following:

```
data = band.ReadAsArray(1400, 6000, 6, 3)
```

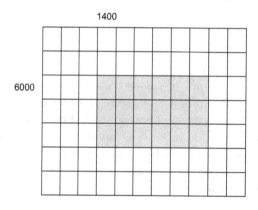

Figure 9.8
Use ReadAsArray(1400, 6000, 6, 3) to read three rows and six columns starting at row 6000 and column 1400.

If you need the pixel values as floating-point instead of byte, you can convert them using NumPy after you've read them in, like this:

```
data = band.ReadAsArray(1400, 6000, 6, 3).astype(float)
```

Or you could have GDAL do the conversion for you as it reads the data. To use this method, you create a floating-point array and then pass it as the buf_obj parameter to ReadAsArray. Make sure you create the array with the same dimensions as the data being read.

```
import numpy as np
data = np.empty((3, 6), dtype=float)
band.ReadAsArray(1400, 6000, 6, 3, buf_obj=data)
```

The NumPy empty function creates an array that hasn't been initialized with any values, so it contains garbage until you fill it somehow. The first parameter to the function is a tuple containing the dimensions of the array to create. If it's a two-dimensional array, the tuple contains the number of rows and then the number of columns. The dtype parameter is optional and specifies the type of data that the array will hold. If not provided, the array will hold floating-point numbers.

To write a data array out to a specific location in other dataset, pass the offsets to WriteArray. It will write out all data in the array you pass to the function, beginning at the offsets you provide.

```
band2.WriteArray(data, 1400, 6000)
```

One important thing to remember when reading partial datasets is that you have to make sure you don't try to read more data than exists, or you'll get an error. For example, if an image has 100 rows, and you ask it to start reading at offset 75 and read in 30 rows, that would go past the end of the image and will fail. A similar problem will occur if you pass an array to WriteArray that's too large to fit in the raster, given your starting offsets.

Access window out of range error messages

The following message means that I tried to read a 30 x 30 array from band 1 of testio.tif, starting at column 0 and row 75. The problem is that testio.tif only has 100 rows and 100 columns, so there aren't 30 rows to read if I start at number 75.

"ERROR 5: testio.tif, band 1: Access window out of range in RasterIO(). Requested (0,75) of size 30x30 on raster of 100x100."

How might you use this information to process a large dataset that won't fit in RAM? Well, one way would be to deal with a single block at a time. Remember that rasters store their data on disk in blocks. Because the data in a block are stored together on disk, it's efficient to process images in these chunks.

The basic idea is shown in figure 9.9. You start with the first block of rows and columns, and then go to the next block in either the x or y direction (this example uses the latter). Each time you jump to the next block, you need to make sure there's really a full block's worth of data to read. For example, if the block size is 64 rows, you need to check that at least 64 rows are left that you haven't read yet. If there aren't, then you can only read in as many as are left, and you'll get an error if you try to access more. Once you've worked your way to the end, you move to the next block of columns and start over, working through the rows. Again, you always need to make sure that you don't try to read more columns than exist in the raster.

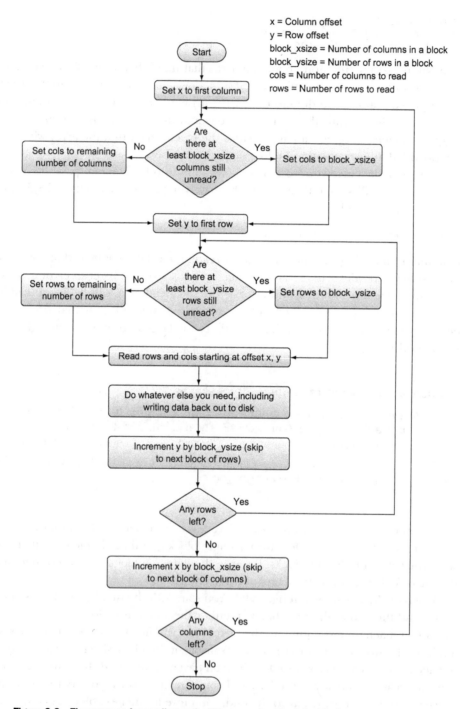

Figure 9.9   The process for reading and writing a raster block by block

Listing 9.2 shows how you might convert a digital elevation model from meters to feet, one block at a time. This is a small dataset, so you probably wouldn't need to break it up like this in the real world, but you would process a large dataset in the same way. This also shows you an example of dealing with NoData values in your raster, which are pixels that are considered to have a null value. Pixels must have a value, but a specific value can be specified as NoData, and therefore ignored.

> **Listing 9.2 Processing a raster by block**

```
import os
import numpy as np
from osgeo import gdal

os.chdir(r'D:\osgeopy-data\Washington\dem')

in_ds = gdal.Open('gt30w140n90.tif')
in_band = in_ds.GetRasterBand(1)
xsize = in_band.XSize
ysize = in_band.YSize
block_xsize, block_ysize = in_band.GetBlockSize()        Get block size and
nodata = in_band.GetNoDataValue()                         NoData value

out_ds = in_ds.GetDriver().Create(
    'dem_feet.tif', xsize, ysize, 1, in_band.DataType)
out_ds.SetProjection(in_ds.GetProjection())
out_ds.SetGeoTransform(in_ds.GetGeoTransform())
out_band = out_ds.GetRasterBand(1)

for x in range(0, xsize, block_xsize):
    if x + block_xsize < xsize:                           Get number of
        cols = block_xsize                                columns to read
    else:
        cols = xsize - x
    for y in range(0, ysize, block_ysize):
        if y + block_ysize < ysize:                       Get number of
            rows = block_ysize                            rows to read
        else:
            rows = ysize - y
        data = in_band.ReadAsArray(x, y, cols, rows)
        data = np.where(data == nodata, nodata, data * 3.28084)
        out_band.WriteArray(data, x, y)

out_band.FlushCache()
out_band.SetNoDataValue(nodata)                           Compute statistics at end,
out_band.ComputeStatistics(False)                         after setting NoData
out_ds.BuildOverviews('average', [2, 4, 8, 16, 32])
del out_ds
```

Read and write one block's worth of data (annotation for the `data = in_band.ReadAsArray` ... `out_band.WriteArray` lines)

You can probably figure out what's happening at the beginning of this example. You open the dataset and get information about the band, including the size of its blocks and its NoData value. After creating the output dataset, you start looping through the blocks in the horizontal (x) direction. You start at column 0 and go up to the last column, represented with index xsize. The twist is that each time through the loop, you increment x by the number of columns in a block (the third argument to range is the amount to increment by), so you skip from the beginning of one block to the beginning of the next. Then you store the number of columns to read in a variable called cols. If there's a full block's worth of columns left to read, this variable is set to the number of columns in a block. But if there aren't enough columns, as would be the case when x is equal to 10 in figure 9.10, the number of remaining columns (three in the figure) is used instead. You need to do this because you'll get an error if you try to read more rows or columns than exist.

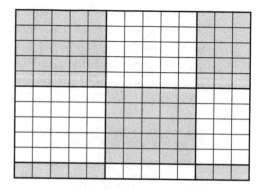

Figure 9.10   A small example image with a block size of five rows and five columns. Alternating blocks are shaded to make it easy to see. The upper-left pixel has offset 0,0.

After computing the number of columns to read, you repeat the process for the number of rows. As shown in figure 9.10, the first two times through that second loop you read five rows, but the third time there's only one row left to read. After processing the first five columns of all rows, you go to the next iteration of the outer loop and process the next five columns, and then the last three columns.

Once you figure out how many rows and columns to read, you pass those numbers, along with the current row and column offsets, to ReadAsArray to get a block's worth of data back:

```
data = in_band.ReadAsArray(x, y, cols, rows)
```

The next step is to convert the values, which come in as meters, to feet. You use the NumPy where function to help with this. This function is like an if-else statement. The first parameter is the condition to check, in this case whether or not the pixel value is equal to the NoData value. The second parameter is the output value if the condition is true. If the incoming pixel is NoData, you output NoData as well. The third parameter is the value to output if the condition is false, so this is where you convert the values to feet by multiplying them by 3.28084:

```
data = np.where(data == nodata, nodata, data * 3.28084)
```

After converting valid pixels to feet, you pass the data to WriteArray using the current row and column offsets before continuing on to the next block:

```
out_band.WriteArray(data, x, y)
```

After processing all of the blocks, you calculate statistics and build the overviews. To exclude the NoData pixels from the statistics calculation, you have to tell the band which value represents `NoData` before calling `ComputeStatistics`. You might be tempted to calculate statistics inside your loop, but you want the statistics to be based on all of the pixels in the band, so you need to wait until all of the pixel values have been calculated.

Obviously this method of looping through blocks is more complicated than reading and writing an entire band at once, but it's invaluable if you're low on RAM.

9.3.1 Using real-world coordinates

Until now, we've only considered pixel offsets when deciding where to start reading or writing data, but most of the time you'll have real-world coordinates instead. Fortunately, converting between the two is easy, as long as your coordinates use the same SRS as the raster. You saw earlier how to calculate coordinates of individual pixels, and now you need to reverse that process. All of the data required, including the origin coordinates, pixel sizes, and rotation values, are stored in the geotransform you've been copying between datasets. The geotransform is a tuple containing the six values shown in table 9.2. The rotation values are usually 0; in fact, I can't recall ever using an image that wasn't north up, but they're certainly out there.

Table 9.2 `GeoTransform` items

Index	Description
0	Origin x coordinate
1	Pixel width
2	x pixel rotation (0° if image is north up)
3	Origin y coordinate
4	y pixel rotation (0° if image is north up)
5	Pixel height (negative)

You could use this information to apply the affine transformation yourself, but GDAL provides a function that does it for you, called `ApplyGeoTransform`, that takes a geotransform, an x value, and a y value. When used with a dataset's geotransform, this function converts image coordinates (offsets) to real-world coordinates. But right now you're interested in going the other direction, so you need to get the inverse of the dataset's geotransform. Fortunately, a function exists for that, but you use it differently depending on the version of GDAL that you're using. If you're using GDAL 1.x, the `InvGeoTransform` function returns a success flag and a new geotransform that can be used to go the other direction:

```
gt = ds.GetGeoTransform()
success, inv_gt = gdal.InvGeoTransform(gt)
```

If all went well, then the success flag will be 1, but if the affine transformation couldn't be inverted, it returns 0 instead. Because of this, you should check the value of the success flag before continuing if you're not certain that the geotransform can be inverted.

If you're using GDAL 2.x, then the InvGeoTransform function only returns one item: a geotransform if one could be calculated, or None if not. In this case, you need to make sure that the returned value isn't equal to None:

```
inv_gt = gdal.InvGeoTransform(gt)
```

Now that you have an inverse geotransform, you can use it to convert real-world coordinates to image coordinates. For example, say you need the pixel value at coordinates 465200, 5296000. The following code would get it, assuming that the raster covers that location:

```
offsets = gdal.ApplyGeoTransform(inv_gt, 465200, 5296000)
xoff, yoff = map(int, offsets)
value = band.ReadAsArray(xoff, yoff, 1, 1)[0,0]
```

The ApplyGeoTransform function returns an x and a y value as floating-point numbers, but you need integer offsets to pass to ReadAsArray. If you forget to convert the offsets to integers, you'll get an error. After getting the integers, you read in one row and one column starting at those offsets. You might think that this would return a number, but not quite. Remember that ReadAsArray returns a two-dimensional array, and it does this even for only one row and/or one column. To get the actual pixel value, you still have to get the value in the first row and first column (position [0,0]) in the array.

This method is extremely inefficient if you need to sample pixel values at many different locations, however. In that case, you're better off reading in the entire band and then pulling the appropriate values from that array. This is because read and write operations are expensive, so doing a new read operation for each point is much slower than doing one large read operation for the whole band. The code to get the same pixel value using this method might look like this:

```
data = band.ReadAsArray()
x, y = map(int, gdal.ApplyGeoTransform(inv_gt, 465200, 5296000))
value = data[yoff, xoff]
```

Obviously, you wouldn't read the whole band in for each point; you'd do that once but then repeat the last two lines for each point. Notice that the row and column offsets are reversed when pulling the pixel value from the NumPy array, because NumPy wants offsets as [row, column], not [x, y] (which is the same as [column, row]).

> **TIP** Use [row, column] offsets for NumPy arrays. This is the reverse of what you're used to using with GDAL.

The ability to convert between real-world coordinates and offsets is also important if you want to extract a spatial subset and save it to a new image, because you need to change the origin coordinates in the geotransform. Say you wanted to extract Vashon Island (figure 9.11) out of the natural color Landsat image you created earlier, and

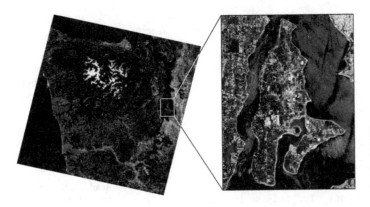

Figure 9.11 The goal of listing 9.3 is to extract Vashon Island out of the natural color Landsat image created earlier.

you're given the upper-left and lower-right coordinates of the area of interest. You need to convert these into pixel offsets so you know what data to read, but it's unlikely that these bounding coordinates correspond exactly to pixel boundaries, so you also need to find the true upper-left coordinates for the subset you extracted. The following listing shows an example of this.

Listing 9.3 Extracting and saving a subset of an image

```python
import os
from osgeo import gdal

vashon_ulx, vashon_uly = 532000, 5262600
vashon_lrx, vashon_lry = 548500, 5241500

os.chdir(r'D:\osgeopy-data\Landsat\Washington')
in_ds = gdal.Open('nat_color.tif')
in_gt = in_ds.GetGeoTransform()

inv_gt = gdal.InvGeoTransform(in_gt)
if gdal.VersionInfo()[0] == '1':
    if inv_gt[0] == 1:
        inv_gt = inv_gt[1]
    else:
        raise RuntimeError('Inverse geotransform failed')
elif inv_gt is None:
    raise RuntimeError('Inverse geotransform failed')

offsets_ul = gdal.ApplyGeoTransform(
    inv_gt, vashon_ulx, vashon_uly)          # Compute upper-left
offsets_lr = gdal.ApplyGeoTransform(         # and lower-right offsets
    inv_gt, vashon_lrx, vashon_lry)
off_ulx, off_uly = map(int, offsets_ul)
off_lrx, off_lry = map(int, offsets_lr)

rows = off_lry - off_uly                     # Compute number of rows
columns = off_lrx - off_ulx                  # and columns to extract
```

```
gtiff_driver = gdal.GetDriverByName('GTiff')
out_ds = gtiff_driver.Create('vashon2.tif', columns, rows, 3)
out_ds.SetProjection(in_ds.GetProjection())
subset_ulx, subset_uly = gdal.ApplyGeoTransform(
    in_gt, off_ulx, off_uly)
out_gt = list(in_gt)                             Put new origin coordinates
out_gt[0] = subset_ulx                           in geotransform
out_gt[3] = subset_uly
out_ds.SetGeoTransform(out_gt)

for i in range(1, 4):
    in_band = in_ds.GetRasterBand(i)
    out_band = out_ds.GetRasterBand(i)
    data = in_band.ReadAsArray(                  Read in data using
        off_ulx, off_uly, columns, rows)         computed values
    out_band.WriteArray(data)       ◄────┐  Write out data
                                         └  starting at the origin
del out_ds
```

You've seen everything in this example before, but you haven't seen it put together quite like this. The important parts are where you compute the offsets for the upper-left and lower-right corners of Vashon Island, based on the coordinates at the top of the script (in real life you probably wouldn't want the coordinates hardcoded in, but it works for the example). Then you subtract the upper-left offsets from the lower-right offsets to get the total numbers of rows and columns to extract.

Once you have that basic information, you create an output image with these new dimensions, rather than the dimensions of the original image. The projection information is copied over unchanged, but you have to alter the geotransform to reflect the upper-left coordinates of the subset. You can't use the upper-left coordinates that you calculated because those probably fall in the middle of a pixel somewhere, but you need the coordinates of the pixel corner. Notice that you use the original geotransform for this, not the inverted one, because you're converting offsets to real-world coordinates. Then, because the geotransform is returned as a tuple, you have to convert it to a list before you can insert the new upper-left coordinates.

After all that housekeeping, you copy the data from the original to the new image. You start reading at the upper-left offsets, and grab the numbers of columns and rows that you computed earlier. There's no reason to provide offsets when writing the data out because the new image will only contain the subset, so you want to start writing at the origin.

You'd probably also compute statistics and build overviews, but those steps aren't absolutely necessary, so I left them out in the interest of space.

9.3.2 *Resampling data*

A nice feature of the ReadAsArray function is that you can use it to resample data as it's read in, either by specifying the output buffer size or passing an existing buffer array. As a reminder, here's what the function signature looks like:

```
band.ReadAsArray([xoff], [yoff], [win_xsize], [win_ysize], [buf_xsize],
                 [buf_ysize], [buf_obj])
```

The win parameters specify the number of rows and columns to read from the band, and the buf parameters specify the size of the array to put those pixel values into. An array with larger dimensions than the original will resample to smaller pixels, while one with smaller dimensions will resample to larger pixels using nearest-neighbor interpolation.

RESAMPLING TO SMALLER PIXELS

To resample data to a finer resolution, provide an array that's larger than the data being read in so that the pixel values need to be repeated to fill the target array. For example, this will create four pixels for every one pixel, essentially cutting the pixel size in half, as shown in figure 9.12:

```
band.ReadAsArray(1400, 6000, 3, 2, 6, 4)
```

This works because you're reading three columns and two rows from the band, but putting that data into an array with six columns and four rows, so each row and column is duplicated to fill the output array.

Figure 9.12 Pixel values are repeated four times each when the numbers of rows and columns are doubled.

This is all well and good, but how do you deal with the new cell size if you need to write the data out to a new image? It's easy, because all you have to do is alter the geo-transform so that it specifies a smaller pixel size. Take a look at the following listing to see how you might resample an entire image to a smaller pixel size.

Listing 9.4   Resample an image to a smaller pixel size

```
import os
from osgeo import gdal

os.chdir(r'D:\osgeopy-data\Landsat\Washington')

in_ds = gdal.Open('p047r027_7t20000730_z10_nn10.tif')
in_band = in_ds.GetRasterBand(1)
out_rows = in_band.YSize * 2          Get number of output
out_columns = in_band.XSize * 2       rows and columns

gtiff_driver = gdal.GetDriverByName('GTiff')
out_ds = gtiff_driver.Create('band1_resampled.tif',      Create output dataset
    out_columns, out_rows)
```

```
out_ds.SetProjection(in_ds.GetProjection())
geotransform = list(in_ds.GetGeoTransform())
geotransform [1] /= 2
geotransform [5] /= 2
out_ds.SetGeoTransform(geotransform)
```

Edit the geotransform so pixels are one-quarter previous size

```
data = in_band.ReadAsArray(
    buf_xsize=out_columns, buf_ysize=out_rows)
out_band = out_ds.GetRasterBand(1)
out_band.WriteArray(data)
```

Specify a larger buffer size when reading data

```
out_band.FlushCache()
out_band.ComputeStatistics(False)
out_ds.BuildOverviews('average', [2, 4, 8, 16, 32, 64])  ◄──┐
del out_ds
```

Build appropriate number of overviews for larger image

This example has a few important things to note. First, you double the number of rows and columns when creating the new dataset, and you pass these same numbers as parameters to ReadAsArray. This ensures that your input data dimensions match your output data dimensions, and also causes the data to be resampled to those larger dimensions. Instead of using the buf_xsize and buf_ysize parameters, you could have used an existing array for the buf_obj parameter and gotten the same results. You could also have provided the win_xsize and win_ysize parameters, but they default to the original numbers of rows and columns, which is what you want.

You also edit the geotransform to reflect the smaller pixel size. The second item in the geotransform is the pixel width, and the sixth is the pixel height, so you divide each of those values by two and overwrite the original values. Because this image still covers the same spatial extent as the original, you don't need to change any of the other values. Once you finish editing, you set the geotransform onto the new dataset. Fortunately, editing the geotransform doesn't alter the geotransform for the original image because the tuple isn't linked to the dataset, so you're not introducing any complications there.

If you hadn't changed the pixel size and instead copied the original geotransform to the new dataset, your output would have looked like the larger image shown in figure 9.13. As you may recall, the spatial extent of a raster is determined from the origin coordinates and the pixel size. The upper-left corner coordinates would be the same, but the incorrect pixel size would cause the image to cover twice the distance in each direction. In this case, a satellite image of northwestern Washington State would appear to extend into eastern Washington and south into Oregon, which is obviously incorrect.

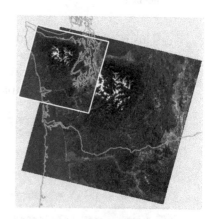

Figure 9.13 This illustrates the result of resampling to a smaller pixel size without changing the size in the geotransform. The smaller image in the upper left is correct. The larger one was created by using the unedited geotransform from the input image.

It should be clear by now how important an accurate geotransform is. If a raster appears to be in the wrong location or the wrong size when opened in a GIS, then an incorrect geotransform is a likely culprit, as is an erroneous spatial reference system.

RESAMPLING TO LARGER PIXELS

You can also resample to a coarser resolution by providing a smaller buffer array when reading the data. In this case, one pixel takes the place of several cells, and nearest-neighbor interpolation is used to determine which value is used (figure 9.14). The following example replaces four pixels with one:

```
data = np.empty((2, 3), np.int)
band.ReadAsArray(1400, 6000, 6, 4, buf_obj=data)
```

Here, an empty integer NumPy array with three columns and two rows is created beforehand and then passed as an argument to ReadAsArray. The six columns and four rows requested from the image are resampled to fit into this smaller array. By the way, you don't need to catch the return value from ReadAsArray in this case, because you already have the data variable. But not only is the data variable filled automatically, it's also returned from the function, so you can grab it that way if you'd like, but it's not necessary.

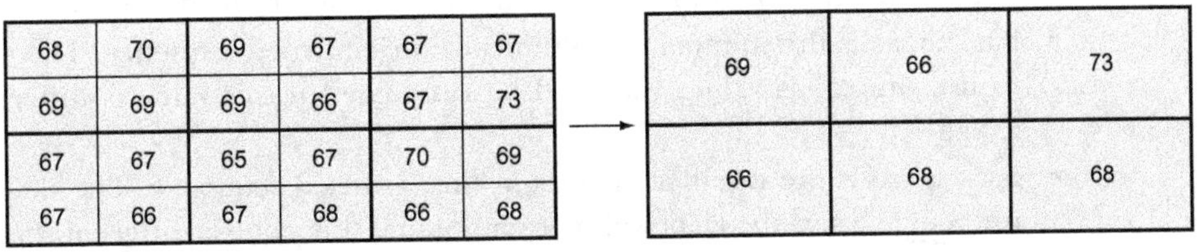

68	70	69	67	67	67
69	69	69	66	67	73
67	67	65	67	70	69
67	66	67	68	66	68

69	66	73
66	68	68

Figure 9.14 Nearest-neighbor interpolation is used to select a pixel value when resampling to smaller dimensions. In this case, the lower-right pixel value for each block of four pixels is used in the output.

Although this technique usually uses nearest-neighbor interpolation to resample, if you have an overview layer of the requested resolution, then that will be used instead. If the appropriate overview was built with average interpolation, then that's what you'd get when using ReadAsArray, rather than nearest-neighbor.

As with resampling to smaller pixels, you need to change the pixel size in the geotransform when writing the data back out to another dataset. The only difference is that in this case, you want to decrease the number of rows and columns and increase the pixel size. You could alter listing 9.4 to resample to coarser pixels by dividing the rows and columns by 2 instead of multiplying, and multiplying instead of dividing the pixel size. You could also get away with building fewer overview layers. Other than that, the technique is the same. If you forget to change the pixel size, then you end up with an image that's compressed into too small of an area instead of the too large area shown back in figure 9.13.

9.4   *Byte sequences*

If you've looked through appendix E, you've probably noticed that ReadAsArray and WriteArray aren't the only ways to read and write data with GDAL. (Appendixes C through E are available online on the Manning Publications website at https://www.manning.com/books/geoprocessing-with-python.) You can also read data into a sequence of Python bytes, which is much like a string made up of the ASCII codes corresponding to the numeric pixel values. Unlike strings, byte sequences can't be modified, although you'll learn how to get around that in this section. This is a bit faster than converting to a NumPy array, but I prefer to get an array so that I can manipulate it mathematically. But if you'd like to use bytes instead, or don't need to manipulate the data, the parameters for ReadRaster are similar to those for ReadAsArray. Here's the signature for the dataset version of ReadRaster:

```
ReadRaster([xoff], [yoff], [xsize], [ysize], [buf_xsize], [buf_ysize],
           [buf_type], [band_list], [buf_pixel_space], [buf_line_space],
           [buf_band_space])
```

- xoff is the column to start reading at. The default value is 0.
- yoff is the row to start reading at. The default value is 0.
- xsize is the number of columns to read. The default is to read them all.
- ysize is the number of rows to read. The default is to read them all.
- buf_xsize is the number of columns in the returned sequence. The default is to use the xsize value. Data will be resampled if this value is different than xsize.
- buf_ysize is the number of rows in the returned sequence. The default is to use the ysize value. Data will be resampled if this value is different than ysize.
- buf_type is the target GDAL data type for the returned sequence. The default is the same as the original data.
- band_list is a list of band indices to read. The default is to read all bands.
- buf_pixel_space is the byte offset between pixels in the sequence. The default is the size of buf_type.
- buf_line_space is the byte offset between lines in the sequence. The default is the size of buf_type multiplied by xsize.
- buf_band_space is the byte offset between bands in the sequence. The default is the size of buf_line_space multiplied by ysize.

The first six parameters are the same as for ReadAsArray. The buf_type parameter is a GDAL data type constant from table 9.1 and is used to specify the data type used for the byte sequence. This can be used to change the data type as it's read in. For example, if the raster is of type byte, but you provide GDT_float32 for this parameter, then the resulting byte string will represent the pixel values as floating-point instead of byte. You can also provide a list of bands to read, and they'll be returned in the order you specify. You can even include a band more than once, although I'm not sure why you'd want to. The last three parameters change the spacing of data in the returned byte string and can be used to work with unusually interleaved datasets, but chances are that you'll never need these. The parameters for the band version of `ReadRaster` are the same, except that `band_list` and `buf_band_space` are missing.

Anyway, if you were to print out the results of a call to `ReadRaster`, the result would be something like `b'\x1c\x1d\x1c\x1e'`, which doesn't mean a whole lot to me. You can access elements by index, however, and those will look more familiar. Byte strings are immutable, which means they can't change, but you can convert them to byte arrays if you need to edit values. The following interactive session shows you an example of this:

```
>>> data = ds.ReadRaster(1400, 6000, 2, 2, band_list=[1])
>>> data
b'\x1c\x1d\x1c\x1e'
>>> data[0]
28
>>> bytearray_data = bytearray(data)
>>> bytearray_data[0] = 50
>>> bytearray_data[0]
50
```

You can also convert a byte string to a tuple using the built-in struct module. Here you need to provide a format string that specifies what type, and how many, elements are contained in the string. In this example, you're using a format string such as "BBBB" to specify four bytes. See the Python struct documentation for other formats.

```
tuple_data = struct.unpack('B' * 4, data)
```

If you want to turn the byte string into a NumPy array, you can do that using the tuple from `unpack`, or by using the NumPy `fromstring` function to convert the byte string directly (although if you want a NumPy array, maybe you should use `ReadAsArray`). As with using `unpack`, you have to provide the data type that the sequence uses when converting it to a NumPy array. Both of these methods return a one-dimensional array, so you'll have to reshape it to multidimensional if needed. Examples of these operations are shown here:

```
numpy_data1 = np.array(tuple_data)
numpy_data2 = np.fromstring(data, np.int8)
reshaped_data = np.reshape(numpy_data2, (2,2))
```

The parameters for writing data from byte strings are similar to those for reading, although the first five arguments are required instead of optional:

```
def WriteRaster(xoff, yoff, xsize, ysize, buf_string, [buf_xsize],
                [buf_ysize], [buf_type], [band_list], [buf_pixel_space],
                [buf_line_space], [buf_band_space])
```

- `xoff` is the column to start writing at.
- `yoff` is the row to start writing at.
- `xsize` is the number of columns to write.
- `ysize` is the number of rows to write.

- buf_string is the byte sequence to write.
- buf_xsize is the number of columns in the byte sequence. The default is to use the xsize value. Data will be resampled if this value is different than xsize.
- buf_ysize is the number of rows in the byte sequence. The default is to use the ysize value. Data will be resampled if this value is different than ysize.
- buf_type is the GDAL data type of the byte sequence. The default is the same as the dataset being written to.
- band_list is a list of band indices to write. The default is to write all bands.
- buf_pixel_space is the byte offset between pixels in the byte sequence. The default is the size of buf_type.
- buf_line_space is the byte offset between lines in the byte sequence. The default is the size of buf_type multiplied by xsize.
- buf_band_space is the byte offset between bands in the byte sequence. The default is the size of buf_line_space multiplied by ysize.

Once again, the band version is the same, except that the band_list and buf_band_space parameters don't exist.

You could write a byte sequence, called data, that contains six columns and four rows out to a dataset like this:

```
ds.WriteRaster(1400, 6000, 6, 4, data, band_list=[1])
```

Let's try resampling an image to a larger pixel size using bytes instead of NumPy arrays.

Listing 9.5 Resample an image to a larger pixel size using byte sequences

```
import os
import numpy as np
from osgeo import gdal

os.chdir(r'D:\osgeopy-data\Landsat\Washington')

in_ds = gdal.Open('nat_color.tif')
out_rows = int(in_ds.RasterYSize / 2)          ⎤ Get number of output
out_columns = int(in_ds.RasterXSize / 2)       ⎦ rows and columns
num_bands = in_ds.RasterCount

gtiff_driver = gdal.GetDriverByName('GTiff')
out_ds = gtiff_driver.Create('nat_color_resampled.tif',
    out_columns, out_rows, num_bands)          ⎤ Create output dataset

out_ds.SetProjection(in_ds.GetProjection())
geotransform = list(in_ds.GetGeoTransform())   ⎤ Edit the geotransform
geotransform[1] *= 2                           ⎥ so pixel sizes are larger
geotransform[5] *= 2
out_ds.SetGeoTransform(geotransform)

data = in_ds.ReadRaster(                        ⎤ Use a smaller buffer
    buf_xsize=out_columns, buf_ysize=out_rows)  ⎦ to read and write data
```

```
out_ds.WriteRaster(0, 0, out_columns, out_rows, data)
out_ds.FlushCache()
for i in range(num_bands):
    out_ds.GetRasterBand(i + 1).ComputeStatistics(False)

out_ds.BuildOverviews('average', [2, 4, 8, 16])
del out_ds
```

Build appropriate number of overviews for smaller image

In many ways this example is similar to listing 9.4, except that the numbers of output rows and columns are halved instead of doubled, and the pixel size is doubled instead of halved. Notice that in this case you need to ensure that the numbers of rows and columns are integers, because the result of the division might be floating-point, and the dataset creation function doesn't like that.

The interesting part is where you read and write the data. Because all rows, columns, and bands are read by default, you didn't have to do anything about those. But because you want the data resampled into half as many rows and columns, you pass these smaller numbers in using the buf_xsize and buf_ysize parameters. This causes the data to be resampled as GDAL reads it into the byte sequence. Then you write the data out to the new dataset starting at the first row and column. You also tell WriteRaster how many rows and columns are contained in the byte sequence, because unlike a NumPy array, this isn't obvious. A byte sequence that is 32 bytes long might contain one 32-bit integer, or it might contain four 8-bit integers. Although WriteRaster can figure out how many bytes are in the sequence, it doesn't know how to convert those to pixel values until you tell it how many values there are supposed to be.

9.5 Subdatasets

Several types of datasets can contain other datasets, which each in turn contain bands (figure 9.15). One example of this is the MODIS imagery distributed by the United States Geological Service, which comes as a hierarchical data format (HDF) file. If your dataset contains subdatasets, you can get a list of them with the GetSubDatasets function and then use that information to open the one you want.

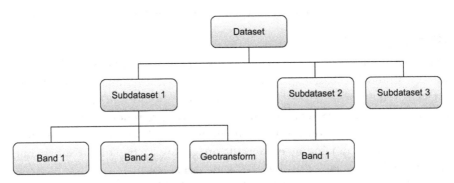

Figure 9.15 Several types of datasets include subdatasets. Each of these subdatasets is structured like a normal dataset and contains its own bands and georeferencing information.

As an example, let's open a subdataset contained in a MODIS file. Note that the HDF driver isn't included in GDAL by default, so this example won't work for you if your GDAL version doesn't include HDF support. Assuming you can work with HDF files, the first step is to open the HDF file as a dataset:

```
ds = gdal.Open('MYD13Q1.A2014313.h20v11.005.2014330092746.hdf')
```

Now you can get the list of subdatasets contained in this open dataset. The GetSub-Datasets method returns a list of tuples, with one tuple per subdataset. Each tuple contains the name and the description of the subdataset, in that order. The following snippet gets this list and then prints out the name and description for each of the sub-datasets:

```
subdatasets = ds.GetSubDatasets()
print('Number of subdatasets: {}'.format(len(subdatasets)))
for sd in subdatasets:
    print('Name: {0}\nDescription:{1}\n'.format(*sd))
```

The first few lines of output look like this, and show the information for the first sub-dataset, which is the NDVI (normalized difference vegetation index), but there are 11 more not shown:

```
Number of subdatasets: 12
Name: HDF4_EOS:EOS_GRID:"MYD13Q1.A2014313.h20v11.005.2014330092746.hdf":
➥ MODIS_Grid_16DAY_250m_500m_VI:250m 16 days NDVI
Description:[4800x4800] 250m 16 days NDVI MODIS_Grid_16DAY_250m_500m_VI
➥ (16-bit integer)
```

To open a subdataset, pass its name to gdalOpen. For example, this gets the tuple corresponding to the first subdataset, gets the first item (the name) from the tuple, and then uses that to open the subdataset:

```
ndvi_ds = gdal.Open(subdatasets[0][0])
```

Similarly, you would use subdatasets[4][0] to open the fifth subdataset. Once you've opened a subdataset like this, it can be treated like any other dataset. For example, you could get the first band in the NDVI subdataset using ndvi_ds.GetRasterBand(1).

9.6   *Web map services*

Let's take a quick look at web map services, which are used to serve images across the web for things like basemaps. We'll try out an OGC web map service that creates an image based on your request, but you have other methods of accessing basemaps. For example, both OpenStreetMap and Google use pre-rendered tiled images. To use those, you need to know the tile that you want, and nothing is rendered on the fly (well, it could be, depending on how the images are cached on the server, but the idea is that the tiles already exist so they provide fast access).

GDAL allows you to use XML files to specify the parameters for a map service, and all of the possibilities are documented at http://www.gdal.org/frmt_wms.html.

The following listing shows the XML describing an imagery basemap from the US National Map.

> **Listing 9.6 XML describing a web map service**

```
<GDAL_WMS>
    <Service name="WMS">
        <Version>1.3.0</Version>
        <ServerURL>http://raster.nationalmap.gov/arcgis/services/
        ➥ Orthoimagery/USGS_EROS_Ortho_1Foot/ImageServer/WMSServer?
        ➥ </ServerURL>
        <CRS>CRS:84</CRS>
        <ImageFormat>image/png</ImageFormat>
        <Layers>0</Layers>
    </Service>
    <DataWindow>
        <UpperLeftX>-74.054444</UpperLeftX>
        <UpperLeftY>40.699167</UpperLeftY>
        <LowerRightX>-74.034444</LowerRightX>
        <LowerRightY>40.679167</LowerRightY>
        <SizeX>300</SizeX>
        <SizeY>300</SizeY>
    </DataWindow>
    <BandsCount>4</BandsCount>
</GDAL_WMS>
```

You need to know certain information about the service to create an XML specification, however. OGC web map services allow you to request information about them using a GetCapabilities request. If you don't know the base URL for the service, you're out of luck, but assuming you do know it, tack "?request=GetCapabilities&service=WMS" onto the end and view the result in a browser. For example, the URL for the service defined in listing 9.6 is http://raster.nationalmap.gov/arcgis/services/Orthoimagery/USGS_EROS_Ortho_1Foot/ImageServer/WMSServer?request=GetCapabilities&service=WMS.

This is a lot of information, but we'll focus on a few parts that are important for the Service section of the XML. Look at the first line of output:

```
<WMS_Capabilities xmlns=http://www.opengis.net/wms
➥ xmlns:xsi="http://www.w3.org/2001/XMLSchema-instance" version="1.3.0"
➥ xsi:schemaLocation="http://www.opengis.net/wms
➥ http://schemas.opengis.net/wms/1.3.0/capabilities_1_3_0.xsd">
```

Part of that line specifies the WMS version as 1.3.0. Add that information to the Version section of your XML. Now look through the GetCapabilities results until you find the GetMap section. The first part of it looks like this:

```
<GetMap>
    <Format>image/tiff</Format>
    <Format>image/png</Format>
    <Format>image/png24</Format>
    <Format>image/png32</Format>
    <Format>image/bmp</Format>
```

```
<Format>image/jpeg</Format>
<Format>image/svg</Format>
<Format>image/bil</Format>
```

These are the formats that the service can provide, and you should include one of them in the ImageFormat section of your XML. Now look for the Layer section in the GetCapabilities output. Here are the first few lines of that section:

```
<Layer>
    <Name>0</Name>
    <Title>USGS_EROS_Ortho_1Foot</Title>
    <Abstract>
        The USGS_EROS_Ortho_1Foot service from The National Map contains 1
    foot orthoimagery, and is viewable at all scales.
    </Abstract>
```

We want to use the layer called USGS_EROS_Ortho_1Foot, but the Name value is the important one. In this case, the name is "0," which isn't too descriptive, but it's what you need to add to the Layer section of the XML. If you keep looking at the Layer section of the capabilities, you'll see a lengthy list of CRS values, which are the coordinate systems supported by the service. Here are the first few:

```
<CRS>CRS:84</CRS>
<CRS>EPSG:4326</CRS>
<CRS>EPSG:3857</CRS>
```

You guessed it. Select one of these for your output and add it to the CRS section of your XML.

Now that the service is defined in your XML, you need to specify the geographic extent that you want to retrieve. You do this with the DataWindow section. The Upper-LeftX, UpperLeftY, LowerRightX, and LowerRightY are the minimum x, maximum y, maximum x, and minimum y values, respectively. The SizeX and SizeY parameters specify the number of columns and rows for the output image.

Once you have your XML saved, pass the filename to the GDAL Open function, and if everything is configured correctly, it will be opened as a dataset. At this point you could get the bands and read the data into an array, or you could save the image to a local file using CreateCopy. For example, this snippet uses the XML from listing 9.6 to save a local image of Liberty Island in New York Harbor (figure 9.16):

Figure 9.16 An image of Liberty Island in New York Harbor obtained using an OGC web map service

```
ds = gdal.Open('listing9_6.xml')
gdal.GetDriverByName('PNG').CreateCopy(r'D:\Temp\liberty.png', ds)
```

If you need to request images with different spatial extents or other parameters that regularly change, it would make sense to create an XML template and format it with the desired values when required.

9.7 Summary

- Raster datasets are ideal for continuous data without sharp boundaries, such as elevation, precipitation, or satellite imagery.
- In the interest of disk space, don't use smaller pixel sizes or larger data types than necessary.
- If you need to use your data for analysis, be sure to use a lossless compression algorithm or no compression at all.
- Use overviews for rapid display of raster data.
- Always use nearest-neighbor resampling for non-continuous raster data because other methods will result in new pixel values that don't correspond to the originals.
- For best performance, make as few read/write calls as possible, but don't try to keep more data in memory than you have RAM.
- Don't forget to edit the geotransform if you change spatial extent or pixel size.
- Don't try to read or write past the edge of an image.
- Use the buffer parameters to resample data while reading or writing.
- Use `ReadAsArray` if you want to use NumPy to manipulate your data in memory, but `ReadRaster` is slightly faster if you only need to copy data.

Working with raster data

10

This chapter covers

- Georeferencing with ground control points
- Working with attributes, histograms, and color tables
- Using the GDAL virtual format
- Reprojecting rasters
- Using GDAL error handling

In the last chapter you learned the basics of raster processing, such as how to read and write data and work with individual bands, and how rasters use geotransforms to orient themselves to the real world. This was a great first step, but what if you have an old aerial photograph or scanned paper map that you'd like to turn into a geographic dataset? You might want to do that because it's fun and interesting, or you might want to do a change analysis using this data along with more current imagery. To do that, you must overlay the old data on the new. You can do this using ground control points, which are essentially a collection of points with known locations. This chapter will teach you how to use these points.

You'll also learn how to work with raster attribute tables. Although most of the raster examples we've looked at so far have been continuous data, such as satellite imagery, raster datasets can also contain thematic data. In this case, each unique pixel value corresponds to a classification of some kind, such as vegetation or soil type. Pixel values are numeric, though, so how do you know what each value stands for? For example, the landcover classification map shown in figure 10.1 has 125 different classes. I certainly can't remember what each value stands for; 76 doesn't mean nearly as much to me as "Inter-Mountain Basins Semi-Desert Grassland" does. Fortunately, it's possible to store information like this in raster attribute tables.

> **NOTE TO PRINT BOOK READERS: COLOR GRAPHICS** Many graphics in this book are best viewed in color. The eBook versions display the color graphics, so they should be referred to as you read. To get your free eBook in PDF, ePub, and Kindle formats, go to https://www.manning.com/books/geoprocessing-with-python to register your print book.

Figure 10.1 A landcover classification map, where each unique pixel value corresponds to a specific landcover classification

Take a look at figure 10.1 again. It uses a constant set of colors to display each landcover class. Water is always blue (or almost black if you're viewing this in black and white) and the playa west of the Great Salt Lake are always a pale yellowish color (or a very light gray in black and white). Although constant colors are certainly not required for data analysis, it's nice to have them when looking at a dataset. You saw earlier how red, green, and blue bands can be used to draw an image, but this dataset has only one band that contains classification values. Instead of the RGB bands, it has what's called a *color table* that specifies what color each unique pixel value should be drawn in.

These are only a few examples of other components of raster datasets. You'll learn how to work with these, and more, in this chapter. You'll also learn tricks for handling errors in GDAL.

10.1 Ground control points

You've learned how geotransforms work to georeference an image, using the upper-left coordinates and pixel size. You don't always have this information, however. For example, if you found an old aerial photo from 1969 and scanned it in, you'd have a digital image, but you couldn't load it into a GIS and see it displayed in the correct location. Your scanner creates a digital image, but it doesn't attach any sort of geo-graphic information to it. All is not lost, however, as long as you know what area the photo is of and can identify a few locations. These locations are called ground control points (GCPs), which are points for which you know the real-world coordinates. If you can associate a number of pixels around the image with actual coordinates, then the

A   B   C

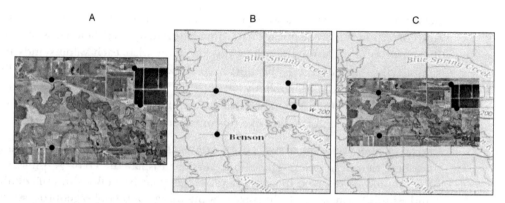

Figure 10.2 An example of using four known locations to warp an image to fit correctly on a map. Figure A shows an aerial photo with the points overlaid on top, figure B shows a topographical map with the same points, and figure C shows the photo stretched so that the points match up with the topo map.

image can be warped to overlay on a map, as shown in figure 10.2. This method isn't used as often as geotransforms, but it's necessary in certain cases. Plus, once an image has been georeferenced this way, a geotransform can be computed so that it can be used instead if desired. You should be aware that because the image will be stretched, warped, and/or rotated so that the GCPs overlay the real coordinates, the pixel size and raster dimensions might be changed during the process.

It should be apparent that fixed landmarks make good GCPs because they're the easiest thing to pinpoint and get real coordinates for. For example, if you have an aerial photograph that includes a freeway, using a car on that freeway won't work because you probably have no way of knowing its location at the exact time the photo was taken (if you do, then by all means, use it). An exit ramp, however, is a good choice because it doesn't move, it will be easily visible in the photo, and it's not difficult to get the coordinates for.

Depending on the type of transformation you use to warp the image, you'll need a different number of GCPs. One commonly used algorithm, a *first-order polynomial*, fits a linear equation to the image's x coordinates so the GCP image coordinates match, as closely as possible, the real GCP coordinates that you provide. The same is done for the y coordinates. This method requires at least three points. If your coordinates are exact, then theoretically you don't need more points, but this is probably not the case, and you'll get better results with a few more points evenly distributed around the image. This algorithm works well if your image needs to be scaled or rotated, as in figure 10.3A. If your image needs to be bent, as in the shape changes (figure 10.3B), then you're better off using a higher-order polynomial, such as a quadratic or cubic equation, with more GCPs.

A polynomial transformation might end up shifting several of your GCPs slightly to minimize error across the image, as in figure 10.4A. If you want to eliminate error

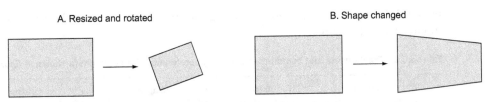

Figure 10.3   A scaled and rotated raster (A) and a raster whose shape has changed (B)

around the GCPs and are willing to accept greater error in other parts of the image, as in figure 10.4B, you can use a *spline* method instead. A spline doesn't use a single equation, but instead uses different equations for different parts of the data, so it can fit the provided points exactly. This might cause other parts of the image to be warped in odd ways, however. You can use various interpolation methods with the gdalwarp utility that comes with GDAL, but you'll only see how to use a first-order (linear) polynomial with Python.

How do you go about using GCPs? The first things you need are the known coordinates for specific pixel offsets. You could get this information the hard way, such as opening your raster in image or photo processing software and using it to determine pixel offsets. An easier method, however, is to use the QGIS georeferencer plugin. This allows you to click on a location in your image and on an already georeferenced map, and it will tell you the pixel offsets and corresponding real-world coordinates. It will even export the necessary gdalwarp command to do the georeferencing for you. But you're here to learn how to do the same job with Python, so let's look at the example back in figure 10.2. This aerial photo of a small area shows a few roads and several large water treatment ponds. I've determined the coordinates for four locations, shown in table 10.1 and as dots in the figure. I chose points that could be identified on both the image in figure 10.2A and the topo map in figure 10.2B. Getting the point coordinates is the hard part of the process, but it's something you'll have to do

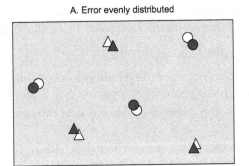
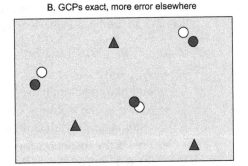

Figure 10.4   Different error distributions. Triangles are GCPs; circles are random points. Solid shapes are the true location; hollow shapes are the location the point ends up in the warped raster.

by hand. In this case, the topo map was georeferenced so I could figure out the coordinates from the map.

Table 10.1   Pixel offsets and coordinates used to georeference the aerial photo shown in figure 10.2

Photo column	Photo row	Longitude	Latitude
1078	648	-111.931075	41.745836
3531	295	-111.901655	41.749269
3722	1334	-111.899180	41.739882
1102	2548	-111.930510	41.728719

The following listing shows how you would attach these ground control points to the photo.

Listing 10.1   Adding ground control points to a raster

```
import shutil
from osgeo import gdal, osr
orig_fn = r'D:\osgeopy-data\Utah\cache_no_gcp.tif'
fn = r'D:\osgeopy-data\Utah\cache.tif'            ◁─── Make a copy of the
shutil.copy(orig_fn, fn)                               file to work with
ds = gdal.Open(fn, gdal.GA_Update)
sr = osr.SpatialReference()
sr.SetWellKnownGeogCS('WGS84')
gcps = [gdal.GCP(-111.931075, 41.745836, 0, 1078, 648),
        gdal.GCP(-111.901655, 41.749269, 0, 3531, 295),
        gdal.GCP(-111.899180, 41.739882, 0, 3722, 1334),
        gdal.GCP(-111.930510, 41.728719, 0, 1102, 2548)]
ds.SetGCPs(gcps, sr.ExportToWkt())
ds = None
```

When adding GCPs to a raster, make sure you open the dataset for updating, as you do here. You also need the spatial reference system of the known coordinates; in this case they use the WGS84 datum but are unprojected (lat/lon). The last thing you need is a list of GCPs, and you can create each of those with the GCP constructor shown here:

```
gdal.GCP([x], [y], [z], [pixel], [line], [info], [id])
```

- x, y, and z are the real-world coordinates corresponding to the point. All are optional and default to 0, although you probably don't want x and y values to be 0.
- pixel is the column offset for the pixel with known coordinates. This is optional and the default is 0.
- line is the row offset for the pixel with known coordinates. This is optional and the default is 0.
- info and id are two optional strings used to identify the GCP, but in my experience they don't carry over to the image. I rarely use GCPs, however, so perhaps there are instances where they do. The default is a blank string.

In listing 10.1 you use the information in table 10.1 to create a list of four GCPs, and then you attach those GCPs to the dataset with `SetGCPs`. This function requires a list of GCPs and a WKT string containing projection information for the real-world coordinates.

Now that you've added GCPs, software that understands them can display your image in its correct location. If you don't need to know what GCPs were used to georeference the image and would rather use the more common geotransform method of georeferencing, you can create a geotransform from the GCPs and set that on the dataset instead of attaching the GCPs. To create a geotransform using a first-order transformation, pass your list of GCPs to `GCPsToGeoTransform`. Then make sure you set both the geotransform and the projection information on your dataset:

```
ds.SetProjection(sr.ExportToWkt())
ds.SetGeoTransform(gdal.GCPsToGeoTransform(gcps))
```

You don't have to convert your GCPs to a geotransform if you don't want to, however.

10.2 *Converting pixel coordinates to another image*

As you learned in the last chapter, functions can help you convert between real-world coordinates and pixel offsets. Also, a `Transformer` class can be used for that or to go between offsets in two different rasters. One example of why you might want to do this is if you're mosaicking rasters together, because each input image goes in a different part of the mosaic. To illustrate this, let's combine a few digital orthophotos of Cape Cod together into one raster.

To combine the images, it's necessary to know the extent of the output mosaic. The only way to find this is to get the extent of each input raster and calculate the overall minimum and maximum coordinates (figure 10.5). To make this a little easier, you'll create a function that gets the extent of a raster. It uses the geotransform to get the upper-left coordinates and then calculates the lower right coordinates using the pixel size and raster dimensions:

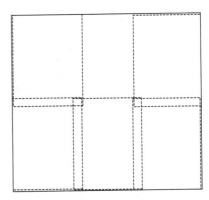

Figure 10.5 Dotted lines show the footprints of six rasters to be mosaicked together. The solid outer line is the footprint for the output raster.

```
def get_extent(fn):
    '''Returns min_x, max_y, max_x, min_y'''
    ds = gdal.Open(fn)
    gt = ds.GetGeoTransform()
    return (gt[0], gt[3], gt[0] + gt[1] * ds.RasterXSize,
        gt[3] + gt[5] * ds.RasterYSize)
```

You can see in the following listing how this function is used to help find the output extent. Once you know the extent, you can calculate the output dimensions and create the raster. Then you can finally start copying data from each file.

Listing 10.2   Mosaic multiple images together

```
os.chdir(r'D:\osgeopy-data\Massachusetts')
in_files = glob.glob('O*.tif')
min_x, max_y, max_x, min_y = get_extent(in_files[0])
for fn in in_files[1:]:                                       Calculate output
    minx, maxy, maxx, miny = get_extent(fn)                   extent from all inputs
    min_x = min(min_x, minx)
    max_y = max(max_y, maxy)
    max_x = max(max_x, maxx)
    min_y = min(min_y, miny)

in_ds = gdal.Open(in_files[0])
gt = in_ds.GetGeoTransform()
rows = math.ceil((max_y - min_y) / -gt[5])                    Calculate dimensions
columns = math.ceil((max_x - min_x) / gt[1])

driver = gdal.GetDriverByName('gtiff')
out_ds = driver.Create('mosaic.tif', columns, rows)
out_ds.SetProjection(in_ds.GetProjection())                  Create output
out_band = out_ds.GetRasterBand(1)

gt = list(in_ds.GetGeoTransform())
gt[0], gt[3] = min_x, max_y                                   Calculate new
out_ds.SetGeoTransform(gt)                                    geotransform

for fn in in_files:
    in_ds = gdal.Open(fn)
    trans = gdal.Transformer(in_ds, out_ds, [])
    success, xyz = trans.TransformPoint(False, 0, 0)         Get output offsets
    x, y, z = map(int, xyz)
    data = in_ds.GetRasterBand(1).ReadAsArray()
    out_band.WriteArray(data, x, y)                           Copy data

del in_ds, out_band, out_ds
```

The first thing you do in listing 10.2 is loop through all of the input files and use their extents to calculate the final mosaic's extent, and then you calculate the numbers of rows and columns for the output. You do this by getting the distance between the min and max values in each direction and dividing by the pixel size. You make sure to not accidentally cut the edges off by using the `ceil` function to round any partial numbers up to the next integer. Then you create a new dataset using these dimensions. You still need to create an appropriate geotransform, but that's easily done by copying one from an input file and changing the upper-left coordinates to the ones you calculated.

By this point you have an empty raster of the appropriate size, so it's time to start copying data. This is where the transformer comes in. For each input dataset, you create a transformer between that dataset and the output mosaic. The third parameter is for transformer options, but you're not using any of them here. Once you have the transformer, you can easily calculate the correct pixel offsets for the mosaic that correspond to the upper-left corner of the input raster using `TransformPoint`:

```
TransformPoint(bDstToSrc, x, y, [z])
```

- bDstToSrc is a flag specifying if you want to compute offsets from the destination raster to the source raster or vice versa. Use True to go from the destination to the source and False to go the other way.
- x, y, and z are the coordinates or offsets that you want to transform. z is optional and defaults to 0.

You want to compute offsets in the destination raster (the second one provided when you created the transformer) based on the source, so you use False for the first parameter. The x and y parameters are both 0 because you want the offsets corresponding to the first row and first column in the input. The function returns a list containing a success flag and a tuple with the requested coordinates, but the coordinates are floating-point so you convert them to integers. Finally, you read the data from the input raster and write it out to the mosaic using the offsets you just calculated. Then you go on to the next input dataset.

The resulting mosaic is shown in figure 10.6. You can see how the color balancing between the images isn't perfect. One other thing to be aware of is that if the input rasters overlap, then pixel values in the overlap area will be overwritten by the last raster that covers the overlap. The order in which you combine the rasters might be important to you so that you get the correct pixel values. Or you could implement a fancier algorithm to average the pixel values or handle them in another way.

You can also transform coordinates between pixel offsets and real-world coordinates by not providing

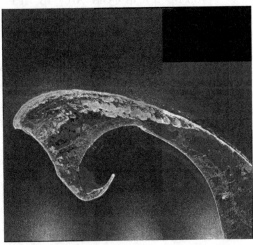

Figure 10.6 A simple mosaic of six aerial photos of Cape Cod, Massachusetts

one of the datasets. For example, this would get the real-world coordinates for the pixel at column 1078 and row 648, assuming that the dataset has a valid geotransform:

```
trans = gdal.Transformer(out_ds, None, [])
success, xyz = trans.TransformPoint(0, 1078, 648)
```

I prefer to use ApplyGeoTransform for this, as you saw in the previous chapter, but you should use whichever one makes the most sense to you.

10.3 Color tables

In thematic rasters the pixel values represent a classification such as vegetation type instead of color information like in a photograph. If you want to control how these datasets are displayed, then you need a color table. The map of Utah vegetation types shown back in figure 10.1 uses a color table so that the image looks the same whether

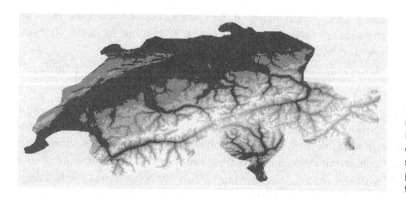

Figure 10.7   Digital elevation model for Switzerland that has been classified into five elevation ranges and then stretched so that the small pixel values appear different from one another

you open it in QGIS, ArcMap, or even the Windows Photo Viewer. Color tables only work for integer-type rasters, and I have only had luck getting the color mapping to work on pixel values of 255 and below (values that fit into a byte).

To see how color tables work, let's create one for an elevation dataset that has been classified into ranges. This file has been created for you and is in the book data's Switzerland folder. It's called dem_class.tif, and the elevation values have been classified into five different ranges, so the pixel values range from 0 to 5, with 0 being set to NoData. If you look at this file in something like Windows Photo Viewer, you'll only see a black rectangle, because that's how it interprets such small pixel values. If you open it up in QGIS or another GIS package, it's likely that the software will automatically stretch the data for you so you'll see something like figure 10.7.

If you add a color map to this image, then it will draw correctly without being stretched, and you can specify the colors that are used for each elevation range. Let's try it.

Listing 10.3   Add a color map to a raster

```
os.chdir(r'D:\osgeopy-data\Switzerland')
original_ds = gdal.Open('dem_class.tif')
driver = gdal.GetDriverByName('gtiff')
ds = driver.CreateCopy('dem_class2.tif', original_ds)      ◄──┐  Make a copy
band = ds.GetRasterBand(1)                                    │  of the dataset

colors = gdal.ColorTable()
colors.SetColorEntry(1, (112, 153, 89))                          Create an RGB
colors.SetColorEntry(2, (242, 238, 162))                         ColorTable and
colors.SetColorEntry(3, (242, 206, 133))                         add colors
colors.SetColorEntry(4, (194, 140, 124))
colors.SetColorEntry(5, (214, 193, 156))

band.SetRasterColorTable(colors)
band.SetRasterColorInterpretation(                           Add the ColorTable
    gdal.GCI_PaletteIndex)                                   to the band

del band, ds
```

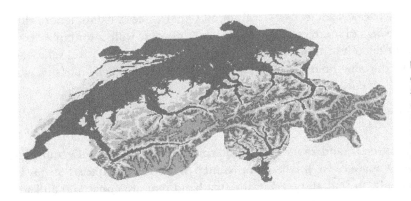

Figure 10.8 Digital elevation model for Switzerland that has been classified into five elevation ranges and then had a color map applied. If you look at the color version online, it will look much different from the automatic symbology shown in figure 10.7.

The first part of this listing doesn't have anything to do with the color table; it's making a copy of the image so that the original doesn't get modified. The interesting part is when you create an empty color table called colors and then add colors to it. The first parameter to SetColorEntry is the pixel value that you want to set the color for, and the second parameter is a tuple or list containing the red, green, and blue values for the color. You set colors for pixel values between 1 and 5, inclusive. Because this is a byte dataset, there are 255 possible pixel values and the color table contains zeros (black) for the values that you didn't change. Finally, you add the color map to the band and tell the band that it's using a color map by setting the color interpretation to paletted, although that second step isn't necessary because GDAL figures it out. Now your image looks like figure 10.8, although software that doesn't understand the NoData setting will draw a black background.

You can also edit existing color tables. Say you want to change the color map you created so that the highest elevation range displays as something closer to white. Grab the color table from the band and change the entry you're interested in, which is the pixel value 5 in this case:

```
os.chdir(r'D:\osgeopy-data\Switzerland')
ds = gdal.Open('dem_class2.tif', gdal.GA_Update)         Open the dataset
band = ds.GetRasterBand(1)                               for updating
colors = band.GetRasterColorTable()
colors.SetColorEntry(5, (250, 250, 250))
band.SetRasterColorTable(colors)                         Set the modified ColorTable
del band, ds                                             onto the band
```

Remember to open the dataset for writing. If you don't, your changes won't take effect and you won't get an error message, either. You also need to add the color map back to the band because the one you're editing is no longer linked to the band.

10.3.1 Transparency

Have you ever seen colors referred to as RGBA instead of plain RGB? The *A* stands for a fourth value called *alpha*, which is used to specify opacity. The higher the alpha value, the more opaque the color. You can add an alpha band to your image and then certain software packages, such as QGIS, will use it. Others, like ArcMap, ignore the alpha band when using color tables. If you want to go this route with color tables, you need to create your dataset with two bands, where the first one is your pixel values as before, and the second one holds alpha values. You also need to specify that this second band is an alpha band at creation time, like this:

```
ds = driver.Create('dem_class4.tif', original_ds.RasterXSize,
    original_ds.RasterYSize, 2, gdal.GDT_Byte, ['ALPHA=YES'])
```

Then add values between 0 and 255 to your alpha band, where 0 means fully transparent and 255 is fully opaque. We'll talk about NumPy in the next chapter, but this is how you'd use NumPy to find all pixels in the first band that are equal to 5 and set them approximately 25% transparent:

```
import numpy as np
data = band1.ReadAsArray()
data = np.where(data == 5, 65, 255)
band2.WriteArray(data)
band2.SetRasterColorInterpretation(gdal.GCI_AlphaBand)
```

Here you use the NumPy where function to create a new array based on the values of the original data array from band 1. It's like an if-else statement, where the condition is whether or not the pixel value is equal to 5. If it is, then the corresponding cell in the output array gets a value of 65, which is roughly a quarter of 255. If the original pixel has a value other than 0, then the output gets a value of 255, which is no transparency. Write that new array to the second band, and make sure you set the color interpretation for that band to alpha.

If you wanted to create an image with transparency that more software will understand, then you could create a four-band image and forego the color map. In this case, you'd put the red value in the first band, green in the second, blue in the third, and the alpha value in the fourth band. The disadvantage to this is that your dataset would be at least twice as large because it would have twice as many bands. It would probably have even another band to hold your original pixel values instead of only color information. Otherwise, you'd lose your information about pixel classifications, such as landcover types.

10.4   Histograms

Sometimes you need a frequency histogram for pixel values. One example of this would be calculating the area of each vegetation type in a vegetation classification. If you know how many pixels are classified as pinyon-juniper, for instance, then you can multiply that number by the area of a pixel (which is pixel width times pixel height) to get the total acreage of pinyon-juniper.

The easiest way to get a histogram is to use the GetHistogram function on a band. You can specify exactly what bins you want to use, but the default is to use 256 of them. The first one includes values between -0.5 and 0.5, the second bin goes from 0.5 to

1.5, and so on. So if you have byte data, this histogram will have one bin per possible pixel value (0, 1, 2, and so on). If no histogram data already exist for the raster, then this function computes an approximate one by default, but you can request an exact one. The function looks like this:

```
GetHistogram([min], [max], [buckets], [include_out_of_range], [approx_ok],
        [callback], [callback_data])
```

- min is the minimum pixel value to include in the histogram. The default value is 0.5.
- max is the maximum pixel value to include in the histogram. The default value is 255.5.
- buckets is the number of bins you want. The size of the bins is determined by taking the difference between max and min and dividing that by buckets. The default value is 256.
- include_out_of_range denotes whether or not to lump pixel values below the minimum value into the minimum bin, and the pixel values larger than the maximum into the maximum bin. The default value is False. Use True if you want to enable this behavior.
- approx_ok denotes whether or not it's okay to use approximate numbers, either by looking at overviews or only sampling a subset of the pixels. The function will run faster this way. The default value is True. Use False if you want exact counts.
- callback is a function that's called periodically while the histogram is being computed. This is useful for showing progress while processing large datasets. The default value is 0, which means you don't want to use a callback function.
- callback_data is data to pass to the callback function if you're using one. The default value is None.

This code snippet shows the difference between approximate and exact values, using the classified elevation raster that we looked at earlier:

```
os.chdir(r'D:\osgeopy-data\Switzerland')
ds = gdal.Open('dem_class2.tif')
band = ds.GetRasterBand(1)
approximate_hist = band.GetHistogram()
exact_hist = band.GetHistogram(approx_ok=False)
print('Approximate:', approximate_hist[:7], sum(approximate_hist))
print('Exact:', exact_hist[:7], sum(exact_hist))
```

The histogram consists of a list of counts, in order by bin. In this case the first count corresponds to pixel value 0, the second to pixel value 1, and so on. Here you only print the first seven entries, because the remaining 249 of them are all 0 for this dataset. The results are shown here and in figure 10.9:

```
Approximate: [0, 6564, 3441, 3531, 2321, 802, 0] 16659
Exact: [0, 27213, 12986, 13642, 10632, 5414, 0] 69887
```

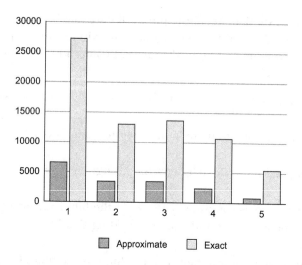

Figure 10.9   **The approximate and exact histograms generated from the classified elevation raster**

Notice that the numbers, including the sum, for the approximate histogram are much smaller than those for the exact. Therefore, the approximate numbers are not the way to go if you need to tabulate area, but they'd probably work well if you want relative frequencies. Also notice that no counts exist for a pixel value of 0. That's because 0 is set to NoData, so it gets ignored.

GDAL stores these histograms in an XML file alongside the raster. If you ran this code, there should now be a file called dem_class2.tif.aux.xml in your Switzerland folder. If you open it and take a look, you'll see both sets of histogram data. As long as you don't delete that file, those specific histograms won't need to be computed again because GDAL can read the information from the XML file.

You can also set a certain binning scheme to be the default for an image. For example, say you want to lump pixel values 1 and 2 together, 3 and 4 together, and leave 5 alone. You can do that like this:

```
hist = band.GetHistogram(0.5, 6.5, 3, approx_ok=False)
band.SetDefaultHistogram(1, 6, hist)
```

In this example, you create a histogram with three bins that include pixel values 1 through 6. Why go up to 6 instead of 5, when the actual data values only go up to 5? The bins are created of equal size, so if there were three bins between 1 and 5, then the breaks would be in the wrong places. The breaks for the example would be at 2.5 and 4.5, but they would be at 2.2 and 3.8 if you used a range of 5 instead of 6. In that case, the pixels with values 4 and 5 would be lumped together and 3 would be alone, which isn't the desired outcome (figure 10.10).

Once you compute the histogram, you set it as the default. The SetDefault-Histogram function wants a minimum pixel value, maximum value, and then the list of counts. Once you've set a default, you can use GetDefaultHistogram to get that particular one. While GetHistogram returns a list of counts, GetDefaultHistogram

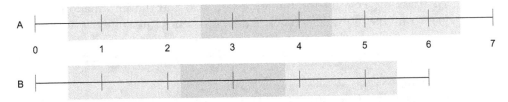

Figure 10.10 The results of creating three bins between 0.5 and 6.5 (A) and 0.5 and 5.5 (B). In case A, values 1 and 2 share a bin, 3 and 4 share a bin, and 5 has a bin to itself (there are no pixels with a value of 6). In case B, however, value 3 is the one that doesn't share a bin.

returns a tuple containing the minimum pixel value, maximum pixel value, number of bins, and a list of counts:

```
min_val, max_val, n, hist = band.GetDefaultHistogram()
print(hist)
[40199, 24274, 5414]
```

When you call `GetHistogram`, you provide the min and max values and the number of bins to use, so that function doesn't need to return that information because you already know it. These values are returned when you call `GetDefaultHistogram` because you might not know what values were used to create the default histogram.

10.5 Attribute tables

Integer raster datasets can have attribute tables, although in my experience they don't have nearly as many fields as vector attribute tables. Instead of a record in the table corresponding to an individual feature, each record of a raster attribute table corresponds to a particular pixel value. For example, all pixels with a value of 56 will share the same record in the attribute table, because each pixel doesn't represent an individual feature, but multiple pixels with the same value should represent the same thing, whether it's a certain color, an elevation, a land use classification, or something else.

An attribute table doesn't even make sense for many rasters. I can't think of an attribute I would want to attach to various pixel values in an aerial photo, for example. In fact, raster attribute tables make the most sense for categorical data such as landcover or soil type, when you'd want attributes containing information about each category.

Let's create an attribute table for the classified elevation raster we've been working with. Table 10.2 shows the elevation classes used to create the dataset, which would be useful information to store.

Table 10.2 Pixel values and the corresponding elevation ranges for the classified elevation raster

Pixel value	Elevation range (meters)
1	0 – 800
2	800 – 1300
3	1300 – 2000
4	2000 – 2600
5	2600+

We'll use the code in the following listing to add the information from table 10.2 along with the histogram counts to the raster's attribute table.

Listing 10.4   Add an attribute table to a raster

```
os.chdir(r'D:\osgeopy-data\Switzerland')
ds = gdal.Open('dem_class2.tif')
band = ds.GetRasterBand(1)
band.SetNoDataValue(-1)                                        Set NoData to a
                                                               nonexisting value

rat = gdal.RasterAttributeTable()
rat.CreateColumn(
    'Value', gdal.GFT_Integer, gdal.GFU_Name)
rat.CreateColumn(
    'Count', gdal.GFT_Integer, gdal.GFU_PixelCount)            Create the
rat.CreateColumn(                                              attribute table
    'Elevation', gdal.GFT_String, gdal.GFU_Generic)
rat.SetRowCount(6)

rat.WriteArray(range(6), 0)
rat.WriteArray(                                                Load in a list of
    band.GetHistogram(-0.5, 5.5, 6, False, False), 1)         data to a column
rat.SetValueAsString(1, 2, '0 - 800')
rat.SetValueAsString(2, 2, '800 - 1300')
rat.SetValueAsString(3, 2, '1300 - 2000')
rat.SetValueAsString(4, 2, '2000 - 2600')                     Set individual values
rat.SetValueAsString(5, 2, '2600 +')

band.SetDefaultRAT(rat)                                        Add the table
band.SetNoDataValue(0)                                         to the band
del band, ds
```

Restore NoData → `band.SetNoDataValue(0)`

When you create a new raster attribute table, you need to define the columns it will have. You provide three pieces of information when you add a column. The first is the name of the column. The second is the data type, which can be one of GFT_Integer, GFT_Real, or GFT_String. The last thing is the purpose of the column using one of the GFU constants from appendix E. (Appendixes C through E are available online on the Manning Publications website at https://www.manning.com/books/geoprocessing-with-python.) The desktop software I use either doesn't support raster attribute tables at all (QGIS) or doesn't see most of them as anything special (ArcGIS), but there's probably software out there that does. Therefore, the only types I'm usually interested in are the ones used here. You used GFU_Name for the column containing the pixel values, GFU_PixelCount for the histogram counts, and GFU_Generic for the description. You also told it that there will be six rows.

The next step is to add the data. You know the pixel values range from 0 to 5, so you use the range function to get a list containing those numbers and then add that list to the first column, which is pixel values. The first parameter to WriteArray is the data to put in the column, and the second parameter is the index of the column that you want to add the data to. Because you already specified that there were six rows, you'd have gotten an error about too many values if you'd provided a list with more than six items. You don't have to write all rows at once, though. If you'd only provided four values, it would have filled the first four rows of the column. An optional third parameter tells it which row to begin writing data on, so you could then add the remaining data by passing a 4 as the last parameter so that it would start writing on the correct row.

You use this same method to add the histogram counts to the second column, except this time the list of values comes from the GetHistogram function. Remember that the 0 values are ignored when computing histograms, but you might want the number of zeros in the attribute table. One way to get a histogram that includes those values is to set NoData to a bogus value and then calculate a histogram that hasn't been calculated before. It needs to be a new histogram because if that particular one has already been calculated, then GDAL will pull the information out of the XML file you saw earlier, and the zeros still won't be counted. That's why you set NoData to -1 before creating the attribute table, and then use a set of parameters that you haven't yet used to retrieve a histogram. Conveniently, you tell it that you want six bins, which is exactly how many rows the table has.

Then you set the elevation ranges. You could create a list holding those descriptions, but instead you add each one individually by specifying the row, then the column, and then the value to put in the table. You don't bother adding a description for the first row with index 0, because there's not much to say about NoData.

Once you add all of your data to the table, you add it to the band with SetDefaultRAT, and then make sure to restore the NoData value to your band. Figure 10.11 shows a screenshot of this raster attribute table in ArcGIS. Unfortunately, you can't view the results with QGIS.

dem_class2.tif

OID	Value	Count	Elevation
0	0	55467	
1	1	27213	0 - 800
2	2	12986	800 - 1300
3	3	13642	1300 - 2000
4	4	10632	2000 - 2600
5	5	5414	2600 +

Figure 10.11 Raster attribute table for the classified elevation raster

10.6 *Virtual raster format*

The GDAL virtual format (VRT) isn't another property you can add to a raster, like an attribute or color table, but it's a useful format that allows you to define a dataset using an XML file. Virtual raster datasets use other datasets to store the data, but the XML describes how to pull the data out of those other files. A VRT can be used to subset the data, modify properties such as the projection, or even combine multiple datasets into one. In these cases, the original datasets aren't changed, but the modifications are made to the data in memory when it's read by the software.

For example, say you had a raster dataset that covered a large spatial area, but you needed to run different analyses on various spatial subsets of the original raster. You could clip out the areas you need, or you could define a VRT for each of these subsets and not have to create the subsetted rasters on disk. For even more information on VRTs than you will see here, check out http://www.gdal.org/gdal_vrttut.html.

Before we look at manipulating data with a VRT, let's look at an extremely simple example (called simple_example.vrt in the Landsat/Washington data folder). This XML defines a VRT dataset with one band, and that band is the blue band from the natural color GeoTIFF you created in the last chapter.

Listing 10.5   XML to define a VRT dataset with one band

```
<VRTDataset rasterXSize="8849" rasterYSize="8023">
    <SRS>
        PROJCS["WGS 84 / UTM zone 10N",GEOGCS["WGS 84",DATUM["WGS_1984",...
    </SRS>
    <GeoTransform>343724.25, 28.5, 0, 5369585.25, 0,-28.5</GeoTransform>
    <VRTRasterBand dataType="Byte" band="1">
        <SimpleSource>
            <SourceFilename relativeToVRT="1">
                nat_color.tif                          Filename and
            </SourceFilename>                           band number
            <SourceBand>3</SourceBand>
            <SourceProperties RasterXSize="8849" RasterYSize="8023"
                DataType="Byte" BlockXSize="8849" BlockYSize="1" />
        </SimpleSource>
    </VRTRasterBand>
</VRTDataset>
```

This XML contains general dataset information such as the numbers of rows and columns, the spatial reference system, and the geotransform. The SRS needs to be in WKT, which would have taken up a lot of space, so I opted to truncate it for the example (the third line). In real life, you'd need the entire SRS string. There's also a VRTRasterBand element for each band in the dataset, which is only one in this case. This contains the data type, band number, rows and columns, and the information required to load the data. This is a simple case, so it only needs a filename and a band number. You want the blue band, which is the third one in nat_color.tif. The `rela-tiveToVRT` attribute tells it whether or not the file path to the data is relative to the location of the VRT file itself. If you want an absolute filename, use a 0 here. In this particular case, the image file and the VRT file are in the same directory, but if you were to move the VRT file without moving the image file, the VRT would be unable to load any data.

Creating VRT datasets from Python can be a bit tricky because you need to supply part of the XML yourself. The most basic example is providing the source filename and band number. You could set up an XML template something like the following and then use it when adding bands to the VRT:

```
xml = '''
<SimpleSource>
  <SourceFilename>{0}</SourceFilename>
  <SourceBand>1</SourceBand>
</SimpleSource>
'''
```

This snippet assumes that you'll always use the first band from the source raster. That's because you're going to use it to define a natural color raster without copying any data around like you did in the last chapter, and each of the input rasters only has one band anyway. Most things work the same way with a VRT as they do for other dataset types, so creating the new dataset is the same as before. Even though no data will be copied, you still need to make sure that you create the dataset with the same dimensions as the originals:

```
os.chdir(r'D:\osgeopy-data\Landsat\Washington')
tmp_ds = gdal.Open('p047r027_7t20000730_z10_nn30.tif')
driver = gdal.GetDriverByName('vrt')
ds = driver.Create('nat_color.vrt', tmp_ds.RasterXSize,
    tmp_ds.RasterYSize, 3)
ds.SetProjection(tmp_ds.GetProjection())
ds.SetGeoTransform(tmp_ds.GetGeoTransform())
```

Now you can go about adding the links to the three input rasters. For each one, you need to create a dictionary with one entry, where the key is `'source_0'` and the value is the XML string containing the filename. Then you add that dictionary as metadata for the band in the `'vrt_sources'` domain. Repeat this process for all three bands.

```
metadata = {'source_0': xml.format('p047r027_7t20000730_z10_nn30.tif')}
ds.GetRasterBand(1).SetMetadata(metadata, 'vrt_sources')

metadata = {'source_0': xml.format('p047r027_7t20000730_z10_nn20.tif')}
ds.GetRasterBand(2).SetMetadata(metadata, 'vrt_sources')

metadata = {'source_0': xml.format('p047r027_7t20000730_z10_nn10.tif')}
ds.GetRasterBand(3).SetMetadata(metadata, 'vrt_sources')
```

Now you can use QGIS to open the VRT dataset, and you'll see a three-band image such as that shown in figure 10.12. Unlike the GeoTIFF you created before, this won't open up in regular image-processing software, however.

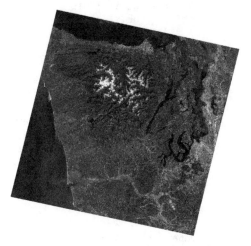

10.6.1 Subsetting

I mentioned earlier that you can use VRTs to subset images without creating another subsetted image. The process of creating the empty dataset is similar to what you did in the previous chapter when subsetting. You still need to figure out the numbers of rows and columns and the new geotransform, and then use that information to create the new dataset. That's what the first part

Figure 10.12 Stacked VRT created from three single-band images. It doesn't look like much when printed in black and white, but you can view the color version online, which will look like natural color (the way our eyes would see it).

of the following listing does, except that it uses a VRT driver. Things change after you create the dataset, however, because in this example you need to create the appropriate XML for each raster band and insert it into the VRT. The process is explained after the code listing.

Listing 10.6 Subset a raster using a VRT

```
import os
from osgeo import gdal

os.chdir(r'D:\osgeopy-data\Landsat\Washington')
tmp_ds = gdal.Open('nat_color.tif')
tmp_gt = tmp_ds.GetGeoTransform()

inv_gt = gdal.InvGeoTransform(tmp_gt)
if gdal.VersionInfo()[0] == '1':
    if inv_gt[0] == 1:
        inv_gt = inv_gt[1]
    else:
        raise RuntimeError('Inverse geotransform failed')
elif inv_gt is None:
    raise RuntimeError('Inverse geotransform failed')

vashon_ul = (532000, 5262600)
vashon_lr = (548500, 5241500)
ulx, uly = map(int, gdal.ApplyGeoTransform(inv_gt, *vashon_ul))
lrx, lry = map(int, gdal.ApplyGeoTransform(inv_gt, *vashon_lr))
rows = lry - uly
columns = lrx - ulx
gt = list(tmp_gt)
gt[0] += gt[1] * ulx
gt[3] += gt[5] * uly

ds = gdal.GetDriverByName('vrt').Create('vashon.vrt', columns, rows, 3)
ds.SetProjection(tmp_ds.GetProjection())
ds.SetGeoTransform(gt)

xml = '''
<SimpleSource>
  <SourceFilename relativeToVRT="1">{fn}</SourceFilename>
  <SourceBand>{band}</SourceBand>
  <SrcRect xOff="{xoff}" yOff="{yoff}"
          xSize="{cols}" ySize="{rows}" />
  <DstRect xOff="0" yOff="0"
          xSize="{cols}" ySize="{rows}" />
</SimpleSource>
'''

data = {'fn': 'nat_color.tif', 'band': 1,
        'xoff': ulx, 'yoff': uly,
        'cols': columns, 'rows': rows}

meta = {'source_0': xml.format(**data)}
ds.GetRasterBand(1).SetMetadata(meta, 'vrt_sources')
```

Account for GDAL version when getting the inverse geotransform

XML describing a band

Data to be inserted into the XML

Insert band 1 into VRT

```
data['band'] = 2
meta = {'source_0': xml.format(**data)}
ds.GetRasterBand(2).SetMetadata(meta, 'vrt_sources')
```
Change to band 2 and insert

```
data['band'] = 3
meta = {'source_0': xml.format(**data)}
ds.GetRasterBand(3).SetMetadata(meta, 'vrt_sources')
```

```
del ds, tmp_ds
```

As mentioned right before the code listing, you have to create XML to subset your raster with a VRT. This XML is slightly more complicated than the XML you used a minute ago, because it also includes elements for the source and destination extents. The numbers of rows and columns are the same size for both the source and destination and are the numbers you calculate. The offsets for the source are the offsets you compute to correspond with the upper-left corner of the area of interest, and the offsets for the destination are both 0 because you fill the entire output image:

```
<SrcRect xOff="{xoff}" yOff="{yoff}" xSize="{cols}" ySize="{rows}" />
<DstRect xOff="0" yOff="0" xSize="{cols}" ySize="{rows}" />
```

This time you use named placeholders in the XML to make it easier to see what goes where. To format this string, you need a dictionary with the same keys as placeholders. You can create this dictionary once and then change the band number (because you're using a different band from the three-band input image each time) when you add a new band to the VRT:

```
data['band'] = 2
```

```
meta = {'source_0': xml.format(**data)}
ds.GetRasterBand(2).SetMetadata(meta, 'vrt_sources')
```

If all went well, the VRT will look like figure 10.13 if you open it in QGIS, and it will overlay perfectly on the original image.

10.6.2 *Creating troublesome formats*

Not all raster formats allow you to create and manipulate multiple-band images. For example, if you'd tried to create a natural color JPEG instead of a TIFF in the previous chapter, you would have run into problems because the JPEG driver doesn't allow you to create a multiband image and then add data to the bands. That's a real problem if you want JPEG output! Fortunately, VRTs can come to your rescue. All you need to do is create a VRT that defines the output you want, and then use the JPEG (or whatever

Figure 10.13 Subsetted VRT

format) driver's `CreateCopy` function. For example, to create a JPEG of Vashon Island, open up the VRT you created in the last section and then copy it to a JPEG:

```
ds = gdal.Open('vashon.vrt')
gdal.GetDriverByName('jpeg').CreateCopy('vashon.jpg', ds)
```

If you'd rather create an intermediate TIFF instead of a VRT, go right ahead, and then copy the TIFF to a JPEG. The advantage to using a VRT is that you're not creating possibly large intermediate files on disk.

10.6.3 *Reprojecting images*

Remember talking about reprojecting vector data in chapter 8? Raster data can also be reprojected, but it's more complicated than with vector data. With vectors, you need the new coordinates for each vertex and you're good to go, but with rasters you need to deal with the fact that cells get bent and moved around, and a one-to-one mapping from old cell locations to new cell locations doesn't exist (see figure 10.14). The easiest way to determine the pixel value for a new cell is use the value from the input cell that gets mapped closest to the output cell. This is

Figure 10.14   Example of how pixels get moved around when the raster is projected. The triangles and circles are pixel center points for two different rasters. The triangles were created from a reprojected version of the raster that the circles came from. Notice that the dimensions are even different.

called nearest-neighbor and is the fastest method, the one you'll usually want for categorical data. All others, except mode, will change your categories, which you definitely don't want for categorical data. Continuous data rasters usually won't look good if you use nearest-neighbor, however. For those, I generally use bilinear interpolation or cubic convolution, which use an average of surrounding pixels. Several other methods are available, however, that might be more appropriate for your particular data.

I think that the easiest way to reproject a raster, other than using the gdalwarp utility that comes with GDAL, is to use a VRT. There's a handy function that creates a reprojected VRT dataset for you when you provide the spatial reference information. It looks like this:

```
AutoCreateWarpedVRT(src_ds, [src_wkt], [dst_wkt], [eResampleAlg],
[maxerror])
```

- src_ds is the dataset you want to reproject.
- src_wkt is the WKT representation of the source spatial reference system. The default is None, in which case it will use the SRS information from the source raster. If this raster doesn't have SRS information, then you need to provide it here. You can also provide it here if using None makes you nervous.

- dst_wkt is the WKT representation of the desired spatial reference system. The default is None, in which case no reprojection will occur.
- eRasampleAlg is one of the resampling methods from table 10.3. The default is GRA_NearestNeighbour.
- maxerror is the maximum amount of error, in pixels, that you want to allow. The default is 0, for an exact calculation.

Table 10.3 Resample methods

Constant	Description
GRA_NearestNeighbour	Closest pixel
GRA_Bilinear	Weighted distance average of 4 pixels
GRA_Cubic	Average of 16 pixels
GRA_CubicSpline	Cubic B-spline of 16 pixels
GRA_Lanczos	Lanczos windowed sinc with 36 pixels
GRA_Average	Average
GRA_Mode	Most common value

The AutoCreateWarpedVRT function doesn't create a VRT file on disk, but returns a dataset object that you can then save to another format using CreateCopy. The following example takes the natural color Landsat image that uses a UTM spatial reference, creates a warped VRT with a destination SRS of unprojected WGS84, and copies the VRT to a GeoTIFF:

```
srs = osr.SpatialReference()
srs.SetWellKnownGeogCS('WGS84')
os.chdir(r'D:\osgeopy-data\Landsat\Washington')
old_ds = gdal.Open('nat_color.tif')
vrt_ds = gdal.AutoCreateWarpedVRT(old_ds, None, srs.ExportToWkt(),
    gdal.GRA_Bilinear)
gdal.GetDriverByName('gtiff').CreateCopy('nat_color_wgs84.tif', vrt_ds)
```

The output from this looks like figure 10.15.

Figure 10.15 The original Landsat image that uses a UTM spatial reference, and the new one that uses unprojected lat/lon coordinates

10.7   *Callback functions*

Often you want an indication of how long your process is going to take or how far along it is. If I'm batch processing multiple files, and each one takes a long time, I'll sometimes have my code print out a message telling me which file it's currently working on. That's not a useful technique if I want to see the progress of a GDAL function, such as computing statistics or warping an image, because that bit of processing is out of my hands once I call the function. Fortunately, the GDAL developers thought of this, and many functions take callback functions as arguments (in fact, you saw this in the signature for Get-Histogram). A callback function is one that gets passed to another function and is then called from the function it was passed to (figure 10.16).

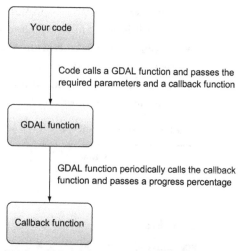

Figure 10.16   Callbacks are functions that get passed as parameters to a second function, where they're later called.

In the case of GDAL, the callback functions are designed to show progress, so they get called often as the process runs. A predefined function is even available for you to use that prints a percentage or a dot at every 2.5% of progress. By the time the process has finished, the output from this function looks like this:

```
0...10...20...30...40...50...60...70...80...90...100 - done.
```

To take advantage of this, just pass gdal.TermProgress_nocb to any function that takes a callback function as a parameter. This example would cause progress information to be printed while calculating statistics:

```
band.ComputeStatistics(False, gdal.TermProgress_nocb)
```

> **TIP**   Several of the methods on OGR layers, such as Intersection and Union, also accept a callback parameter. To use it, import GDAL and pass gdal.TermProgress_nocb as you do with GDAL functions. You could also use callbacks to track the progress of your vector-processing functions using the techniques shown here.

You can also use this function to print out the progress of your own functions. Instead of passing the TermProgress_nocb function to another function, call it yourself with the appropriate percentage. For example, if I wanted to use this instead of print file-names during my batch processing, I could do something like this:

```
for i in range(len(list_of_files)):
    process_file(list_of_files[i])
```

```
        gdal.TermProgress_nocb(i / float(len(list_of_files)))
gdal.TermProgress_nocb(100)
```

This assumes that the list_of_files variable is a list of all of the files to process, and that the process_file function does something with the file. Each time I start processing a new file, I figure out how far I am based on the total number of files and pass that to TermProgress_nocb so I can get a visual indication of my progress. The progress function is also called after finishing the loop in order to tidy things up. Otherwise, if the last percentage passed wasn't quite 100, you'd end up with output like this, where the last bit is missing:

```
0...10...20...30...40...50...60...70...80...90..
```

You might not care about that, but if other people are running your code, they might prefer to know that things finished.

> **NOTE** It's possible that with your version of GDAL, the progress function is called TermProgress instead.

You can also write your own callback function if you'd like your progress information to look different. The function you define needs to have three different parameters. The first is the progress percentage between 0 and 1, the second is a string, and the third is whatever you want. If one of the GDAL functions invokes your callback, it will pass a string specifying what it's doing as the second parameter. You provide the third parameter when you pass in the callback function. The best way to explain how this works is by example, so the following listing is an example that allows the user to specify how often to print a progress indicator dot by passing in another number between 0 and 1 as the progressArg parameter. The message string is also printed out once at the beginning.

Listing 10.7 Use a callback function

```
import sys
def my_progress(complete, message, progressArg=0.02):
    if not hasattr(my_progress, 'last_progress'):
        sys.stdout.write(message)                          This runs the first
        my_progress.last_progress = 0                      time only
    if complete >= 1:
        sys.stdout.write('done\n')
        del my_progress.last_progress       Clear out
    else:                                   progress
        progress = divmod(complete, progressArg)[0]   when done
        while my_progress.last_progress < progress:
            sys.stdout.write('.')
            sys.stdout.flush()                             Show current progress
            my_progress.last_progress += 1
```

There's a trick here that you may not have seen before. Normally when you declare a variable inside a function, it disappears when the function finishes, right? Here you attach the variable to the function as an attribute, so it sticks around and can be used the next time the function is called. The Python hasattr function checks to see if an object has an attribute with a given name, and you check to see if the my_progress function has an attribute called last_progress. If there isn't one, then you assume that this is the first time the function has been called and print the message parameter and create the last_progress attribute. The next time the function is called, that attribute will exist, so the message won't be printed. You'll also use that attribute to keep track of how many dots have been printed so far.

Next you check to see if the process is done. If so, then you print done and delete the last_progress attribute. If you don't delete the attribute, then you can't use this function again in the same script because it will always think that it's done and won't do anything.

If the process isn't finished, which is the case most of the time, you use divmod (which returns a quotient and a remainder as a tuple) to figure out how many dots should be printed. Because multiple dots might need to be printed if the process runs quickly, you keep printing and incrementing last_progress until it equals the required number of dots.

This function uses sys.stdout.write instead of print because print works slightly differently in Python 2 and 3, and so you need to call it in different ways to get the dot to print without a newline after it. Using write solves the problem because it doesn't print a newline unless you request it. You do need to call flush to make sure that the dot shows up immediately, however. The progress function isn't of much use if the dots aren't printed until the processing is finished.

Now that you have your progress function, how do you use it? Exactly the same way that you used TermProgress_nocb, except that you need to include an additional parameter specifying how often you want the indicators printed (the default value of 0.02 is only honored if you call the function yourself, not if it's called by something else). Here's an example:

```
band.ComputeStatistics(False, my_progress, 0.05)
Compute Statistics..................done
```

This is a simple example, but the same concepts apply if you want to do something more complicated. For example, if you had an exceptionally long-running process and wanted to be notified by email at certain points in the process, you could do that.

10.8 *Exceptions and error handlers*

As with OGR, you can have GDAL throw an exception when it runs into a problem. To do this, simply call UseExceptions, and you can turn exceptions off by calling Dont-UseExceptions. Normally, you need to make sure that operations that might have failed did work, such as opening a file. If you don't check and the file wasn't opened, then your script will crash when it tries to use the dataset. This might be fine, depending on what you're doing, but it might not be. Take this simple example of batch-computing statistics:

```
file_list = ['dem_class.tif', 'dem_class2.tiff', 'dem_class3.tif']
for fn in file_list:
    ds = gdal.Open(fn)
    ds.GetRasterBand(1).ComputeStatistics(False)
```

This is great, except for the extra "f" at the end of the second filename. The script will spit out the following error and crash when it tries to get the band, and the last file will never be looked at:

```
ERROR 4: `dem_class2.tiff' does not exist in the file system,
and is not recognised as a supported dataset name.

Traceback (most recent call last):
  File "D:\ errors.py", line 28, in <module>
    ds.GetRasterBand(1).ComputeStatistics(False)
AttributeError: 'NoneType' object has no attribute 'GetRasterBand'
```

You have a few ways you can solve this problem. You might have your code check that the dataset was successfully opened, and if not, print a message so that the user knows that the file was skipped. This way nothing will crash and statistics will be computed for the last file:

```
for fn in file_list:
    ds = gdal.Open(fn)
    if ds is None:
        print('Could not compute stats for ' + fn)
    else:
        ds.GetRasterBand(1).ComputeStatistics(False)
```

A cleaner way to handle errors would be to use a try/except block. If multiple possible points of failure exist, you don't have to check that each one succeeded. Instead, wrap the whole thing in one try block and handle all errors in the except clause:

```
gdal.UseExceptions()
for fn in file_list:
    try:
        ds = gdal.Open(fn)
        ds.GetRasterBand(1).ComputeStatistics(False)
    except:
        print('Could not compute stats for ' + fn)
        print(gdal.GetLastErrorMsg())
```

Although the error was encountered in this case, the error message would not print automatically. If you need it, you can get access to the error message using the Get-LastErrorMsg function. After the except clause is processed, the loop continues with the next filename in the list.

GDAL also has the concept of error handlers that get called whenever a GDAL function runs into an error. The default error handler prints out error messages like the one you saw a second ago. If you don't want these messages printed for some reason, you can shut them up with the built-in quiet error handler. To do this, enable the handler before running the code that you'd like to be quiet. The `PushErrorHandler` function will make a handler the active one until you call `PopErrorHandler`, which will restore the original handler:

```
gdal.PushErrorHandler('CPLQuietErrorHandler')
# do stuff
gdal.PopErrorHandler()
```

You can also use `SetErrorHandler` to enable a handler, but then it's in effect until you pass a different handler to `SetErrorHandler`:

```
gdal.SetErrorHandler('CPLQuietErrorHandler')
# do stuff
gdal.SetErrorHandler('CPLDefaultErrorHandler')
```

Error handlers aren't limited to the GDAL functions, though. You can call them yourself if you'd like. Say you have a function that takes two datasets, but they need to share a spatial reference system for your logic to work. You can call whatever error handler happens to be in effect by using the `Error` function, which takes three parameters. The first is an error class and the second is an error number, both from table 10.4. The third argument is the error message.

```
def do_something(ds1, ds2):
    if ds1.GetProjection() != ds2.GetProjection():
        gdal.Error(gdal.CE_Failure, gdal.CPLE_AppDefined,
            'Datasets must have the same SRS')
        return False
    # now do your stuff
```

You might be wondering why you would do this instead of print your error message and return from the function. You certainly could do that, but this gives you more flexibility in the future. If this function is part of a module you're reusing in different situations, you might want to handle errors in different ways, depending on your application. This gives you that ability, because all you have to do is change your error handler instead of finding and changing all of your print statements (which you might have to change back for another application). This also makes your function treat errors the same way that GDAL does. If `UseExceptions` is in effect, then instead of printing the error message, the call to `Error` will raise an exception that can be caught in a try/except block.

Table 10.4 Possible error classes and numbers

Error classes	Error numbers (types, if you prefer)
CE_None	CPLE_None
CE_Debug	CPLE_AppDefined
CE_Warning	CPLE_OutOfMemory

Table 10.4 Possible error classes and numbers *(continued)*

Error classes	Error numbers (types, if you prefer)
CE_Failure	CPLE_FileIO
CE_Fatal	CPLE_OpenFailed
	CPLE_IllegalArg
	CPLE_NotSupported
	CPLE_AssertionFailed
	CPLE_NoWriteAccess
	CPLE_UserInterrupt

You can write your own error handlers, too. You might do this so you could log error messages to a file or database, or I suppose you could also try to solve the error somehow. If you do choose to write your own function, it needs to accept the same three arguments that get passed to the Error function. Here's a simple example of a handler that logs the error class, reason, and message using the Python logging module:

```
def log_error_handler(err_class, err_no, msg):
    logging.error('{} - {}: {}'.format(
        pb.get_gdal_constant_name('CE', err_class),
        pb.get_gdal_constant_name('CPLE', err_no),
        msg))
```

Get constant names with the ospybook module

The ospybook module contains a function that helps you get the human-readable forms of the GDAL constants. Pass it the case-sensitive prefix corresponding to the type of GDAL constant you want to look up and the numeric value to find. The function returns the name of the constant as a string.

```
>>> import ospybook as pb
>>> print(pb.get_gdal_constant_name('GDT', 5))
GDT_Int32
```

You could use a function like this to easily send your error messages to different places. If you wanted to see the messages on the screen, all you'd have to do is import the logging module and set your error handler:

```
import logging
gdal.PushErrorHandler(log_error_handler)
```

To send the messages to a file instead, you need to add the step of configuring the logger, like this:

```
import logging
logging.basicConfig(filename='d:/temp/log.txt')
```

Now if you called your do_something function on two datasets with different spatial references, a line like this would get added to log.txt:

```
ERROR:root:CE_Failure - CPLE_AppDefined: Datasets must have the same SRS
```

Normally, when you turn on exceptions by calling UseExceptions, the error handlers are turned off and an exception is raised when an error is encountered. This is why the error messages don't automatically print when exceptions are enabled. If you wanted your error message logged using your new log_error_handler function, but you also wanted to use exceptions, you could enable the error handler after enabling exceptions, and then you should get both behaviors.

10.9   Summary

- Use known locations, called ground control points, if you don't have geotransform information for a raster dataset. You can create a geotransform from the GCPs.

- Add a raster attribute table to your thematic datasets so that you know what each pixel value means.

- Add a color table to a thematic dataset if you want it to draw with the same colors all of the time.

- You can manipulate data with virtual raster files without ever creating new files on disk.

- The easiest way to reproject a raster is to use a VRT and then copy it to the desired format.

- It's a good idea to use callback functions to provide progress information for long-running processes.

Map algebra with
NumPy and SciPy

This chapter covers

- Manipulating data with NumPy
- Using NumPy and SciPy for local, focal, and zonal map algebra calculations
- Using GDAL for global map algebra calculations
- Resampling data

You've seen how to read and write raster data, but you still don't know how to manipulate pixel values to do any analysis. Aerial photos make nice basemaps, but many types of raster datasets are used for scientific data analysis. For example, you'll see several examples of landcover classification in the next chapter. If you wanted to create your own landcover model, you might collect satellite imagery, elevation data, and climate data such as average precipitation or temperature, all of which are generally raster datasets. If you wanted to use vector data in the model, such as soil types, you'd convert it to raster first so that you could use it with your raster datasets. You could then use techniques from this chapter to derive slope and aspect from your elevation data and to combine all of your datasets to create a landcover model.

You'll learn several techniques for manipulating raster data in this chapter. For example, you'll learn how to apply calculations on a pixel-by-pixel basis to two or more rasters. You'll also see how to use a small set of neighboring pixels to come up with new values. This is what happens when you smooth or sharpen a digital photo. Other calculations use all pixels in a raster, or divide them up based on some common value. Each of these has different uses, and you'll see examples of all of them.

If you plan on working with large raster datasets with Python, you need to be familiar with the SciPy project, which is a collection of Python modules designed for scientific computing. The NumPy, SciPy, and matplotlib modules are part of this. NumPy was designed to handle large arrays of data, which is perfect for raster data because a band is essentially a two-dimensional array of pixel values. SciPy contains routines for several kinds of scientific analysis, and it uses NumPy to hold the data. We'll look at both of these modules, along with a couple of others, in the next two chapters. Matplotlib is a plotting module that's also part of the SciPy project, and we'll look at it in the last chapter.

11.1 Introduction to NumPy

Entire books have been written about NumPy, but we'll take a brief look at how to create arrays and access specific values. When you use the GDAL ReadAsArray function, the data are put into a NumPy array for you. Once there, you can manipulate your data in many different ways. The bulk of this chapter discusses map algebra, which involves calculations on one or more arrays, and many of the examples won't make much sense if you don't understand the basics of working with NumPy arrays. For more in-depth information, please look at another book or refer to the excellent documentation online at http://www.numpy.org.

Accessing individual cell values in NumPy arrays is much the same as accessing values in a Python list, except that they have an index for each dimension of the array. For example, if you have a two-dimensional array, you need to provide both the row and the column offsets to specify a particular cell. Let's look at this and other basics from inside a Python interactive session.

> **TIP** By convention, the numpy module is renamed to np when importing it into a Python script. You don't have to follow suit, but you'll find that many examples, including those on the NumPy website, do this.

First create an example array using the arange function, which returns an array containing a sequence of numbers. Because this array only has one dimension, you can access elements with a single index. You can also get a slice, or section, of an array by providing starting and ending indices separated by a colon.

```
>>> import numpy as np
>>> a = np.arange(12)
>>> a
array([ 0,  1,  2,  3,  4,  5,  6,  7,  8,  9, 10, 11])
>>> a[1]
1
>>> a[1:5]
array([1, 2, 3, 4])
```

As long as the total number of elements in an array doesn't change, you can reshape it to different dimensions. For example, the array you created contains 12 elements, so it can be reshaped into a two-dimensional array with three rows and four columns because that also contains 12 elements. It couldn't be reshaped into an array with four rows and four columns, however, because that would require 16 elements. A two-dimensional array requires a row and a column index, in that order, to access its elements:

```
>>> a = np.reshape(a, (3,4))
>>> a
array([[ 0,  1,  2,  3],
       [ 4,  5,  6,  7],
       [ 8,  9, 10, 11]])
>>> a[1,2]
6
```

TIP When specifying the shape of an array in a function, be sure to pass the dimensions in a tuple instead of as individual values.

If you only provide one index, n, it returns the entire n^{th} row. You can get an entire column by using a colon as the row index, which is the same as 0:n, where n is the number of rows. You could retrieve the entire second row or third column like this:

```
>>> a[1]
array([4, 5, 6, 7])
>>> a[:,2]
array([ 2,  6, 10])
```

You can access a two-dimensional slice by providing starting and ending indices for both dimensions. Again, not providing a starting index on the left of the colon is the same as using 0, and if you don't provide an ending index, then you get the rest of the values in that dimension. You can also use negative numbers to leave rows or columns off the end.

```
>>> a[1:,1:3]
array([[ 5,  6],
       [ 9, 10]])
>>> a[2,:-1]
array([ 8,  9, 10])
```

Working with NumPy arrays is more than accessing cell values, though. You need to use multiple arrays together to implement many types of analyses. If two or more arrays have the same dimensions, you can perform mathematical and logical operations on them. These work on a cell-by-cell basis, so, for example, if you add two arrays, the [n, m] cell in the first array is added to the [n, m] cell in the second array. The same rule applies to operations such as multiplication; if you want mathematical matrix algebra behavior instead, use the numpy.linalg submodule.

```
>>> a = np.array([[1, 3, 4], [2, 7, 6]])
>>> b = np.array([[5, 2, 9], [3, 6, 4]])
>>> a
array([[1, 3, 4],
       [2, 7, 6]])
```

```
>>> b
array([[5, 2, 9],
       [3, 6, 4]])
>>> a + b
array([[ 6,  5, 13],
       [ 5, 13, 10]])
>>> a > b
array([[False,  True, False],
       [False,  True,  True]], dtype=bool)
```

Many different functions exist for working with arrays, including one that works much like an if-else statement. For example, you could create an array with certain values based on the comparison of two existing arrays like this:

```
>>> np.where(a > b, 10, 5)
array([[ 5, 10,  5],
       [ 5, 10, 10]])
```

The first parameter to the where function is the condition to check, the second parameter is the value to use if the condition is true, and the third is the value to use otherwise. These values can also be arrays, as long as they're the same size as the condition array. For example, you could get the larger of the two values at each location like this:

```
>>> np.where(a > b, a, b)
array([[5, 3, 9],
       [3, 7, 6]])
```

Now that you've seen these examples, let's look at another way to extract data from arrays. You aren't limited to one value or slices of contiguous data, because you can also use a list of indices. As an example, create an array of 12 random integers between 0 and 20 and then extract the ninth, first, and fourth values, in that order:

```
>>> a = np.random.randint(0, 20, 12)
>>> a
array([16, 16, 18,  1, 14,  2, 18, 19,  2, 16, 10,  8])
>>> a[[8, 0, 3]]
array([ 2, 16,  1])
```

If the array is multidimensional, you need to provide a list of lists, with an inner list for each dimension. If you want to extract three values from a two-dimensional array, you would provide a list containing two other lists. The first of these would contain the three row offsets, and the second would contain the three column offsets. You'll use this technique to easily sample pixel values at a list of locations in the next chapter. Try converting the random number array into two-dimensions and look at how it works:

```
>>> a = np.reshape(a, (3, 4))
>>> a
array([[16, 16, 18,  1],
       [14,  2, 18, 19],
       [ 2, 16, 10,  8]])
>>> a[[2, 0, 0], [0, 0, 3]]
array([ 2, 16,  1])
```

You can also use an array of Boolean values that's the same size as your data array, and the returned array will contain only the values that correspond to `True`. Here's an example of this, using the same array a:

```
>>> b
array([[False, False,  True, False],
       [False,  True, False, False],
       [ True,  True,  True, False]], dtype=bool)
>>> a[b]
array([18,  2,  2, 16, 10])
```

Why is this useful? Say you wanted to get the mean pixel value, but only for pixels that had a value greater than five. Using `where` to select the values of interest wouldn't work, because you'd still have to set the nonmatching ones to some value, which would mess up your mean calculation. Using Boolean indexing solves your problem.

```
>>> np.mean(a[a>5])
15.0
```

Sometimes you need to create a new array from scratch. If the cells need to be initialized to a certain value, you can use the `zeros` or `ones` functions. These return floating-point arrays by default, but you can specify the data type if needed. If you need a different number, you can create an array of ones and multiply that by the number you need:

```
>>> np.zeros((3,2))
array([[ 0.,   0.],
       [ 0.,   0.],
       [ 0.,   0.]])
>>> np.ones((2,3), np.int)
array([[1, 1, 1],
       [1, 1, 1]])
>>> np.ones((2,3), np.int) * 5
array([[5, 5, 5],
       [5, 5, 5]])
```

You might have noticed that `np.int` was provided as a second parameter to the `ones` examples. Arrays are created as floating-point by default, but you can specify a different data type if you need. This example didn't specify if it should be a 32-bit or 64-bit integer, and the result is system-dependent. To ensure that you get a 64-bit integer, use `np.int64`. A list of the available NumPy data types can be found at http://docs.scipy.org/doc/numpy/user/basics.types.html.

> **TIP** NumPy data types and GDAL data types aren't the same thing, and you can't use them interchangeably.

If you need an empty array that doesn't have to be initialized to a certain value, you can use the `empty` function. This is faster than initializing an array, but be sure to fill all cells with real data eventually, because ones that you don't fill will contain garbage, like that shown here:

```
>>> np.empty((2,2))
array([[ 2.50516998e-315,   2.50377043e-315],
       [ 1.53313748e-316,   0.00000000e+000]])
```

You'll see examples of several more NumPy functions and techniques for working with arrays as you read through this chapter.

11.2 Map algebra

Map algebra is way of manipulating raster datasets using algebraic operations that you're already familiar with, such as addition and subtraction. In this case, however, two or more rasters are used instead of numbers. You can use these techniques to process raster data in many different ways, from simple to complex. You could touch up a dataset to make it look better on a map, or you could create entirely new datasets derived from one or more others.

Four main types of map algebra exist, all useful for different types of analyses. Several of the array examples in the previous section showed local analysis, where each operation works on individual pixels. This is what happens when you add two arrays together. Focal analyses use a few surrounding pixels, zonal operations work on pixels with the same value, and global analyses work on the entire array. We'll look at examples of all of these.

Before diving into the details, though, let's write a function that will save typing later, as shown in listing 11.1. It will create an output GeoTIFF with the same dimensions, geotransform, and projection as an existing dataset. The function will require five parameters: the existing dataset, the filename for the new dataset, a NumPy array containing data to write to the new image, an output data type, and an optional NoData value.

Listing 11.1 Function to save a new raster

```
def make_raster(in_ds, fn, data, data_type, nodata=None):
    """Create a one-band GeoTIFF.

    in_ds     - datasource to copy projection and geotransform from
    fn        - path to the file to create
    data      - NumPy array containing data to write
    data_type - output data type
    nodata    - optional NoData value
    """
    driver = gdal.GetDriverByName('GTiff')
    out_ds = driver.Create(
        fn, in_ds.RasterXSize, in_ds.RasterYSize, 1, data_type)
    out_ds.SetProjection(in_ds.GetProjection())
    out_ds.SetGeoTransform(in_ds.GetGeoTransform())
    out_band = out_ds.GetRasterBand(1)
    if nodata is not None:
        out_band.SetNoDataValue(nodata)
    out_band.WriteArray(data)
    out_band.FlushCache()
    out_band.ComputeStatistics(False)
    return out_ds
```

This code has nothing new. All it does is create a new raster using information from the existing dataset and the provided data type, write the data into this new raster, and compute statistics. It then returns the new dataset. All of this code would need to be in the rest of the chapter listings, but this function will reduce it to one line. For the sake of convenience, it's already in the ospybook module.

11.2.1 *Local analyses*

Local map algebraic operations are probably the simplest to both understand and perform. They work on two or more arrays that are the same size, and an algebraic equation is applied to each set of pixel locations. Figure 11.1 shows an example of a local calculation that adds two 2D arrays together. This is a simple example, but the operations can be much more involved if required.

Adding two rasters together may not seem useful at first, but it can be. For example, I remember helping with a project many years ago that used this technique to rank land for conservation efforts. A few input rasters were created that ranked locations by individual variables, such as distance to riparian areas. Areas within a certain distance to water had the highest rank, and other distance intervals had different ranks. Another input raster ranked areas on biodiversity, and another on distance to existing developments. There were six or seven of these datasets, all with a small number of rank categories. They were added together to find the locations with the highest overall rank, and therefore the most important for conservation efforts. This simple model was then turned into an interactive online tool that allowed people to change their rankings of the different variables. If a user selected a different ranking structure for a variable, a new raster to reflect those priorities was created and the appropriate rasters were added together to get a new overall importance map. This gave planners a simple tool for exploring different scenarios without knowing anything about GIS. I'm sure much more sophisticated models exist online today, but this was in the days before online mapping was common.

Local analysis can be used for plenty of other things as well. Another common task using multispectral imagery is to compute various indices for tasks such as distinguishing between burned and unburned land or measuring nitrogen contained in vegetation. Let's look at an index used to measure "greenness," the normalized difference vegetation index (NDVI). The NDVI is a simple index that uses red and near-infrared wavelengths to produce a number that ranges from -1 to 1. Growing plants use red wavelengths for photosynthesis, but reflect near-infrared radiation, so a high ratio of these two measurements can indicate photosynthetic activity and healthy vegetation.

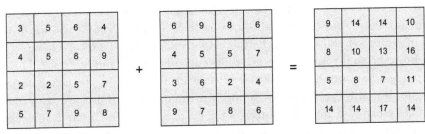

Figure 11.1 Local map algebra calculations work on a pixel-by-pixel basis, so the equation applies to pixels that fall in the same spatial location.

NOTE TO PRINT BOOK READERS: COLOR GRAPHICS Many graphics in this book are best viewed in color. The eBook versions display the color graphics, so they should be referred to as you read. To get your free eBook in PDF, ePub, and Kindle formats, go to https://www.manning.com/books/geoprocessing-with-python to register your print book.

Remember the near-infrared color composite from chapter 9 that clearly showed that stadium turf was artificial? Let's revisit that briefly in figure 11.2. Here you can see the natural color, red band, near-infrared band, near-infrared composite, and NDVI images. The red, near-infrared, and NDVI images are of single bands, where brighter areas have higher values. Notice that the vegetation is dark in the red band image but bright in the near-infrared one. Vegetation is absorbing red light and reflecting near-infrared, and these pixel values measure the amount being reflected back to the sensor, the same way our eyes see what's reflected back. The color infrared composite is a visual image only. Our eyes can see that vegetation appears red (unless you're reading a black-and-white copy of this book, in which case it's gray and doesn't look much different from the natural color image), but that's not useful if you want to use the data in an analysis. This is where the NDVI comes in. The practice fields in the NDVI image are bright, meaning they have high values and represent growing vegetation. The stadium field is dark, making it easy to determine that it's artificial.

In the following example, you'll calculate the NDVI that's shown in figure 11.2E. The formula is simple:

$$NDVI = \frac{NIR - RED}{NIR + RED}$$

You can use NumPy to apply this equation to two arrays, where one holds the red values and the other the near-infrared. You already know how to read data into NumPy arrays, because you did it in chapter 9. One potential problem exists, however. It's possible for both the red and near-infrared pixels to have 0 values, in which case the denominator is also 0, and we all know that you can't divide by 0. This situation probably doesn't exist in the example data we're using, but if you're using a satellite image you may have a large number of 0 values around the edges, and this becomes important.

You have several ways of dealing with this problem, and you might get different advice depending on who you talk to. The first is to proceed as if there were no problem, although I don't suggest this approach. By default, NumPy will warn you that it ran into errors, but the rest of the calculations will be fine (you can change this behavior, however, so if your settings are different, then things might crash). However, the output will have invalid numbers for the pixels that couldn't be calculated, so you'll have to deal with that somehow. If both the numerator and denominator of the equation are equal to 0, then the output is the np.nan value, which stands for *not a number*. If only the denominator is 0, then the output is set to the np.inf value, which stands for *infinity*. If you leave these values in your dataset, not only will it affect your statistics calculations, but different software will treat the values differently. For these reasons,

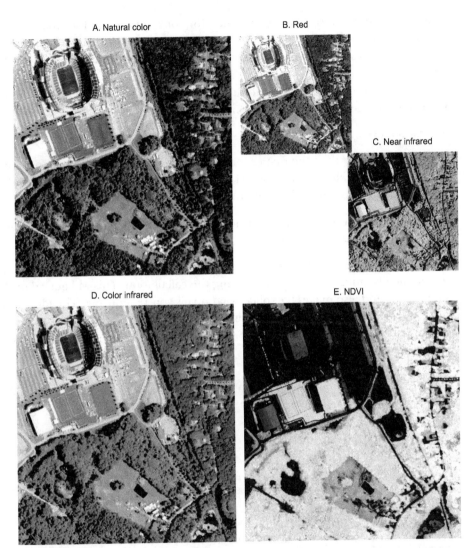

Figure 11.2 Different examples of looking at the same image in different ways. Image A is the natural color image composed of the red, green, and blue wavelengths, and B and C are single bands, where bright areas have higher pixel values. Image D is a visual representation that allows people to see what's vegetation and what isn't, but doesn't help with data analysis. Image E, however, is a single NDVI band in which higher values represent growing vegetation. Not all differences are readily apparent in black and white, but a color version of this figure is available online.

you'll probably want to set the invalid pixels to NoData so that things are standardized. You could do that by checking for pixels equal to either of these values and replacing them with another number, like this:

```
ndvi = (nir - red) / (nir + red)
ndvi = np.where(np.isnan(ndvi), -99, ndvi)
ndvi = np.where(np.isinf(ndvi), -99, ndvi)
```

Then you would also set the `NoData` value for your output band to whatever number you used, in this case -99. I like to use -99 because I know it isn't a valid number for my use cases and it's easy for me to remember, but software packages tend to use much larger numbers.

You might think you could deal with the problem by doing the calculation on only the pixels that don't have 0 as a denominator, like this:

```
ndvi = np.where(nir + red > 0, (nir - red) / (nir + red), -99)
```

This will run faster than the first example, but you'll still get division errors and risk a crash. At least you don't have to check for `nan` or `inf`, because everywhere with a division by 0 will be assigned -99 during the calculation.

> **TIP** Change NumPy's behavior when it encounters floating-point errors with the `numpy.seterr` function.

A better solution to the division by 0 problem is to use masked arrays, which allow you to completely ignore certain pixels during the calculation. This will get rid of the division errors, and also makes it explicit what pixels you're ignoring. The idea is that you mask out the pixels that you want to ignore, do your calculations, and then fill in the missing pixels with your `NoData` value. Check out the following listing for an example of this in action.

Listing 11.2   Compute NDVI for a NAIP image

```
import os
import numpy as np
from osgeo import gdal
import ospybook as pb

os.chdir(r'D:\osgeopy-data\Massachusetts')
in_fn = 'm_4207162_ne_19_1_20140718_20140923_clip.tif'
out_fn = 'ndvi.tif'

ds = gdal.Open(in_fn)
red = ds.GetRasterBand(1).ReadAsArray().astype(np.float)
nir = ds.GetRasterBand(4).ReadAsArray()
red = np.ma.masked_where(nir + red == 0, red)          ← Mask the red band
ndvi = (nir - red) / (nir + red)
ndvi = ndvi.filled(-99)                                 ← Fill the empty cells

out_ds = pb.make_raster(
    ds, out_fn, ndvi, gdal.GDT_Float32, -99)            ← Set NoData to the fill value
overviews = pb.compute_overview_levels(out_ds.GetRasterBand(1))
out_ds.BuildOverviews('average', overviews)
del ds, out_ds
```

This example uses the same input image as that shown in figure 11.2. This dataset is from the National Agriculture Imagery Program (NAIP), part of the United States Department of Agriculture. These aerial images are acquired periodically, with different states processed in different years. Although the visible red, green, and blue bands are always collected so that the images are natural color, sometimes a fourth, near-infrared band is also collected. That's the case with the image used here. The first band is red light, and the fourth is near-infrared, which is why you read these two bands in at the beginning. Because the output is floating-point, you want to ensure that floating point math is used, so you convert the red band from byte to floating-point as you read it into the NumPy array.

Once you have the data in memory, you mask out the red array in all locations where the sum of the two arrays is 0. Although you could also mask the near-infrared data, you don't need to because if one array has a masked value, then no computations happen on that pixel, so it doesn't matter what value the other input arrays have. If you'd rather create a separate mask because you want to apply it to multiple arrays, you could do something like this instead:

```
mask = np.ma.equal(nir + red, 0)
red = np.ma.masked_array(red, mask)
```

Once you mask out the bad pixels, you apply the NDVI equation to the two arrays to create a third one with valid NDVI values in most pixels, but with bad pixels masked out and containing no value. You want your output image to have NoData in those locations, so you fill those pixels in with -99 and then make sure to set -99 as the band's NoData value later. All that's left to do is save your new NDVI array to a new dataset with the same projection and geotransform as the original NAIP image.

11.2.2 Focal analyses

Focal analyses use the pixels that surround the target pixel in order to calculate a value. For a given cell in the output, the value is calculated based on the corresponding cell and its neighbors in the input dataset. This is also called a *moving window* analysis because you can think of it as "window" of cells centered on each pixel in turn. Once the value for the target pixel is calculated, the window moves to the next pixel. Figure 11.3 shows how a 3 x 3 window would "move" across an image. The output values of the dark pixels are calculated using the nine surrounding lightly-shaded

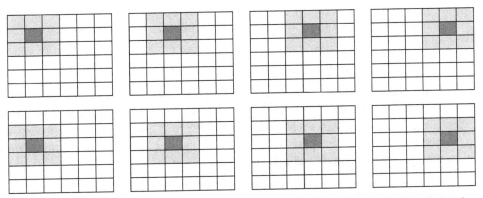

Figure 11.3 In a 3 x 3 moving window analysis, the output value for each dark pixel is calculated using the nine surrounding lightly-shaded input pixels.

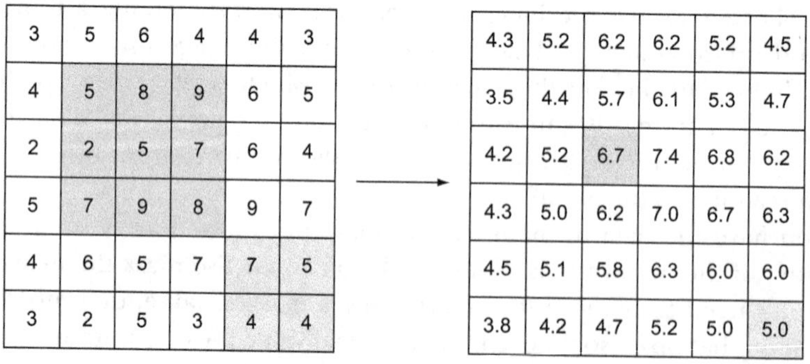

3	5	6	4	4	3
4	5	8	9	6	5
2	2	5	7	6	4
5	7	9	8	9	7
4	6	5	7	7	5
3	2	5	3	4	4

4.3	5.2	6.2	6.2	5.2	4.5
3.5	4.4	5.7	6.1	5.3	4.7
4.2	5.2	6.7	7.4	6.8	6.2
4.3	5.0	6.2	7.0	6.7	6.3
4.5	5.1	5.8	6.3	6.0	6.0
3.8	4.2	4.7	5.2	5.0	5.0

Figure 11.4   A 3 x 3 moving window that calculates the average value of the nine surrounding pixels (or less if the target pixel is on the edge). The shaded areas correspond to the window of input pixels that produce one output pixel value.

input pixels. These types of operations are common for smoothing data and removing random noise. In fact, you've probably used similar filters to touch up your own digital photos. Focal analyses can also be used for anything else that requires input from surrounding pixels, such as computing slope and aspect for an elevation dataset.

Figure 11.4 shows an example of a smoothing filter that computes the mean of a 3 x 3 moving window. The value of each output pixel is the average value of the nine surrounding pixels in the input. The exception to this is if the target pixel is on the edge, so that there aren't a full nine surrounding pixels. In this particular example, the average of the available pixels is used, but you have many ways of dealing with the edge problem. In the figure, the shaded regions in the input (left) raster show the cells that are used to compute the output value for the corresponding shaded cells in the raster on the right.

If the input data array from figure 11.4 is called indata and the result is called outdata, then the upper shaded output pixel in the figure is computed like this:

```
outdata[2,2] = (indata[1,1] + indata[1,2] + indata[1,3] +
                indata[2,1] + indata[2,2] + indata[2,3] +
                indata[3,1] + indata[3,2] + indata[3,3]) / 9
```

This is the average of the nine surrounding pixels. Thankfully, you have a shorter way to write the same thing:

```
outdata[2,2] = np.mean(indata[1:4, 1:4])
```

Using this information, you might be tempted to loop through the rows and columns of a raster to implement a moving window like this one, especially if you have background in a language such as C. To simplify and eliminate special cases where you don't have nine input pixels, you might throw out the outer rows and columns, and then your code would be similar to this:

```
outdata = np.zeros(indata.shape, np.float32)
for i in range(1, rows-1):
    for j in range(1, cols-1):
        outdata[i,j] = np.mean(indata[i-1:i+2, j-1:j+2])
```

Don't try this at home!

This example would run, but it would be excruciatingly slow, and unless your raster was small, you'd wait a long time for the output. It's a bad idea to implement looping like this on a NumPy array if you don't absolutely have to. You're much better off using array slices, and then your processing speed will be closer to the speed you could get with C. To do this, you need to create nine slices, each one corresponding to one of the nine input pixels, as shown in figures 11.5 and 11.6. The first figure shows a small raster with six rows and columns, and the lightly shaded cells correspond to the slice specified with the text below each example. The dark outline defines the 3 x 3 window around the cell at index [2, 2].

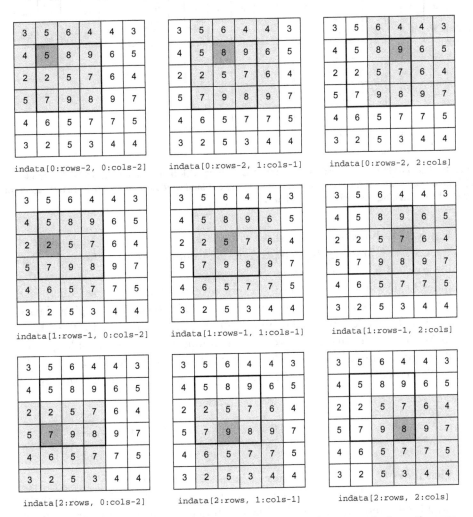

Figure 11.5 **The slices that are used in a 3 x 3 moving window. Each example shows the same input data, but the lightly shaded cells are the slices defined by the text below the example. The dark outline defines the window around pixel [2, 2]. The darker shaded cells are all at index [1, 1] inside the corresponding slice.**

3	5	6	4
4	5	8	9
2	2	5	7
5	7	9	8

5	6	4	4
5	8	9	6
2	5	7	6
7	9	8	9

6	4	4	3
8	9	6	5
5	7	6	4
9	8	9	7

4	5	8	9
2	2	5	7
5	7	9	8
4	6	5	7

5	8	9	6
2	5	7	6
7	9	8	9
6	5	7	7

8	9	6	5
5	7	6	4
9	8	9	7
5	7	7	5

Sum

40	51	55	48
47	60	67	61
45	56	63	60
46	52	57	54

Average

4.4	5.7	6.1	5.3
5.2	6.7	7.4	6.8
5.0	6.2	7.0	6.7
5.1	5.8	6.3	6.0

2	2	5	7
5	7	9	8
4	6	5	7
3	2	5	3

2	5	7	6
7	9	8	9
6	5	7	7
2	5	3	4

5	7	6	4
9	8	9	7
5	7	7	5
5	3	4	4

Figure 11.6   The individual slices created in figure 11.5, along with their sum and average. The shaded cells at index [1, 1] in each slice are the same cells as those in the outlined window in figure 11.5, so averaging the slices is equivalent to averaging the cells in the window.

Figure 11.6 shows these same slices with the cell at index [1, 1] highlighted. Compare the values of these highlighted pixels to the values of the pixels inside the dark outline in figure 11.5. They're the same, so if you take the average of the slices, the value at index [1, 1] will be the average of the nine pixels outlined in figure 11.5. In fact, the output contains the average of all complete 3 x 3 windows in the original dataset. Again, this leaves the edge rows and columns out of the resulting data, for the sake of simplicity. You'd need to cut off more rows and columns from the edges for larger moving windows. For example, a 5 x 5 window would cut off two on each side instead of one.

You could create all nine slices, add them together, and then divide by 9, like this:

```
outdata = np.zeros(indata.shape, np.float32)
outdata[1:rows-1, 1:cols-1] = (
    indata[0:-2, 0:-2] + indata[0:-2, 1:-1] + indata[0:-2, 2:] +
    indata[1:-1, 0:-2] + indata[1:-1, 1:-1] + indata[1:-1, 2:] +
    indata[2:  , 0:-2] + indata[2:  , 1:-1] + indata[2:  , 2:]) / 9
```

That looks like a pain, but again, I have an easier way to do it. If the slices are all stacked into a three-dimensional array, then you can use the mean function, which would definitely be simpler. The dstack function will stack the slices on top of each other, which is what you need. But you still need to get all of the slices so you can pass them to dstack. You could type everything out again, but that isn't any easier than before. Instead, you could use a loop to get each slice and add it to a list. To do this, you need to loop through three rows and three columns. Assuming that you use loop indices 0-2, the current index can be used as the row or column to start the slice. You know that when a slice starts at row 0, then it needs to end at 2 less than the number of rows. If you add 1 to the starting index, then you need to add 1 to the ending index as well. Therefore, you can find the ending index by adding the starting index to the number rows minus 2:

```
slices = []
for i in range(3):
    for j in range(3):
        slices.append(indata[i:rows-2+i, j:cols-2+j])
```

Not only does this require less typing, but it scales much easier to larger windows, as you'll see in a bit. But now that you have a list of slices, you can stack them in the third dimension with the dstack function, which returns a three-dimensional array that can be used to compute means:

```
stacked = np.dstack(slices)
outdata = np.zeros(indata.shape, np.float32)
outdata[1:-1, 1:-1] = np.mean(stacked, 2)
```

By default, the mean function returns the mean of all pixels in the array, but you want the mean calculated for each set of pixels in a single spatial location, like a local analysis. To do this, tell the function which axis you'd like the mean calculated on. The third dimension is axis 2, so if you specify that, you'll get an array with the same numbers of rows and columns as the stacked array, with the value of each cell being the mean of the nine slices in that location. The slices have two less rows and columns than the original data set, however, so you create a zero-filled array the same size as the original data, and then insert the array containing the means into the middle of it, cutting off a row and column on each side.

This method of getting the nine slices can be easily generalized into a function that will return slices of any size you want (well, as long as the dimensions are odd numbers—even numbers don't work well because there's no middle cell). This function, which is in the ospybook module, is shown in the next listing.

Listing 11.3 Function to get slices of any size from an array

```
def make_slices(data, win_size):
    """Return a list of slices given a window size.

    data     - two-dimensional array to get slices from
    win_size - tuple of (rows, columns) for the moving window
    """
    rows = data.shape[0] - win_size[0] + 1       Calculate slice size
    cols = data.shape[1] - win_size[1] + 1
    slices = []
    for i in range(win_size[0]):
        for j in range(win_size[1]):              Create the slices
            slices.append(data[i:rows+i, j:cols+j])
    return slices
```

Original

Smoothed

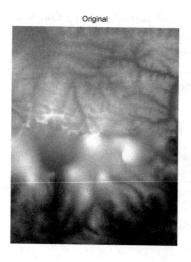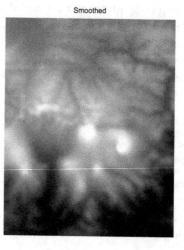

Figure 11.7   A digital elevation model of the area surrounding Mt. Everest, along with a version that has been smoothed using a 3 x 3 moving average filter

Now you can use everything you've learned so far to run an average smoothing filter on an elevation dataset. Figure 11.7 shows a DEM of the area surrounding Mt. Everest. For some reason, a seamline runs right through the middle, and the northern half looks better than the southern half. I thought that perhaps smoothing the dataset would make the seamline less obvious. Whether it did or not is open to debate, but the smoothed image does look different from the original. This is especially obvious in the northern part of the image, where the contours are less distinct in the smoothed version. The following listing shows the code to apply the filter.

Listing 11.4   Smooth an elevation dataset

```
import os
import numpy as np
from osgeo import gdal
import ospybook as pb

in_fn = r"D:\osgeopy-data\Nepal\everest.tif"
out_fn = r'D:\Temp\everest_smoothed_edges.tif'
in_ds = gdal.Open(in_fn)
in_band = in_ds.GetRasterBand(1)
in_data = in_band.ReadAsArray()
slices = pb.make_slices(in_data, (3, 3))
stacked_data = np.ma.dstack(slices)                    Stack the slices

rows, cols = in_band.YSize, in_band.XSize
out_data = np.ones((rows, cols), np.int32) * -99
out_data[1:-1, 1:-1] = np.mean(stacked_data, 2)        Initialize output
                                                        to NoData
pb.make_raster(in_ds, out_fn, out_data, gdal.GDT_Int32, -99)
del in_ds
```

Put result in middle of output

Although it took a while to work up to it, the filtering code turned out to be simple. You use your `make_slices` function to create the nine slices, stack them into a three-dimensional array, and then use the mean function to calculate the average across the third dimension. Because the slices are smaller than the original data, you put the result into the middle of an array of the correct size that has already been initialized to the NoData value. This ensures that the ignored edges are set to NoData, as long as you remember to pass that value to the make_raster function.

Nothing is stopping you from applying much more complicated functions to the cells that make up the moving window. In fact, this is exactly what you'd want to do for many analyses. One example is computing slope from an elevation model. Several algorithms calculate slope, and one of them is shown in figure 11.8.

The next listing shows code for calculating the slope of the Mt. Everest DEM using these equations. Note that for this algorithm to work properly, the elevation units must be the same as the horizontal

$$\frac{dz}{dx} = \frac{(c + 2f + i) - (a + 2d + g)}{8*cell\_width}$$

$$\frac{dz}{dy} = \frac{(g + 2h + i) - (a + 2b + c)}{8*cell\_height}$$

a	b	c
d	e	f
g	h	i

$$dist = \sqrt{\left[\frac{dz}{dx}\right]^2 + \left[\frac{dz}{dy}\right]^2}$$

$$slope(e) = \frac{180*\tan^{-1}(dist)}{\pi}$$

Figure 11.8   The algorithm for computing the slope of cell e from elevation values in the surrounding cells

ones. For example, if your dataset uses a UTM projection, then the coordinates are expressed in meters, so the elevation values must also be meters.

Listing 11.5   Compute slope from DEM

```
import os
import numpy as np
from osgeo import gdal
import ospybook as pb

in_fn = r"D:\osgeopy-data\Nepal\everest_utm.tif"
out_fn = r'D:\Temp\everest_slope.tif'

in_ds = gdal.Open(in_fn)
cell_width = in_ds.GetGeoTransform()[1]
cell_height = in_ds.GetGeoTransform()[5]
band = in_ds.GetRasterBand(1)
in_data = band.ReadAsArray().astype(np.float)
out_data = np.ones((band.YSize, band.XSize)) * -99      ◄── Initialize output
                                                            array with -99
slices = pb.make_slices(in_data, (3, 3))
rise = ((slices[6] + (2 * slices[7]) + slices[8]) -
        (slices[0] + (2 * slices[1]) + slices[2])) / \
       (8 * cell_height)
run = ((slices[2] + (2 * slices[5]) + slices[8]) -
       (slices[0] + (2 * slices[3]) + slices[6])) / \
      (8 * cell_width)

dist = np.sqrt(np.square(rise) + np.square(run))        ┐ Output edges don't
out_data[1:-1, 1:-1] = np.arctan(dist) * 180 / np.pi   ◄── get slope data

pb.make_raster(in_ds, out_fn, out_data, gdal.GDT_Float32, -99)
del in_ds
```

Elevation   Slope

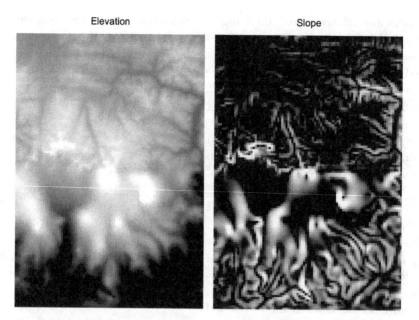

Figure 11.9   **The original Mt. Everest DEM and a slope raster derived from it**

This is more complicated than the smoothing example, but it's still not too bad. Most of it is made up of calculating the slope. You do need to know the order that the slices are stored in the `slices` list to use the correct one in each part of the equation, however. The `make_slices` function returns them in the same order as if you were reading left to right and down, or in other words, in alphabetical order if you refer to figure 11.8. Unlike the smoothing example, in this case you don't stack the slices into a three-dimensional array because you need to reference them individually in the slope equations. Again, you make sure that the edges are set to the `NoData` value. The output looks like that shown in figure 11.9.

USING SCIPY FOR FOCAL ANALYSIS

SciPy is a versatile Python module designed for scientific data analysis, and it uses NumPy arrays to store large amounts of data. It has submodules for interpolation, Fourier transforms, linear algebra, statistics, signal processing, and image processing, among others. The multidimensional image processing submodule contains filtering functions that can be used to perform the same operations you did with NumPy. It's probably easier to use SciPy than NumPy, but hopefully now you understand enough about working with NumPy arrays that you can figure out how to solve other problems that you might run into.

One advantage to using SciPy is that it will handle the edge problems for you by filling in extra cells around the edges so that the calculations can be performed on all cells. It has several different ways of populating these extra pixels, and you can decide which one you want to use. The default is the "reflect" mode, which repeats the values near the edge, but in the opposite order. You can also use the nearest value, a constant value of your own choosing, or a few other value-repeating methods.

One of the built-in filters in SciPy is a uniform filter, which is a smoothing filter that works the same as the smoothing filter from listing 11.4. This next listing shows how you use it.

> **Listing 11.6   Smoothing filter using SciPy**

```
import os
import scipy.ndimage
from osgeo import gdal
import ospybook as pb

in_fn = r"D:\osgeopy-data\Nepal\everest.tif"
out_fn = r'D:\Temp\everest_smoothed.tif'

in_ds = gdal.Open(in_fn)
in_data = in_ds.GetRasterBand(1).ReadAsArray()

out_data = scipy.ndimage.filters.uniform_filter(          Run the filter
    in_data, size=3, mode='nearest')

pb.make_raster(in_ds, out_fn, out_data, gdal.GDT_Int32)
del in_ds
```

As you can see, running the actual filter only requires one line of code, and the rest of it is dealing with reading and writing the data. The only required argument to `uniform_filter` is the NumPy array containing the data to smooth, but you have several optional parameters. You use two of them here. The `size` parameter specifies the size of the moving window to use, and you don't really need to use it here because the default value is 3 anyway. You also use the `mode` parameter to change the method of dealing with the edges so that the closest pixel values are used to fill in the edges.

Other built-in filters exist, including minimum, maximum, and median values. But what about more complicated situations such as the slope calculation? All you need to do is create a function that performs the calculation that you want and then pass that to a generic filter function, as shown in the following listing.

> **Listing 11.7   Calculate slope using SciPy**

```
import os
import numpy as np
import scipy.ndimage
from osgeo import gdal
import ospybook as pb

in_fn = r"D:\osgeopy-data\Nepal\everest_utm.tif"
out_fn = r'D:\Temp\everest_slope_scipy2.tif'

def slope(data, cell_width, cell_height):
    """Calculates slope using a 3x3 window.
```

```
    data         - 1D array containing the 9 pixel values, starting
                   in the upper left and going left to right and down
    cell_width   - pixel width in the same units as the data
    cell_height  - pixel height in the same units as the data
    """
    rise = ((data[6] + (2 * data[7]) + data[8]) -
            (data[0] + (2 * data[1]) + data[2])) / \
           (8 * cell_height)
    run  = ((data[2] + (2 * data[5]) + data[8]) -
            (data[0] + (2 * data[3]) + data[6])) / \
           (8 * cell_width)
    dist = np.sqrt(np.square(rise) + np.square(run))
    return np.arctan(dist) * 180 / np.pi

in_ds = gdal.Open(in_fn)
in_band = in_ds.GetRasterBand(1)
in_data = in_band.ReadAsArray().astype(np.float32)        ◄──┐ Use floating-point

cell_width = in_ds.GetGeoTransform()[1]
cell_height = in_ds.GetGeoTransform()[5]
out_data = scipy.ndimage.filters.generic_filter(
    in_data, slope, size=3, mode='nearest',                   Run the filter
    extra_arguments=(cell_width, cell_height))

pb.make_raster(in_ds, out_fn, out_data, gdal.GDT_Float32)
del in_ds
```

The first thing you do is write a custom filter function called slope that contains the exact same math as before, so it should look familiar. The first argument to your filter function must be a one-dimensional array of data that will be used for the calculation. Conveniently, the cell values will be in the same order that you used earlier with your make_slices function. The first value is the upper-left pixel, the second is the upper-middle pixel, and so on, until ending with the lower-right pixel. If you need it to, your function can also take additional parameters, but this isn't a requirement for custom filters. In this case, you need to know the pixel dimensions for the slope calculation, so your function also requires pixel width and height.

Once you have your custom filter function, you provide it as a parameter when calling the SciPy generic_filter function. The first argument to generic_filter is the NumPy array containing data to filter and the second is the filter function to use. These are the only required parameters, but once again, you can use optional ones. In this case, you specify a 3 x 3 moving window, but it's possible to use different sizes, or even use a Boolean array to indicate exactly which cell values to pass to your filter. You can read more about this in the SciPy documentation. Once again, you change the default method of dealing with the array edges, and finally, you provide a tuple containing the extra arguments that your function requires. The generic_filter function passes the appropriate pixel values to your function to calculate each output cell value.

BREAKING UP FOCAL ANALYSES

What if you want to do a moving window analysis but don't have enough RAM to hold everything in memory? You can break the image up into chunks, but instead of

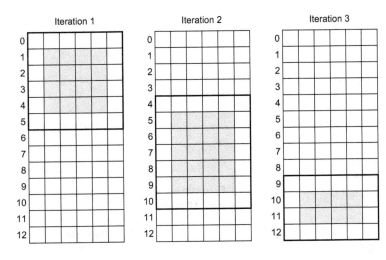

Figure 11.10 Breaking up an image into overlapping chunks. The thick outlines show the cells read from disk, and the shaded cells are the ones that get valid data and are written to the output.

processing discrete sets of data, have them overlap each other. Figure 11.10 shows an example of reading in multiple rows at a time using a step parameter of 5. More than five rows get read in each time, though, because of the overlap. The dark outlines show the rows that are processed, and the shaded areas are the cells that get valid data in that iteration. Because this is a 3 x 3 window, it has one empty row and column on each side. The idea is to tack one row on the top and one on the bottom of each chunk so that every row ends up with valid data. The first time through, one extra row is added at the bottom so six rows get processed instead of five. The second time through, two extra rows are processed. To get an extra row at the top, the starting offset is moved to an earlier row. For example, with a step value of 5, the second iteration would normally start reading at row 5, but instead it starts at row 4 in this example. The third time through is similar, except that the available rows are limited so only four are read in. You can see that all of the rows except the top and bottom end up with valid data.

Although the Everest dataset is small, pretend for a moment that it's too large to process all at once, but you have enough RAM to process approximately 100 rows at a time. The following listing shows how you could do this.

Listing 11.8 Focal analysis broken into chunks

```
import os
import numpy as np
from osgeo import gdal
import ospybook as pb

in_fn = r"D:\osgeopy-data\Nepal\everest.tif"
out_fn = r'D:\Temp\everest_smoothed_chunks.tif'

in_ds = gdal.Open(in_fn)
in_band = in_ds.GetRasterBand(1)
```

```
xsize = in_band.XSize
ysize = in_band.YSize

driver = gdal.GetDriverByName('GTiff')
out_ds = driver.Create(out_fn, xsize, ysize, 1, gdal.GDT_Int32)
out_ds.SetProjection(in_ds.GetProjection())
out_ds.SetGeoTransform(in_ds.GetGeoTransform())
out_band = out_ds.GetRasterBand(1)
out_band.SetNoDataValue(-99)

n = 100
for i in range(0, ysize, n):
    if i + n + 1 < ysize:
        rows = n + 2                    Read two extra
    else:                               rows if possible
        rows = ysize - i
    yoff = max(0, i - 1)                            ◄───  Don't start
                                                         before row 0
    in_data = in_band.ReadAsArray(0, yoff, xsize, rows)
    slices = pb.make_slices(in_data, (3, 3))
    stacked_data = np.ma.dstack(slices)
    out_data = np.ones(in_data.shape, np.int32) * -99
    out_data[1:-1, 1:-1] = np.mean(stacked_data, 2)

    if yoff == 0:
        out_band.WriteArray(out_data)
    else:                                       Don't overwrite good data
        out_band.WriteArray(out_data[1:], 0, yoff + 1)   from previous chunk

out_band.FlushCache()
out_band.ComputeStatistics(False)
del out_ds, in_ds
```

Much of this resembles code you've written before. Remember that you're going to read in two extra rows if possible, so you need to take those extra rows into account when checking if enough rows exist to read an entire chunk. You also make sure you don't try to use -1 as a row offset in the first iteration. Once you determine how many rows to grab, you read them in and process them as before. For all but the first chunk, though, you make sure not to write the first row of the processed data to the output. If you had, it would've overwritten the good data written during the previous iteration. Because you're ignoring this first row, you also have to increase the row offset you use for writing.

11.2.3 *Zonal analyses*

Zonal analyses work on cells that share a certain value, or belong to the same zone. The zones are usually defined by one raster and the analysis performed using values from a second one. For example, if you have a raster showing land ownership categories such as federal, state, and private, and a second raster showing landcover, you could use the ownership categories as zones to determine the acreage of each landcover type within each ownership category, as shown in figure 11.11.

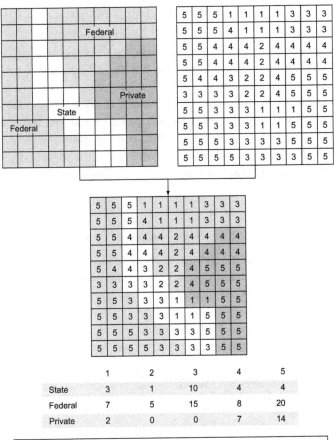

	1	2	3	4	5
State	3	1	10	4	4
Federal	7	5	15	8	20
Private	2	0	0	7	14

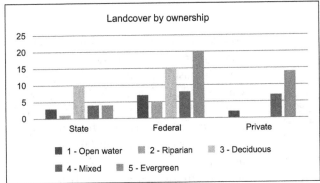

Figure 11.11 An example of a zonal analysis using ownership and landcover type. The number of pixels for each landcover are counted per ownership zone.

First let's look at how you could do this with NumPy and then a more flexible method using SciPy. What you want here is basically a two-dimensional histogram. A regular histogram gives you the number of items in each bin, but in this case, you want to treat zones as one set of bins and landcover as another set of bins, and then get the count for each combination of zone and landcover bin. There are several ways you can define your bins, including letting NumPy do it for you, but here you'll see how to do it yourself. A set of bins is defined by an array of bin edges. The first number in the array is the lower bound for the first bin, the second number is the upper, non-inclusive bound for the first bin and also the lower, inclusive bound for the second bin, and so on. The last number in the array is the upper bound of the last bin. One easy way to get the bins for this particular use case is to get the unique values in the dataset. The NumPy unique function returns these values in sorted order. Because the lower bound is inclusive, this list of numbers would create the lower bound for bins corresponding to the dataset values. All that's left is to add a larger number to the end to form the upper bound for the last bin. The following function creates this array of bin edges for you:

```
def get_bins(data):
    """Return bin edges for all unique values in data. """
    bins = np.unique(data)
    return np.append(bins[~np.isnan(bins)], max(bins) + 1)
```

Now that you know how to define bins, let's see how to use them with the NumPy histogram2d function to get the counts. The two required parameters for this function are the two arrays containing values to bin, and one of the optional arguments lets you specify the bins you want to use. If the values from the upper-left dataset shown in figure 11.11 are in an array called zones, and the values from the upper-right dataset are in an array called landcover, then you can get the two-way histogram like this:

```
>>> hist, zone_bins, landcover_bins = np.histogram2d(
...     zones.flatten(), landcover.flatten(),
...     [get_bins(zones), get_bins(landcover)])
>>> hist
array([[  3.,   1.,  10.,   4.,   4.],
       [  7.,   5.,  15.,   8.,  20.],
       [  2.,   0.,   0.,   7.,  14.]])
```

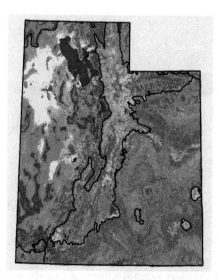

Notice a couple of things here. First, the histogram2d function wants the data arrays to be one-dimensional, and the flatten function takes care of that detail. The histogram2d function returns three values: a two-dimensional histogram and two sets of bins, one for each input array. The histogram rows correspond to the bins from the first array passed in, and the columns correspond to the second. In this case, the two bin outputs will be exactly what you pass in, but if you don't explicitly define your bins, then these two return values would tell you what bins were used for the calculation.

Figure 11.12 SWReGAP landcover classification with ecoregion boundaries drawn on top

If you have SciPy, you can accomplish this same thing with a more general SciPy function called `stats.binned_statistic_2d`. The following listing shows you how to use this to count up the number of pixels with each landcover class in each ecoregion zone shown in figure 11.12.

Listing 11.9   Zonal analysis with SciPy

```
import numpy as np
import scipy.stats
from osgeo import gdal

def get_bins(data):
    """Return bin edges for all unique values in data."""
    bins = np.unique(data)
    return np.append(bins, max(bins) + 1)

landcover_fn = r'D:\osgeopy-data\Utah\landcover60.tif'
ecoregion_fn = r'D:\osgeopy-data\Utah\utah_ecoIII60.tif'
out_fn = r'D:\Temp\histogram.csv'

eco_ds = gdal.Open(ecoregion_fn)
eco_band = eco_ds.GetRasterBand(1)
eco_data = eco_band.ReadAsArray().flatten()
eco_bins = get_bins(eco_data)

lc_ds = gdal.Open(landcover_fn)
lc_band = lc_ds.GetRasterBand(1)
lc_data = lc_band.ReadAsArray().flatten()
lc_bins = get_bins(lc_data)

hist, eco_bins2, lc_bins2, bn = \
    scipy.stats.binned_statistic_2d(
        eco_data, lc_data, lc_data, 'count',      ← Compute histogram
        [eco_bins, lc_bins])
hist = np.insert(hist, 0, lc_bins[:-1], 0)
row_labels = np.insert(eco_bins[:-1], 0, 0)       ← Add bin info to histogram
hist = np.insert(hist, 0, row_labels, 1)
np.savetxt(out_fn, hist, fmt='%1.0f', delimiter=',')   ← Save output
```

The first thing you do is read in the ecoregion and landcover datasets as flattened, one-dimensional arrays (because this function, like `histogram2d`, requires single-dimension arrays), and also calculate the required bins for each dataset. Then you create your histogram using the `binned_statistic_2d` function. The first two parameters are the same as the `histogram2d` function, namely, the two datasets that will be binned. Unlike `histogram2d`, this function can not only count occurrences but also calculate statistics, so it also requires a third array containing the values to calculate statistics on. Because you count the number of pixels, it doesn't matter if you use landcover or ecoregions in this case, but you use landcover. The next parameter to the function specifies which statistic you want to calculate, which is "count" in this case (other options are mean, median, sum, or a custom function that you provide). But you could do something like calculate the mean elevation in each combination of ecoregion and landcover by passing an elevation dataset as the third parameter and "mean" as the fourth. Anyway, you provide the bin boundaries as the last argument, the way you did before. This returns the same outputs as the histogram function, along with one extra one that indicates which bin the data value fell into.

This time you also get ambitious and add several bin labels to your histogram. Because the different landcovers are the columns, you use those bins as the column labels. Remember that the last item in the bins array is the upper bound and isn't needed to label your bins. You insert all but the last number in the landcover bins array as the first row of the histogram. The insert function wants the data being inserting into, the index to insert at, the data to insert, and the axis, where 0 means a row. You use the same idea to insert the ecoregion bins as row labels, except that you have to add a placeholder for the first row, which is now landcover labels. You insert 0 at the beginning and then use axis 1 to insert a column at the beginning of your histogram.

If you wanted to know the area instead of pixel count, you could multiply the histogram array by the area of a pixel before adding the label row and column.

Once your table is complete, you write it out to a text file. The fmt parameter to savetxt specifies how you want the numbers formatted in the output. Without this, you'd probably get scientific notation; nothing is wrong with that, but here you specify integers instead. The % is the first character of a format string, the 1 means you want it to print out at least one digit, the .0 means no numbers after the decimal point, and f means that it will be getting a floating-point number to work with. For more information on format strings, please see the NumPy documentation for savetxt at http://docs.scipy.org/doc/numpy/reference/generated/numpy.savetxt.html.

Your output should look something like figure 11.13.

What if you wanted to know the most common landcover type in each ecoregion but didn't care about counts? In this case, you could use the one-dimensional binned_statistic function instead, because you only need to bin one ecoregion. Unfortunately, mode isn't one of the supported statistics types, but you can pro-

⊿	A	B	C	D	E	F	G
1	0	0	1	2	4	5	8
2	1	8	0	0	0	0	21819
3	2	0	0	0	0	969	0
4	3	0	0	2796	0	10505	84088
5	4	0	0	553	0	157349	0
6	5	0	0	0	0	0	101
7	6	0	5867	209925	49015	235654	3
8	7	0	0	13174	0	2314	0
9	255	12487181	0	0	0	2873	1285

Figure 11.13 The first few columns of the histogram output. The first column specifies the ecoregion (255 is NoData cells), and the top row specifies the landcover category. All other values are cell counts.

vide your own statistical function. All you need to do is write a simple function that returns the mode and then pass it to binned_statistic, like this:

```
def my_mode(data):
    return scipy.stats.mode(data)[0]

modes, bins, bn = scipy.stats.binned_statistic(
    eco_data, lc_data, my_mode, eco_bins)
```

You can use this technique to calculate whatever information you want, as long as it can be computed from a one-dimensional array.

11.2.4 *Global analyses*

Global functions work on the entire image, such as proximity analysis or cost distance. A *proximity analysis* determines the Euclidean distance of each cell to the nearest cell that's marked as a source, while *cost distance* determines the least cost of traveling between each cell and the nearest source, as determined by a cost surface. For example, if you're walking between two points in the mountains, the easiest, and therefore least costly path, may not be the shortest. In this case, the cost surface might be derived from elevation and slope rasters, and cells on a mountain pass would have a lower cost than cells on steep ridges.

GDAL has global analysis functions built in, and you're about to see how to use several of them to determine the distance to the nearest road for areas in the Frank Church—River of No Return Wilderness in Idaho (figure 11.14). You'll start off with a state-wide roads shapefile and a shapefile showing wilderness boundaries. Several polygons make up this wilderness area, so you'll select them out, get the extent of the selected polygons, and use that rectangle to select the roads you're interested in.

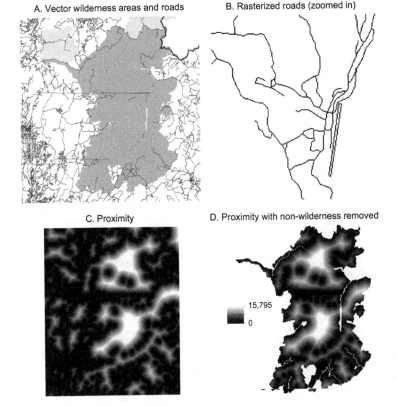

A. Vector wilderness areas and roads

B. Rasterized roads (zoomed in)

C. Proximity

D. Proximity with non-wilderness removed

15,795

0

Figure 11.14 Data used for calculating average distance to roads within a wilderness area. A: The dark shaded area is the wilderness area of interest, and the lines are roads (the roads dataset has artifacts, but those will be treated as real data in the example). B: Roads are no longer smooth lines when rasterized. C: Results of the proximity analysis, where bright areas are further from roads. D: The proximity dataset with non-wilderness removed.

Then you can use GDAL to create a raster that has ones where there are roads and zeros everywhere else. This will be used as a source layer in order to determine distances from roads to every other pixel. The following listing shows how to accomplish all of this. The listing is long, but that's because the data haven't been preprocessed.

Listing 11.10 Proximity analysis

```python
import os
import sys
from osgeo import gdal, ogr

folder = r'D:\osgeopy-data\Idaho'
roads_ln = 'allroads'
wilderness_ln = 'wilderness'
road_raster_fn = 'church_roads.tif'
proximity_fn = 'proximity.tif'
cellsize = 10
```
◀── **Set the cell size for the analysis**

```python
shp_ds = ogr.Open(folder)
wild_lyr = shp_ds.GetLayerByName(wilderness_ln)
wild_lyr.SetAttributeFilter(
    "WILD_NM = 'Frank Church - RONR'")
envelopes = \
    [row.geometry().GetEnvelope() for row in wild_lyr]
coords = list(zip(*envelopes))
minx, maxx = min(coords[0]), max(coords[1])
miny, maxy = min(coords[2]), max(coords[3])
```
Get extent of wilderness area

```python
road_lyr = shp_ds.GetLayerByName(roads_ln)
road_lyr.SetSpatialFilterRect(minx, miny, maxx, maxy)
```
Select roads in wilderness extent

```python
os.chdir(folder)
tif_driver = gdal.GetDriverByName('GTiff')
cols = int((maxx - minx) / cellsize)
rows = int((maxy - miny) / cellsize)

road_ds = tif_driver.Create(road_raster_fn, cols, rows)
road_ds.SetProjection(
    road_lyr.GetSpatialRef().ExportToWkt())
road_ds.SetGeoTransform(
    (minx, cellsize, 0, maxy, 0, -cellsize))
```
Create empty raster for roads

```python
gdal.RasterizeLayer(
    road_ds, [1], road_lyr, burn_values=[1],
    callback=gdal.TermProgress)
```
Burn roads into raster

```python
prox_ds = tif_driver.Create(
    proximity_fn, cols, rows, 1, gdal.GDT_Int32)
prox_ds.SetProjection(road_ds.GetProjection())
prox_ds.SetGeoTransform(road_ds.GetGeoTransform())
gdal.ComputeProximity(
    road_ds.GetRasterBand(1), prox_ds.GetRasterBand(1),
    ['DISTUNITS=GEO'], gdal.TermProgress)
```
Burn proximity data into new raster

```
wild_ds = gdal.GetDriverByName('MEM').Create(
    'tmp', cols, rows)
wild_ds.SetProjection(prox_ds.GetProjection())
wild_ds.SetGeoTransform(prox_ds.GetGeoTransform())
gdal.RasterizeLayer(
    wild_ds, [1], wild_lyr, burn_values=[1],
    callback=gdal.TermProgress)
```

Burn wilderness into
temporary raster

```
wild_data = wild_ds.ReadAsArray()
prox_data = prox_ds.ReadAsArray()
prox_data[wild_data == 0] = -99
prox_ds.GetRasterBand(1).WriteArray(prox_data)
prox_ds.GetRasterBand(1).SetNoDataValue(-99)
prox_ds.FlushCache()
```

Set NoData
outside wilderness

```
stats = prox_ds.GetRasterBand(1).ComputeStatistics(
    False, gdal.TermProgress)
print('Mean distance from roads is', stats[2])

del prox_ds, road_ds, shp_ds
```

Because you're using statewide datasets, the first thing you do is find the extent of the wilderness area that you're interested in. To do this, you open the wilderness shapefile and set an attribute filter that selects out all records where the WILD_NM attribute value was equal to 'Frank Church - RONR'. Because the layer's GetExtent function returns the extent of the entire layer, even when a filter has been applied, you have to come up with a different method to get bounding coordinates. The solution is to make a list of the extents for each of the selected polygons and then find the minimum and maximum coordinates from that. Creating the list of polygon extents is easy enough:

```
envelopes = [row.geometry().GetEnvelope() for row in wild_lyr]
```

Each tuple in this list contains the minimum and maximum x value, and the minimum and maximum y, in that order. Now if you zip these tuples together, you end up with four lists, one each for minimum and maximum x and y. From there, it's a simple matter of extracting the most extreme values in each list:

```
coords = list(zip(*envelopes))
minx, maxx = min(coords[0]), max(coords[1])
miny, maxy = min(coords[2]), max(coords[3])
```

Now that you have the bounding coordinates for the wilderness extent, you set a spatial filter on the roads layer to select only the roads that fall in that rectangle:

```
road_lyr.SetSpatialFilterRect(minx, miny, maxx, maxy)
```

After getting your roads of interest, you turn them into a raster band that you use for your proximity analysis. The raster band must already exist, and then the vector features are burned into it. You figure out how many rows and columns will fit in the extent, given the cell size you chose early in the script. Smaller cell sizes result in more-precise distances, but they also greatly increase processing time. Ten meters is a reasonable size for this example.

```
cols = int((maxx - minx) / cellsize)
rows = int((maxy - miny) / cellsize)
```

Now you have all of the information necessary to create a new raster dataset, which you do. It needs a geotransform so that the roads can be burned into the correct locations, so you construct one from your bounding coordinates and numbers of rows and columns. You also copy the spatial reference from the roads layer. Remember that a layer's GetSpatialRef function returns a spatial reference object, but a raster dataset's SetProjection function requires a WKT string, which is why you have to get the layer's spatial reference as a string:

```
road_ds.SetProjection(road_lyr.GetSpatialRef().ExportToWkt())
road_ds.SetGeoTransform((minx, cellsize, 0, maxy, 0, -cellsize))
```

Now you can finally burn the roads into a raster band using the following function:

```
RasterizeLayer(dataset, bands, layer, [transformer], [transformArg],
               [burn_values], [options], [callback], [callback_data])
```

- dataset is the raster dataset containing the band(s) to burn into.
- bands is the list of bands to burn the data into, where the first one has index 1.
- layer is the OGR layer whose features will be burned into the raster bands.
- transformer is a GDAL transformer object to convert map coordinates into pixel offsets. If not provided, then the function will create its own using the geotransform.
- transformArg is the callback data for the transformer.
- burn_values is the list of values to burn into the raster wherever there are features. If this parameter is provided, it must be the same length as bands. The default for a byte array is 255.
- options is a list of key=value strings. See appendix E for a list of possibilities. (Appendixes C through E are available online on the Manning Publications website at https://www.manning.com/books/geoprocessing-with-python.)
- callback is a callback function for reporting burn progress.
- callback_data is data to be passed to the callback function.

You use this function to burn the value of 1 everywhere there was a road:

```
gdal.RasterizeLayer(
    road_ds, [1], road_lyr, burn_values=[1], callback=gdal.TermProgress)
```

If you load this raster into a GIS, it won't look like much until you start to zoom in. This is because the pixel size is so small that you'll need to zoom in to see many of them. When you do zoom in, you'll see that the roads are blocky, as in figure 11.15. This is a result of them now being represented as pixels instead of smooth vector lines.

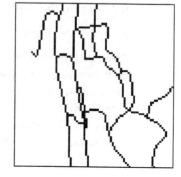

Figure 11.15   Rasterized roads

Once you have a raster representation of the roads, you're almost ready to compute proximity to them using the ComputeProximity function:

```
ComputeProximity(
    srcBand, proximityBand, [options], [callback], [callback_data])
```

- srcBand is the raster band containing the features to compute proximity to. By default, any non-zero pixels are considered features.
- proximityBand is the raster band to store the proximity measurements into.
- options is a list of key=value strings. See appendix E for a list of possibilities.
- callback is a callback function for reporting progress.
- callback_data is data to be passed to the callback function.

Like the RasterizeLayer function, the ComputeProximity function requires that the output raster band already exist. You create a new dataset and copy the spatial reference information and geotransform from your roads raster, and then calculate proximity using map distances instead of the default pixel distances:

```
gdal.ComputeProximity(
    road_ds.GetRasterBand(1), prox_ds.GetRasterBand(1),
    ['DISTUNITS=GEO'], gdal.TermProgress)
```

Although you now have a proximity raster, you only want statistics for the areas within the wilderness area. You could use a zonal analysis to calculate that information, or you could get rid of the non-wilderness data altogether. Either case requires rasterizing the wilderness polygons, though, so you do that in a similar manner as the roads, except you use the MEM driver to store the dataset in memory instead of writing it to disk. Then you read both the wilderness and proximity data into NumPy arrays and change all proximity values to -99 if the wilderness value is 0, which signifies that the pixel doesn't fall in a wilderness polygon.

```
prox_data[wild_data == 0] = -99
```

Once you save the data back to disk, you're able to correctly calculate statistics, and you also have a nice proximity dataset for use later. Figure 11.16 shows part of this dataset, zoomed in far enough to get an idea of how it works. The brighter the pixel, the longer the distance from a road.

11.3 Resampling data

Back in chapter 9 you learned how to resample your data to different cell sizes by changing the size of the arrays used to hold the data. Other ways to

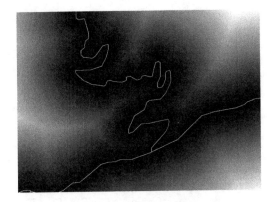

Figure 11.16 A small section of the proximity dataset with roads drawn on top. Brighter areas indicate longer distances from roads.

resample give you more control over the outcome, however. One simple approach is to use slices to keep pixel values at a specific interval and throw out everything in between. To do this, provide a step value when specifying your slice. A step value of 2 tells NumPy to keep every second value, 3 means keep every third value, and so on. This example shows you how to keep every other cell, reducing the rows and columns by half:

```
>>> data = np.reshape(np.arange(24), (4, 6))
>>> data
array([[ 0,  1,  2,  3,  4,  5],
       [ 6,  7,  8,  9, 10, 11],
       [12, 13, 14, 15, 16, 17],
       [18, 19, 20, 21, 22, 23]])
>>> data[::2, ::2]
array([[ 0,  2,  4],
       [12, 14, 16]])
```

For this example, data[0:4:2, 0:6:2] provides the same results as data[::2, ::2]. Not providing starting and stopping indices around the first colon means that you want to start at the beginning and go to the end. The third number, after the second colon, is the step index. If you want to start at the second row and column instead of the first, you can do this:

```
>>> data[1::2, 1::2]
array([[ 7,  9, 11],
       [19, 21, 23]])
```

This should look familiar, because the results are similar to the automatic resampling that happens when you read data from a file into a differently sized array.

You can also increase the size of the array, which is how you'd decrease pixel sizes. To do this, use the NumPy repeat function, which wants an array of data, the number of times to repeat each value, and the axis to use. If you don't provide an axis, then the array is flattened to one dimension. An axis of 0 repeats the rows, and a value of 1 repeats columns, like this:

```
>>> np.repeat(data, 2, 1)
array([[ 0,  0,  1,  1,  2,  2,  3,  3,  4,  4,  5,  5],
       [ 6,  6,  7,  7,  8,  8,  9,  9, 10, 10, 11, 11],
       [12, 12, 13, 13, 14, 14, 15, 15, 16, 16, 17, 17],
       [18, 18, 19, 19, 20, 20, 21, 21, 22, 22, 23, 23]])
```

Notice how each column is repeated twice? To end up with each value repeated four times (so the rows and columns are both doubled), call repeat once on rows and once on columns:

```
>>> np.repeat(np.repeat(data, 2, 0), 2, 1)
array([[ 0,  0,  1,  1,  2,  2,  3,  3,  4,  4,  5,  5],
       [ 0,  0,  1,  1,  2,  2,  3,  3,  4,  4,  5,  5],
       [ 6,  6,  7,  7,  8,  8,  9,  9, 10, 10, 11, 11],
       [ 6,  6,  7,  7,  8,  8,  9,  9, 10, 10, 11, 11],
       [12, 12, 13, 13, 14, 14, 15, 15, 16, 16, 17, 17],
       [12, 12, 13, 13, 14, 14, 15, 15, 16, 16, 17, 17],
       [18, 18, 19, 19, 20, 20, 21, 21, 22, 22, 23, 23],
       [18, 18, 19, 19, 20, 20, 21, 21, 22, 22, 23, 23]])
```

But let's look at something more interesting. You can also use multiple slices to apply custom algorithms instead of using a single pixel value. For example, if you want to resample to pixels that are four times the original size (twice the length and twice the width), you could take the average of those four pixel values and use that as the new value. Figure 11.17 shows an example of this.

3	5	6	4	4	3
4	5	8	9	6	5
2	2	5	7	6	4
5	7	9	8	9	7
4	6	5	7	7	5
3	2	5	3	4	4

4.25	6.75	4.50
4.00	7.25	6.50
3.75	5.00	5.00

Figure 11.17 Increasing cell size and using the average value of the input pixels as the output value

In the case of figure 11.17, you need four numbers to calculate the output value. To accomplish the same thing with slices, you'd need four slices, with each one corresponding to one of the four input values. Unlike the slices you used for moving windows, however, these would each be much smaller than the original array. Instead, they'd be the same size as the output array, and each one would contain one value per output cell, as shown in figure 11.18. The figure shows the original data, but the cells

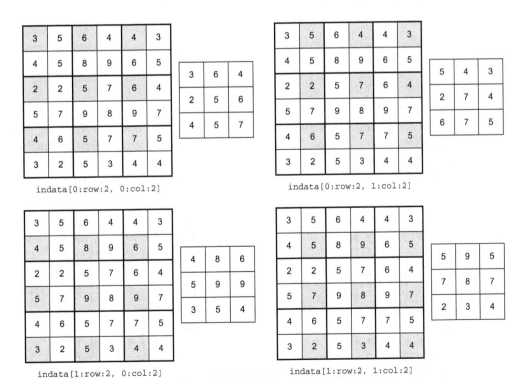

Figure 11.18 Slices used to resample using an average of the input values. The shaded cells are the ones used to create each smaller slice. The smaller arrays are averaged together to get the final result.

that would make up each slice are highlighted. One slice would contain the upper-left pixel from every set of four pixels used to calculate the output. Another slice would contain each upper-right pixel, and so on. Each of the slices in this example has three rows and three columns, which is the same size as the output. If you take the average of these slices on a pixel-by-pixel basis, you end up with the values shown in figure 11.17. For example, the upper-left corner would have a value of $(3 + 5 + 4 + 5) / 4 = 17 / 4 = 4.25$.

Let's look at how to implement this with code. Once again, your life will be easier if you write a function to create the required slices. The following listing shows one that returns the slices in a list, given the original data and the window size (2 x 2 in the example from figure 11.18).

Listing 11.11   Function to make stepped slices

```
def make_resample_slices(data, win_size):
    """Return a list of resampled slices given a window size.

    data      - two-dimensional array to get slices from
    win_size - tuple of (rows, columns) for the input window
    """
    row = int(data.shape[0] / win_size[0]) * win_size[0]
    col = int(data.shape[1] / win_size[1]) * win_size[1]
    slices = []

    for i in range(win_size[0]):
        for j in range(win_size[1]):
            slices.append(data[i:row:win_size[0], j:col:win_size[1]])
    return slices
```

The first thing this function does is calculate the last row and column that will be used. This is necessary because the original data might not be divisible by the size of the window you want. For example, in figure 11.19 the input array has five rows and five columns. The figure shows two of the slices needed to take an average of four pixels, but the second one is smaller than the first. If you try to use these two slices

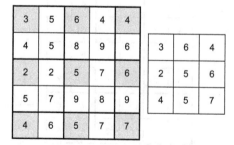
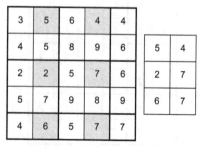

Figure 11.19   When the dimensions of the array to be resampled are not divisible by the dimensions of the window (2 x 2 in this case), the slices are different sizes. This must be accounted for so that the slices have the same size.

together, you'll get an error because they're different sizes. The function in listing 11.11 handles this by cutting off the fifth row and column, so that the slices are created from four rows and columns instead. To do this, the total number of rows or columns is divided by the number of rows or columns in a window. For the example in figure 11.19, this would be 5 / 2 = 2 for both, so the data can fit two full windows in each direction. Multiply that by the window size to get the total number of rows or columns to use, which is four in the example. This number is used to put an upper bound on the slices.

Once the numbers of rows and columns are known, the function creates one slice for each input location, using the window size as the step parameter so only one input pixel per window is extracted. The slices are all returned as a list, and once you have that you can apply any algorithm you want, as long as you can code it up. To get an average, you could stack the slices and then use the NumPy mean function, as you did for moving windows.

> **TIP** Don't forget to change the geotransform to reflect the new cell size when resampling your data.

Techniques like this are great if your output cell size is a multiple of the original, but they don't work in other cases. Let's look at a way to extract specific pixels that can't be specified with a step parameter. To do this, you need to know the original pixel size, the new pixel size, and the numbers of rows and columns in the original image. Get a scaling factor for width by dividing the new pixel width by the original width, and do the same for pixel height. For example, if your original image has a pixel width of 10 but your target width is 25, then the scaling factor is 2.5. The new pixels are 2.5 old pixels wide.

Divide the scaling factor in half to determine that the center of a new pixel is 1.25 old pixels from the edge. This center point is what you want because you use the center of new pixels to determine which nearby old pixels to use when resampling. To get the center x values for the new pixels in terms of the original cell offsets, create an array that starts at the center (1.25) and increments by the scaling factor (2.5). You need to make sure this array goes up to, but not past, the total number of columns in the original array. Do the same for rows, so that you have two arrays containing x and y offsets. Figure 11.20 shows an example of a few pixels. The alternating shaded areas

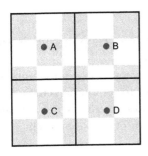

Indexes for each point
in terms of the small cells

A: [1.25, 1.25]

B: [1.25, 3.75]

C: [3.75, 1.25]

D: [3.75, 3.75]

Figure 11.20   Resampling to a larger pixel size that's not a multiple of the original size. The alternating shaded areas are the original pixels and the thick outlines show the new ones. The dots are the center points for the new pixels.

are the original pixels (of size 10) and the thick outlines denote the pixels with size 25. The dots are the center points of the large pixels, and the text shows the coordinates for these points in terms of the smaller pixels.

Once you have the lists of x and y offsets, you can use the NumPy `meshgrid` function to get two new arrays that contain all possible coordinates obtained from these values. For example, if your row offsets are (3, 5) and column offsets are (2, 4), then the possible combinations are [(3, 2), (3, 4), (5, 2), (5, 4)], and `meshgrid` would return two four-element arrays, one for the row offsets and one for the columns.

The following listing shows a function that computes the scaling factors, makes the offset arrays, and then creates and returns the coordinate arrays.

Listing 11.12 Function to get new pixel offsets in terms of old pixels

```
def get_indices(source_ds, target_width, target_height):
    """Returns x, y lists of all possible resampling offsets.

    source_ds     - dataset to get offsets from
    target_width  - target pixel width
    target_height - target pixel height (negative)
    """
    source_geotransform = source_ds.GetGeoTransform()
    source_width = source_geotransform[1]
    source_height = source_geotransform[5]
    dx = target_width / source_width
    dy = target_height / source_height
    target_x = np.arange(dx / 2, source_ds.RasterXSize, dx)
    target_y = np.arange(dy / 2, source_ds.RasterYSize, dy)
    return np.meshgrid(target_x, target_y)
```

Once you have coordinates, you can take advantage of the fact that offset lists can be used to extract values from NumPy arrays to get the values of the original pixels that fall directly under the center of the new pixels. This is nearest-neighbor resampling, which uses the value of the closest pixel in the original array and doesn't do any other processing. To do this, you'd extract the values at the calculated indices and be done with it, like this:

```
ds = gdal.Open(fn)
data = ds.ReadAsArray()
x, y = get_indices(ds, 25, -25)
new_data = data[y.astype(int), x.astype(int)]
```

The only trick is that you need to convert the indices to integers or NumPy will complain when you attempt to use them to index an array.

Nearest-neighbor is simple, fast, and one of the few appropriate resampling methods for categorical data, but it's not the greatest choice for continuous data. For these types of data, you usually want to use several surrounding pixels to calculate your new value. You could use an average as you did earlier, or one of several other common resampling methods. Two examples of these are *bilinear interpolation*, which takes a weighted average of the four closest pixels, and *cubic convolution*, which fits a smooth curve through the 16 nearest pixels and uses that to calculate a new value. You're going to write a function that uses the output from your get_indices function to perform bilinear interpolation. The hatched areas in figure 11.21 show which four original pixels are used for each center point.

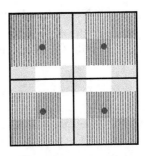

Figure 11.21 The hatched areas show the four original pixels closest to the new pixel center point. These are the ones used to calculate a value for the new pixel.

To get the values of these four pixels, the first thing you need to do is subtract 0.5 from the calculated indices so that they correspond to the center, instead of the edge, of the input pixels. Then you need to determine the integers on either side of these coordinates, which gives you the offsets to use. For example, if the row coordinate is 4.25, then you'd use rows 4 and 5. If you do that for column offsets as well, you have two rows and two columns and can use those to get the four input pixels surrounding the target pixel.

Once you have an input pixel value, multiply it by the distance in both directions from that pixel to the target pixel. This is the part that weights the closer pixels heavier than further pixels. Then add the four weighted values together to get the final output value. If you'd like a more detailed explanation of the algorithm, you can find many sources online.

The following listing shows a function that performs bilinear interpolation, given the original data and the center offsets for the new pixels.

Listing 11.13 Function for bilinear interpolation

```
def bilinear(in_data, x, y):
    """Performs bilinear interpolation.

    in_data - the input dataset to be resampled
    x       - an array of x coordinates for output pixel centers
    y       - an array of y coordinates for output pixel centers
    """
    x -= 0.5
    y -= 0.5
    x0 = np.floor(x).astype(int)     First and second
    x1 = x0 + 1                      column offsets
    y0 = np.floor(y).astype(int)     First and second
    y1 = y0 + 1                      row offsets

    ul = in_data[y0, x0] * (y1 - y) * (x1 - x)
    ur = in_data[y0, x1] * (y1 - y) * (x - x0)
    ll = in_data[y1, x0] * (y - y0) * (x1 - x)
    lr = in_data[y1, x1] * (y - y0) * (x - x0)

    return ul + ur + ll + lr
```

Now to use bilinear interpolation to resample a raster, you can use your `get_indices` function to get offsets, which you then pass to the `bilinear` function. Don't forget to edit the geotransform when saving the output, as shown in the following listing.

Listing 11.14   Bilinear interpolation

```
in_fn = r"D:\osgeopy-data\Nepal\everest.tif"
out_fn = r'D:\Temp\everest_bilinear.tif'
cell_size = (0.02, -0.02)

in_ds = gdal.Open(in_fn)
x, y = get_indices(in_ds, *cell_size)            Resample
outdata = bilinear(in_ds.ReadAsArray(), x, y)

driver = gdal.GetDriverByName('GTiff')
rows, cols = outdata.shape
out_ds = driver.Create(                          New image is same
    out_fn, cols, rows, 1, gdal.GDT_Int32)       size as outdata
out_ds.SetProjection(in_ds.GetProjection())

gt = list(in_ds.GetGeoTransform())
gt[1] = cell_size[0]
gt[5] = cell_size[1]                             Change the geotransform
out_ds.SetGeoTransform(gt)

out_band = out_ds.GetRasterBand(1)
out_band.WriteArray(outdata)
out_band.FlushCache()
out_band.ComputeStatistics(False)
```

If you'd like to try other types of interpolation, `scipy.ndimage` contains several interpolation methods. See http://docs.scipy.org/doc/scipy-0.16.1/reference/ndimage .html#module-scipy.ndimage.interpolation for more information.

Resampling with GDAL command-line utilities

This might be a good time to mention the GDAL command-line utilities. There are currently about 30 of them, and new ones get added occasionally. These aren't Python tools; you need to run them from a command prompt or terminal window. Let's see how to use gdalwarp to resample an image. This utility is designed for transforming rasters between spatial reference systems, but you can also use it to resample without changing the spatial reference. The command line looks like this:

```
gdalwarp -tr 0.02 0.02 -r bilinear everest.tif everest_resampled.tif
```

The -tr option is for target resolution, in this case indicating that both cell width and height should be 0.02. As you've probably guessed, -r stands for resampling method, and this specifies bilinear. Other options include but aren't limited to nearest-neighbor, average, cubic convolution, and mode. The input file is everest.tif, and the new file will be called everest_resampled.tif. Many more options are available, and they're all documented at http://www.gdal.org/gdalwarp.html.

(continued)

If you have GDAL version 2.x, you can also use the same options with the gdal_translate utility, which is designed to convert data between different formats. (I wish I knew how many times I've pointed people to this tool over the years!)

Although these aren't Python tools, you can call them from Python using the subprocess module, which sends commands out to the operating system:

```
import subprocess

args = [
    'gdalwarp',
    '-tr', '0.02', '0.02',
    '-r', 'bilinear',
    'everest.tif', 'everest_resample.tif']
result = subprocess.call(args)
```

The `result` variable will hold 0 if the process completed successfully, and 1 if not. It's preferred that you break your command up into a list of arguments like the example so that Python can handle special cases such as spaces in filenames, but you can also pass a string instead, like this:

```
result = subprocess.call('gdalwarp -tr 0.02 0.02 -r bilinear everest.tif
         everest_resample.tif')
```

11.4 Summary

- If you need to work with large arrays of data in Python, the NumPy module is your answer.
- Use the SciPy module to perform many different scientific data analyses on NumPy arrays.
- Local map algebra computations work on a pixel-by-pixel basis, such as calculating NDVI for a pixel.
- Focal map algebra computations involve a moving window that uses surrounding pixels to calculate the output value, such as calculating slope.
- Zonal calculations work on pixels that are all in the same zone, such as calculating the histogram of landcover types based on land ownership.
- Global calculations, such as proximity analysis, involve the entire raster dataset.

Map classification

One common use for raster data is map classification, which involves categorizing the pixels into groups. For example, say you wanted to create a vegetation land-cover dataset. You might use satellite imagery, elevation, slope, geology, precipitation, or other input data in order to create your classifications. The techniques we've looked at so far will help you prepare your datasets, but you need something else in order to classify pixels. Many different classification techniques exist, and which one you use will probably depend on your use case and available resources. This section is by no means a comprehensive introduction to map classification, and you should consult a remote sensing book if you want to learn more, but at least you'll get an idea of what's possible.

You'll use four bands from a Landsat 8 image to see how well you can replicate the landcover classification from the SWReGAP project that you saw earlier. These classifications include groupings such as "Great Basin Pinyon-Juniper Woodland"

and "Inter-Mountain Basins Playa." Although this project covered five states in the southwestern United States, you'll only look at the area covered by one Landsat scene (figure 12.1). Landsat scenes contain more than four bands, but you'll only use the three visible bands (red, green, and blue, which make up the natural color image) and a thermal band. You'll also use versions of these bands that have been resampled from 30-meter to 60-meter pixels so that the examples run faster and your computer is less likely to run into memory issues.

You'll use the SWReGAP field data when necessary. Don't expect your results to rival theirs, though, because that project involved many years of work, with thousands of locations visited in person to collect data, a much more comprehensive set of predictor variables at 30-meter resolution, and more-sophisticated modeling methods. In addition, the SWReGAP dataset consists of more than 100 distinct landcover classifications, but these examples won't produce nearly so many classes. You'll see that your simpler models can replicate several of the same general patterns, however.

The examples in this section use a few new Python modules: Spectral Python, SciKit-Learn, and SciKit-Learn Laboratory. Please see appendix A for installation instructions.

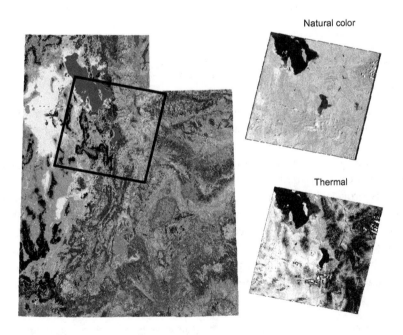

Figure 12.1 The SWReGAP landcover dataset for Utah with the Landsat scene footprint drawn on top. The red, green, and blue bands of the Landsat dataset make up the natural color image, and the thermal band is shown alone.

12.1   *Unsupervised classification*

Unsupervised classification methods group pixels together based on their similarities, with no information from the user about which ones belong together. The user selects the independent, or predictor, variables of interest, and the chosen algorithm does the rest. This doesn't mean that you don't need to know what you're classifying, however. Once a classification is produced, it's up to the user to interpret it and decide which types of features correspond to which generated classes, or if they even do correspond nicely.

The Spectral Python module is designed for working with hyperspectral image data, of which Landsat data is an example. You'll use a k-means clustering algorithm to group the pixels into clusters and then visually compare your results to the SWReGAP classification. But first, let's write a function that takes a list of filenames as a parameter, reads in all bands from all files, and returns the data as a three-dimensional NumPy array. We'll use this function in the next few listings, and to make things easier, it's in the ospybook module.

Listing 12.1   Function to stack raster bands

```
def stack_bands(filenames):
    """Returns a 3D array containing all band data from all files."""
    bands = []
    for fn in filenames:
        ds = gdal.Open(fn)
        for i in range(1, ds.RasterCount + 1):
            bands.append(ds.GetRasterBand(i).ReadAsArray())
    return np.dstack(bands)
```

Now back to the classification problem. A k-means algorithm begins with an initial set of cluster centers and then assigns each pixel to a cluster based on distance. This distance is computed as if the pixel values were coordinates. For example, if two pixel values were 25 and 42, the distance would be 17, no matter where the pixels were in relation to each other spatially.

After this process has completed, the centroids of the clusters are then used as starting points, and the process repeats until the maximum number of iterations or a user-defined stopping condition is reached.

Running the default classification is quite easy, as you'll see in the following listing. In fact, it's only one line of code, and the bulk of the example consists of setting things up and saving the output. That code has been shortened, also, by using custom functions in the ospybook module.

Listing 12.2   K-means clustering with Spectral Python

```
import os
import numpy as np
import spectral                              ◄─┐  Import spectral
from osgeo import gdal
import ospybook as pb
```

```
folder = r'D:\osgeopy-data\Utah'
raster_fns = \
    ['LE70380322000181EDC02_60m.tif', 'LE70380322000181EDC02_TIR_60m.tif']
out_fn = 'kmeans_prediction_60m2.tif'

os.chdir(folder)

data = pb.stack_bands(raster_fns)                          Run the model
classes, centers = spectral.kmeans(data)

ds = gdal.Open(raster_fns[0])
out_ds = pb.make_raster(ds, out_fn, classes, gdal.GDT_Byte)
levels = pb.compute_overview_levels(out_ds.GetRasterBand(1))
out_ds.BuildOverviews('NEAREST', levels)
out_ds.FlushCache()
out_ds.GetRasterBand(1).ComputeStatistics(False)

del out_ds, ds
```

The only required parameter to the kmeans function is an array containing the predictor variables, which is in the three-dimensional array returned by stack_bands. You could also specify the number of output clusters desired, maximum number of iterations, several initial clusters, or a few other things. The default 10 clusters and 20 iterations are sufficient for the example, however. Feel free to consult the Spectral Python online documentation for more information.

Assuming you got the same results I did, the algorithm only created nine classes instead of ten, but it would have created ten if it could resolve them with the given data. I went to the trouble to try to match the resulting classes with SWReGAP classes so that you can see a visual comparison, although admittedly, this works best if you're looking at a color version of figure 12.2. The classification is definitely different, but at least the mountains in the east are clearly separated from the flats and playas to the

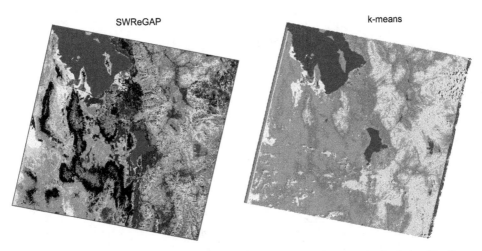

Figure 12.2   The SWReGAP landcover dataset and one created using unsupervised classification

west. It's possible that the match would be better if more clusters had been requested, because the SWReGAP data contains a much larger number of classes. If you were to run a classification like this, you'd have to determine what each cluster represented, as I tried to match up clusters with existing classifications.

12.2 *Supervised classification*

Supervised classification, unlike unsupervised techniques, requires input from the user in the form of training data. A training dataset contains all of the independent variables that correspond to a known value for the dependent variable. For example, if you knew for a fact that a particular pixel was an agricultural field, then you could sample your input datasets, such as satellite imagery, at that location and include those pixel values as the independent variables. The model is fitted using these input data and then it can be applied to your full datasets to get a spatial representation of the model results.

It used to be that training data had to be collected by visiting locations in person and documenting first-hand what the actual classification should be. In this age of high-resolution online imagery, however, in certain cases researchers can determine these values from imagery without leaving their desks. This is definitely a more cost-effective solution, although it certainly isn't appropriate or possible for every situation. Because accurate training datasets are essential for supervised classification, consider collecting data in the field if possible. Even if the truth can be determined by looking at imagery, the modeling process is still necessary, unless you want to manually classify every pixel.

We'll take a look at one example of supervised classification using a decision tree. This type of model consists of a hierarchical set of conditions based on the model's independent variables, and has at least one pathway that leads to each possible outcome. Figure 12.3 shows a simple, if not accurate, example of a decision tree.

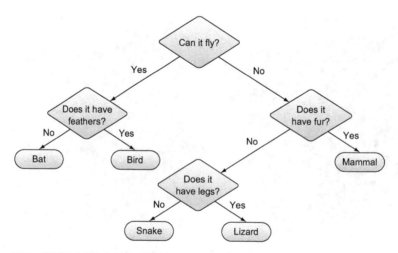

Figure 12.3 A simple example of a decision tree

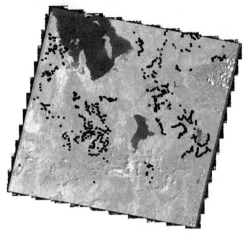

Figure 12.4   The locations of the ground-truthed data points used in listing 12.3

Listing 12.3 uses the `scikit-learn` module to create a decision tree that predicts landcover type based on four bands from a Landsat 8 image and actual field data from the SWReGAP project. The locations of these ground-truthed points are shown in figure 12.4.

You have a text file that contains coordinates for the points in figure 12.4 and an integer value signifying the landcover class, which looks something like this:

```
x                 y                 class
377455.171684     4447157.33631     82
372685.109412     4443741.27817     119
372823.111316     4443875.28023     48
```

That's a good start, but you still need the independent variables. You'll sample the Landsat bands at the coordinates in the text file to get a dataset that looks more like this:

```
band1 band2 band3 band4 class
136   116   92    233   82
156   129   112   253   119
150   127   109   239   48
```

These data will be used to build a model that you'll then apply to the entire extent of the Landsat bands to get a spatial dataset containing predictions. This process is shown in the following listing.

Listing 12.3   Map classification using CART

```
import csv
import os
import numpy as np
from sklearn import tree
from osgeo import gdal
import ospybook as pb
```
◄─┐ **Import sci-kit learn**

```
folder = r'D:\osgeopy-data\Utah'
raster_fns = \
    ['LE70380322000181EDC02_60m.tif', 'LE70380322000181EDC02_TIR_60m.tif']
out_fn = 'tree_prediction60.tif'
train_fn = r'D:\osgeopy-data\Utah\training_data.csv'
gap_fn = r'D:\osgeopy-data\Utah\landcover.img'

os.chdir(folder)

xys = []
classes = []
with open(train_fn) as fp:
    reader = csv.reader(fp)
    next(reader)
    for row in reader:
        xys.append([float(n) for n in row[:2]])
        classes.append(int(row[2]))
```

Read coordinates
and class from csv

```
ds = gdal.Open(raster_fns[0])
pixel_trans = gdal.Transformer(ds, None, [])
offset, ok = pixel_trans.TransformPoints(True, xys)
cols, rows, z = zip(*offset)
```

Get pixel offsets

```
data = pb.stack_bands(raster_fns)

sample = data[rows, cols, :]
```

Sample satellite
data at pixel offsets

```
clf = tree.DecisionTreeClassifier(max_depth=5)
clf = clf.fit(sample, classes)
```

Fit the
classification tree

```
rows, cols, bands = data.shape
data2d = np.reshape(data, (rows * cols, bands))
prediction = clf.predict(data2d)
prediction = np.reshape(prediction, (rows, cols))
```

Apply model to
satellite data

Set pixels
with no
satellite
data to 0

```
prediction[np.sum(data, 2) == 0] = 0

predict_ds = pb.make_raster(ds, out_fn, prediction, gdal.GDT_Byte, 0)
predict_ds.FlushCache()
levels = pb.compute_overview_levels(predict_ds.GetRasterBand(1))
predict_ds.BuildOverviews('NEAREST', levels)

gap_ds = gdal.Open(gap_fn)
colors = gap_ds.GetRasterBand(1).GetRasterColorTable()
predict_ds.GetRasterBand(1).SetRasterColorTable(colors)
```

Copy colormap from
SWReGAP dataset

```
del ds
```

This is a little more complicated than the unsupervised example, but it's still not that bad. The first task is to read in the coordinates and landcover class from the text file. You skip the header line, and then convert the first two values to floating-point (because they're read in as strings) and put them in a list. When finished, this list contains a list of lists, with each inner list containing the x and y coordinates. You need the coordinates in this format later. You also put the landcover class integer in another list for later use.

Then you open one of the raster datasets so you can create a transformer object to convert between map coordinates and pixel offsets. You use this with your list of coordinates to get pixel offsets in two lists called cols and rows.

After reading the four satellite bands into a three-dimensional array, you take advantage of the fact that you can pass lists of coordinates as indices to pull data out of an array, and sample all of the points in one line of code:

```
sample = data[rows, cols, :]
```

This samples the 3D array at each of the provided row and column offsets, and returns every value in the third dimension, which is the four different satellite bands. The result is a two-dimensional array, where each row contains the four pixel values from the four bands.

Now you have all of the data required in order to fit the model, so you create a new decision tree classifier using default parameters (see the scikit-learn documentation to read the nitty-gritty details of the optional parameters) and then pass the fit method your independent variables and known landcover classifications at those same points. Make sure you don't change the order of any of the lists; otherwise, the satellite pixel values won't match up with the appropriate landcover value and your model won't be fitted correctly.

```
clf = tree.DecisionTreeClassifier(max_depth=5)
clf = clf.fit(sample, classes)
```

All that's left is to apply your fitted model to the full set of pixel values. Unfortunately, the predictor variables need to be in a two-dimensional array for this to work, so you reshape the array so that it has a large number of rows (rows * cols) and four columns, one for each band. You pass this to the predict function, and then reshape the resulting one-dimensional array back into two dimensions:

```
rows, cols, bands = data.shape
data2d = np.reshape(data, (rows * cols, bands))
prediction = clf.predict(data2d)
prediction = np.reshape(prediction, (rows, cols))
```

Another way to handle the prediction is to loop through the rows and process one at a time. An added advantage to this method is that it uses less memory. For example, my laptop crashed when I tried to run the prediction on the entire 30-meter dataset at once, but it did it row by row without a problem. You'd do it something like this:

```
prediction = np.empty(data.shape[0:2])
for i in range(data.shape[0]):
    prediction[i,:] = clf.predict(data[i,:,:])
```

Landsat bands have 0 values around the edges of the image, but those pixels are still assigned a value with the model. If all four Landsat bands contain 0 at a location, then you know that there's no data for that cell, so you change those to 0 in the prediction data as well. You could have used any number that wasn't a valid landcover classification, as long as you set it as the NoData value. After saving the prediction as a GeoTIFF,

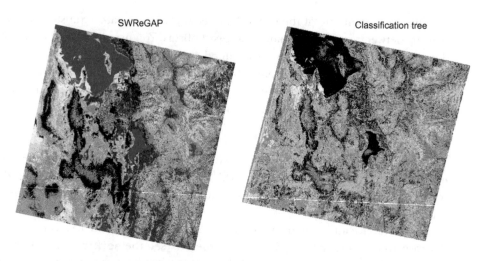

Figure 12.5   The SWReGAP landcover dataset and one created using a decision tree

you copy the color table from the real SWReGAP landcover classification so you can visually compare the results. Again, you can see in figure 12.5 that this model predicts some of the same general patterns, but the results are still different. If you're viewing this in color, you'll see that it even failed to predict water correctly! This is a strong indication that the model needs more work. A better set of training data or independent variables would probably help.

12.2.1 *Accuracy assessments*

Accuracy assessments are usually performed on models such as this to get an idea of how good they are. Because the model should do a good job of predicting the values that were used to build it, accuracy assessments are usually performed using a separate set of data to test the model on different values. I've provided a separate dataset for this, but if you need to split your data into training and assessment groups, you may want to look into the cross-validation tools in scikit-learn. One easy accuracy assessment method is to use a confusion matrix, which breaks out the results by predicted and observed values so you can see how well each classification was predicted. Although you can figure out the total percentage of correct classifications from the confusion matrix, better measures of accuracy exist. One of these is Cohen's kappa coefficient, which ranges from -1 to 1, where the higher the number, the better the predictions. The following listing shows you how to use the scikit-learn module to construct a confusion matrix and SciKit-Learn Laboratory to compute the kappa statistic.

Listing 12.4   Confusion matrix and kappa statistic

```
import csv
import os
import numpy as np
```

```
from sklearn import metrics          Import sci-kit
import skll                          modules
from osgeo import gdal

folder = r'D:\osgeopy-data\Utah'
accuracy_fn = 'accuracy_data.csv'
matrix_fn = 'confusion_matrix.csv'
prediction_fn = r'D:\osgeopy-data\Landsat\Utah\tree_prediction60.tif'

os.chdir(folder)

xys = []
classes = []
with open(accuracy_fn) as fp:
    reader = csv.reader(fp)
    next(reader)
    for row in reader:
        xys.append([float(n) for n in row[:2]])
        classes.append(int(row[2]))

ds = gdal.Open(prediction_fn)
pixel_trans = gdal.Transformer(ds, None, [])
offset, ok = pixel_trans.TransformPoints(True, xys)
cols, rows, z = zip(*offset)

data = ds.GetRasterBand(1).ReadAsArray()
sample = data[rows, cols]
del ds
                                         ◄──  Compute kappa
print('Kappa:', skll.kappa(classes, sample))

labels = np.unique(np.concatenate((classes, sample)))        Create the
matrix = metrics.confusion_matrix(classes, sample, labels)   confusion matrix

matrix = np.insert(matrix, 0, labels, 0)                     Add labels and
matrix = np.insert(matrix, 0, np.insert(labels, 0, 0), 1)    save the matrix
np.savetxt(matrix_fn, matrix, fmt='%1.0f', delimiter=',')
```

Most of this code should look familiar because obtaining the data points needed for the accuracy assessment is similar to collecting the model training data. The difference is that instead of sampling the satellite imagery, you sample the prediction output and compare those results to the known classifications.

Once you have the known and predicted values for each location, computing kappa is easy. All you need to do is pass an array containing the true values and one containing the predicted values to the kappa function. Again, the order of the values is important, because your results will be extremely inaccurate if the known values are compared to predicted values from other locations. The kappa statistic for this model is 0.24, so the classification is slightly better than random, but it's certainly nothing to brag about, either. In fact, a number that low indicates a poor classification.

Technically, you only need the same inputs that you use for the kappa statistic to create the confusion matrix, but you also create a list of unique classification values to

0	2	5	14	22	23	26	28	30	32	36	37
2	1	0	0	0	0	0	0	0	0	0	0
5	1	0	0	0	0	0	0	0	1	0	0
14	0	0	1	1	0	0	0	0	0	0	0
22	0	2	0	16	0	0	1	0	0	0	0
23	0	0	0	3	0	0	0	0	0	0	0
26	0	0	0	0	0	0	1	0	1	0	0
28	0	0	0	0	0	0	4	0	1	0	0
30	0	0	0	0	0	0	1	0	0	0	0
32	0	0	0	2	0	0	1	0	3	0	0
36	0	0	0	0	0	0	0	0	0	0	0
37	2	1	0	0	0	0	0	0	0	0	3

Figure 12.6 The first few rows and columns of the confusion matrix for the classification tree model. Rows are predictions, and columns are actual values, so two pixels were predicted as class 22 but were class 5.

use as labels. The classes will be listed in this order in the resulting matrix. After creating the matrix, you add the labels in much the same way as you added labels to your two-way histogram earlier. The matrix looks something like figure 12.6, where the rows correspond to predicted values and columns to known values. For example, 16 pixels that were predicted as class 22 were predicted correctly, but two were really class 5 and one was actually class 28.

12.3 Summary

- Unsupervised classification algorithms group pixels based on how alike they are.
- Supervised classification algorithms use ground-truthed data to predict which set of conditions results in each class.

Visualizing data

13

This chapter covers

- Making quick plots of vector data using matplotlib
- Plotting raster data with matplotlib
- Creating maps with Mapnik

As you no doubt have noticed, the ability to view your data is essential. While you can use desktop GIS software, such as QGIS, sometimes it's nice to see your data as you work, without needing to open it up in other software. This is the idea behind the VectorPlotter class in the ospybook module. Other times you might need to create a picture of your data, such as a quick-and-dirty plot to show a colleague, or perhaps a much nicer map to post online or give to a client. This isn't a book on cartography (which is good, because I'm cartographically challenged), so this chapter will show you the basics of displaying data in a few different ways, but won't focus on techniques for making the data look pretty. You'll see how to use both the matplotlib and Mapnik modules to plot your data. If you want something pretty, you'll want to go with Mapnik, but matplotlib is great for quick visualizations.

13.1   *Matplotlib*

Matplotlib is a general-purpose plotting library for Python and can be used for any kind of graphic you can think up. This module is extensive, and like NumPy and SciPy, entire books have been written on it. If you're interested in seeing an overview of what can be done, check out the examples in the matplotlib gallery at http://matplotlib.org/gallery.html. The gallery contains many impressive examples for making charts and graphs, but we're more interested in spatial data, so this section will concentrate on quick-and-crude plots of geographical datasets. In fact, the VectorPlotter class uses matplotlib, and you'll learn the basics of how that class plots vector data.

Matplotlib has several parts, but the one that you interact with the most to plot data is pyplot, and that's what we'll use here. It's convention to rename this as plt when importing it:

```
import matplotlib.pyplot as plt
```

You can use pyplot in interactive or non-interactive mode. Back in chapters 3 through 7, you used a VectorPlotter from an interactive console and saw the changes to your plots immediately. This was matplotlib at work in interactive mode. This mode is extremely handy for playing with matplotlib and learning how it works. It's also useful for interactively exploring data.

Plotting isn't interactive by default, however. This makes sense, because interactivity wouldn't be helpful for a script that creates a graphic and saves it to disk with no input from the user. Exceptions exist to every rule, though, and you may find that if you're using IPython in pylab mode or an IDE such as Spyder, then interactive mode will be on by default. When in interactive mode, the plot is automatically shown to the user, but if you want to show the plot when using non-interactive mode, then you must call the plt.show() method after adding all of the graphics to your plot. This will stop the script's execution until the user closes the plot window. You might be tempted to use interactive mode so that the user can see the plot as it's created, but you'll probably have bad luck with that because the plot window disappears when the script ends. The user might see parts of the plot as it's created, but if the script ends as soon as the plot is finished, then the user may never get a chance to see the final product.

If you want to turn interactive mode on, either from a script or the console, use this:

```
plt.ion()
```

You can turn interactive mode back off with plt.ioff().

13.1.1   *Plotting vector data*

You might be surprised to learn that plotting vector data isn't that difficult. The data's made up of x and y coordinates, after all. First you'll see how to use the plot function to draw points, lines, and polygons in general, and then you'll graduate to plotting shapefiles. Once you can do that, you'll learn how to create holes in the special case of donut polygons, so that other data can show through if needed. The plot function has many options, most of which will be ignored here, but you can read about them all in the online documentation found at http://matplotlib.org/api/pyplot_api.html #matplotlib.pyplot.plot. This function wants, at the minimum, lists of x and y coordinates. If that's all you provide, then a line is plotted using those coordinates and a color from the matplotlib color cycle. For example, the following code plots the line $y = x^2$, shown in figure 13.1:

```
import matplotlib.pyplot as plt

x = range(10)
y = [i * i for i in x]
plt.plot(x, y)
plt.show()
```

You can specify a color and change this from a line to a series of points simply by providing a marker specification, as in the following example. In this case, 'ro' means that it should draw red circles instead of the default line. The markersize parameter makes the points a bit larger than they would have been by default. (Don't forget to call plt.show() to draw each of these plots.)

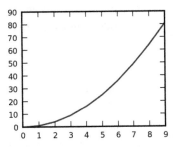

Figure 13.1   A simple line plot

```
plt.plot(x, y, 'ro', markersize=10)
```

The result of this code is shown in figure 13.2. You can also plot a single point by passing in an x and a y value instead of lists of values. You might think that the coordinates would be enough, but you have to provide a marker symbol such as 'ro' or else it still tries to draw a line. Because one point isn't enough information to draw a line, you end up with a blank plot.

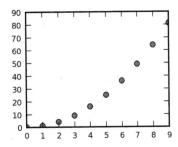

Figure 13.2   A simple point plot

Because polygons are closed lines, you can draw a hollow polygon exactly the same way as a line. Make sure that the first and last sets of coordinates are the same so that the polygon is closed. For example, the following code snippet adds a 0 to the end of each list so that a line from figure 13.1 is drawn back to the origin. In addition, the lw named parameter is used to change the line thickness (lw is short for linewidth, which you could also use). The results are shown in figure 13.3.

```
x = list(range(10))
y = [i * i for i in x]
x.append(0)
y.append(0)
plt.plot(x, y, lw=5)
```

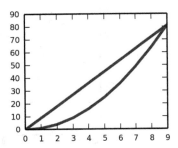

Figure 13.3   A simple closed line plot

Believe it or not, you now know pretty much everything you need to know to make simple plots of vector data, assuming you remember what you learned in earlier chapters. To draw the features in a layer, open it and, for each feature, get the geometry coordinates and plot them as you did here. Let's try it out with the global landmass shapefile. This particular dataset is convenient because all of the geometries are simple polygons, and you don't need to worry about multipolygons. You have one donut polygon in the mix, but you can ignore that for now and plot the outer ring. For each feature, get the first ring from its geometry, and then get the coordinates from that. Remember that the coordinates come in a list of pairs, so the zip function comes in handy because you can use it to create two separate lists of x and y coordinates. The following listing demonstrates this pattern and results in a plot like figure 13.4A.

Listing 13.1 Plotting simple polygons

```
import matplotlib.pyplot as plt
from osgeo import ogr

ds = ogr.Open(r'D:\osgeopy-data\global\ne_110m_land.shp')
lyr = ds.GetLayer(0)
for row in lyr:
    geom = row.geometry()
    ring = geom.GetGeometryRef(0)
    coords = ring.GetPoints()          ◄─── List of (x, y) tuples
    x, y = zip(*coords)
    plt.plot(x, y, 'k')                ◄─── 'k' means black
plt.axis('equal')                      ◄─── Equalize axis units
plt.show()
```

One little detail that the listing takes care of has not been mentioned yet. For your spatial plots to look right, you need to set the axis units equal to each other. If you comment this line out, you'll end up with a plot more like figure 13.4B. By default the data are fitted into the available space so that the data fill it all up. The distance

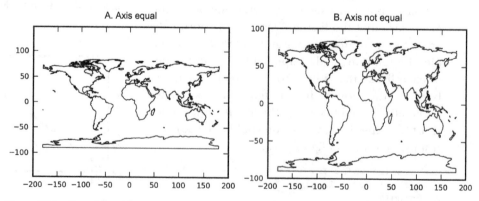

Figure 13.4 Two plots of the continents using closed lines for polygons. Plot A sets the axes equal to each other and the proportions are correct, unlike plot B in which the default axis limits are used.

covered by a single unit might be different on each axis. If you look closely at part B of the figure, you'll see that the horizontal axis ranges from -200 to 200 but the vertical one from -100 to 100, and yet they both use up the same amount of space on paper. Setting the axes units equal fixes this distortion.

As you've seen, drawing simple polygons isn't difficult. Dealing with multipolygons and donut polygons adds a little more complexity to the code, but it's still the exact same process. In the case of a multipolygon, you need to loop through each polygon in the multipolygon, and then for each polygon (whether from a multipolygon or not), loop through the rings and plot each one. The following listing shows this process for the countries shapefile, which gives you a plot like figure 13.5.

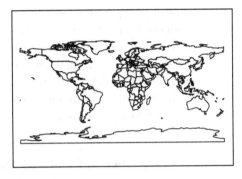

Figure 13.5   A plot of countries using closed lines but accounting for multipolygons and holes

---

Listing 13.2   Plotting polygons

```python
import matplotlib.pyplot as plt
from osgeo import ogr

def plot_polygon(poly, symbol='k-', **kwargs):
    """Plots a polygon using the given symbol."""
    for i in range(poly.GetGeometryCount()):           # Loop through rings
        subgeom = poly.GetGeometryRef(i)
        x, y = zip(*subgeom.GetPoints())
        plt.plot(x, y, symbol, **kwargs)

def plot_layer(filename, symbol, layer_index=0, **kwargs):
    """Plots an OGR polygon layer using the given symbol."""
    ds = ogr.Open(filename)
    for row in ds.GetLayer(layer_index):
        geom = row.geometry()
        geom_type = geom.GetGeometryType()
        if geom_type == ogr.wkbPolygon:
            plot_polygon(geom, symbol, **kwargs)
        elif geom_type == ogr.wkbMultiPolygon:
            for i in range(geom.GetGeometryCount()):    # Loop through
                subgeom = geom.GetGeometryRef(i)        # subpolygons
                plot_polygon(subgeom, symbol, **kwargs)

plot_layer(r'D:\osgeopy-data\global\ne_110m_admin_0_countries.shp', 'k-')
plt.axis('equal')
plt.gca().get_xaxis().set_ticks([])                     # Turn off tick marks
plt.gca().get_yaxis().set_ticks([])
plt.show()
```

This example breaks things up into a few functions to make things easier. The plot_polygon function loops through the rings in a polygon and plots each one. The other function, plot_layer, opens a data source, gets the layer indicated by the optional layer_index parameter, and loops through all of the features and plots their geometries. If the geometry is a polygon, it passes it along to plot_polygon, but if it's a multipolygon, it passes each polygon part to plot_polygon separately. Both of these functions allow you to use **kwargs to pass optional parameters that are used by the matplotlib plot function (see the sidebar).

These functions make it easy to plot a shapefile, because all you have to do is pass the filename and symbol to plot_layer, set your axes to be equal, and then show the plot. This listing also shows you how to turn tick marks off if you don't want them drawing alongside the axes.

Using **kwargs in functions

The same way you've used a single asterisk to explode a list into individual values that can be passed as ordered arguments to a function, you can use double asterisks to explode a dictionary for use as named arguments. For example, if a function can accept a variety of optional parameters, you could create a dictionary containing the ones you want to use, with the parameter names as the keys, and then pass that to the function instead of each argument individually. This behavior is useful for passing arguments through your function to another one.

For example, the matplotlib plot function accepts a large number of optional parameters that control the output. It would be nice to use these with the plot_polygon and plot_layer functions in listing 13.2, but those functions have no reason to worry about the optional parameters. They only need to pass them along to plot when the time comes. To do this, add a variable prefixed with ** as the last parameter to your function. This variable is called kwargs by convention, but you can call it whatever you want. It does have to be the last parameter, however. Then you can pass it along to other functions, and the parameters that the user provided eventually arrive in the intended function.

You probably want to plot lines and points in addition to polygons, so create two more simple functions to plot those geometry types and add a few more conditional statements to plot_layer. This additional code is shown in the following listing, and an example of the output is shown in figure 13.6.

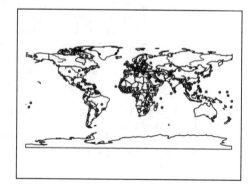

Figure 13.6   A plot of countries, rivers, and cities using basic lines and points

Listing 13.3   Plotting lines and points

```python
import os
import matplotlib.pyplot as plt
from osgeo import ogr

def plot_polygon(poly, symbol='k-', **kwargs):
    """Plots a polygon using the given symbol."""
    for i in range(poly.GetGeometryCount()):
        subgeom = poly.GetGeometryRef(i)
        x, y = zip(*subgeom.GetPoints())
        plt.plot(x, y, symbol, **kwargs)

def plot_line(line, symbol='k-', **kwargs):
    """Plots a line using the given symbol."""
    x, y = zip(*line.GetPoints())
    plt.plot(x, y, symbol, **kwargs)

def plot_point(point, symbol='ko', **kwargs):
    """Plots a point using the given symbol."""
    x, y = point.GetX(), point.GetY()
    plt.plot(x, y, symbol, **kwargs)

def plot_layer(filename, symbol, layer_index=0, **kwargs):
    """Plots an OGR layer using the given symbol."""
    ds = ogr.Open(filename)
    for row in ds.GetLayer(layer_index):
        geom = row.geometry()
        geom_type = geom.GetGeometryType()
        if geom_type == ogr.wkbPolygon:
            plot_polygon(geom, symbol, **kwargs)
        elif geom_type == ogr.wkbMultiPolygon:
            for i in range(geom.GetGeometryCount()):
                subgeom = geom.GetGeometryRef(i)
                plot_polygon(subgeom, symbol, **kwargs)
        elif geom_type == ogr.wkbLineString:
            plot_line(geom, symbol, **kwargs)
        elif geom_type == ogr.wkbMultiLineString:
            for i in range(geom.GetGeometryCount()):
                subgeom = geom.GetGeometryRef(i)
                plot_line(subgeom, symbol, **kwargs)
        elif geom_type == ogr.wkbPoint:
            plot_point(geom, symbol, **kwargs)
        elif geom_type == ogr.wkbMultiPoint:
            for i in range(geom.GetGeometryCount()):
                subgeom = geom.GetGeometryRef(i)
                plot_point(subgeom, symbol, **kwargs)

os.chdir(r'D:\osgeopy-data\global')
plot_layer('ne_110m_admin_0_countries.shp', 'k-')
plot_layer('ne_110m_rivers_lake_centerlines.shp', 'b-')
plot_layer(
    'ne_110m_populated_places_simple.shp', 'ro', ms=3)
plt.axis('equal')
plt.gca().get_xaxis().set_ticks([])
plt.gca().get_yaxis().set_ticks([])
plt.show()
```

New function (plot_line)

New function (plot_point)

New code in plot_layer

Plot new layers

This listing doesn't contain new concepts, only new code. You extend the `plot_layer` function so it calls the correct functions for lines, multilines, points, and multipoints. Then at the end of the listing, you use the updated function to plot country outlines again, but you also add major rivers and large cities. You also take advantage of `**kwargs` to pass a marker size for the city points so that they don't draw so big as to hide other features in the plot.

Until now you've treated polygons as closed lines when plotting them. What if you want to fill them with a color? You can do this by changing your `plot_polygon` function to use the matplotlib `fill` function instead of `plot`, like this:

```
def plot_polygon(poly, symbol='w', **kwargs):
    """Plots a polygon using the given symbol."""
    for i in range(poly.GetGeometryCount()):
        x, y = zip(*poly.GetGeometryRef(i).GetPoints())
        plt.fill(x, y, symbol, **kwargs)
```

Now the `symbol` parameter should be a color to use for the fill, so using `y` for yellow would result in figure 13.7 with the continents filled in.

The only problem with this method is that polygons with holes in them will be plotted incorrectly, because the holes will be plotted using the same fill color. You could fix this by only plotting the first ring with the fill color and using white for the later rings, but that wouldn't create a hole because nothing underneath would show through. If you need real holes, you can use matplotlib `PathPatches`, but it's a little more complicated than what you've done so far. To draw a polygon, you not only need the vertex coordinates, but also a set of codes denoting whether to draw a line or move the pen to that location. You use this information to create a `Path`, and then create a `PathPatch` from that. The

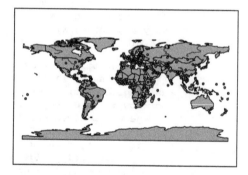

Figure 13.7 A repeat of figure 13.6, but the closed lines are filled with a color

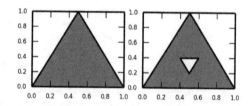

Figure 13.8 Simple patch polygons

`PathPatch` is the object that you add a fill color to. Once you have that, then you need to add it to the plot. For example, this bit of code draws the solid red triangle shown in figure 13.8:

```
import matplotlib.pyplot as  plt
from matplotlib.path import Path
import matplotlib.patches as patches
coords = [(0, 0), (0.5, 1), (1, 0), (0, 0)]
codes = [Path.MOVETO, Path.LINETO, Path.LINETO, Path.LINETO]
```

```
path = Path(coords, codes)
patch = patches.PathPatch(path, facecolor='red')
plt.axes().add_patch(patch)
plt.show()
```

The first code is MOVETO, meaning that the pen should move to the first set of coordinates without drawing anything. This makes sense if you've already drawn something else and don't want a line connecting the last point in the previous path to the first point in this path. The LINETO code corresponds to the rest of your coordinates, meaning that the points will be connected. Once you've created the path, then you can use it to create a patch, which can be filled. You need to add the patch to the drawing area of the plot, which is called the *axes* (which in turn contains the x and y axis).

To put a hole in a patch, create a path as before, but use a MOVETO code to move to the first set of coordinates for the hole, and then add the vertices in the opposite direction as the outer set in order to indicate that this should create a hole. If the coordinates for the outer ring are in clockwise order, then the coordinates for the holes must be in counterclockwise order. For example, you can put a hole in your earlier triangle like this:

```
outer_coords = [(0, 0), (0.5, 1), (1, 0), (0, 0)]
outer_codes = [Path.MOVETO, Path.LINETO,
               Path.LINETO, Path.LINETO]
inner_coords = [(0.4, 0.4), (0.5, 0.2),
                (0.6, 0.4), (0.4, 0.4)]
inner_codes = [Path.MOVETO, Path.LINETO,
               Path.LINETO, Path.LINETO]
coords = np.concatenate((outer_coords, inner_coords))
codes = np.concatenate((outer_codes, inner_codes))
path = Path(coords, codes)
patch = patches.PathPatch(path, facecolor='red')
```

Clockwise

Counterclockwise

Once you have all of your coordinates and codes in two lists or NumPy arrays, then you can use them as before to create the patch with a hole that is shown in figure 13.8. The following listing applies this process to spatial data to make a plot of world countries like that in figure 13.9.

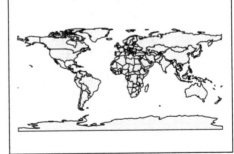

Figure 13.9 Countries drawn with patches instead of lines

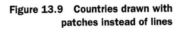

Listing 13.4 Draw world countries as patches

```
import numpy as np
import matplotlib.pyplot as  plt
from matplotlib.path import Path
import matplotlib.patches as patches
from osgeo import ogr
```

```
def order_coords(coords, clockwise):
    """Orders coordinates."""
    total = 0
    x1, y1 = coords[0]
    for x, y in coords[1:]:
        total += (x - x1) * (y + y1)
        x1, y1 = x, y
    x, y = coords[0]
    total += (x - x1) * (y + y1)
    is_clockwise = total > 0
    if clockwise != is_clockwise:
        coords.reverse()
    return coords
```

> **Determine order of coordinates**
>
> **Reorder if needed**

```
def make_codes(n):
    """Makes a list of path codes."""
    codes = [Path.LINETO] * n
    codes[0] = Path.MOVETO
    return codes
```

```
def plot_polygon_patch(poly, color):
    """Plots a polygon as a patch."""
    coords = poly.GetGeometryRef(0).GetPoints()
    coords = order_coords(coords, True)
    codes = make_codes(len(coords))
    for i in range(1, poly.GetGeometryCount()):
        coords2 = poly.GetGeometryRef(i).GetPoints()
        coords2 = order_coords(coords2, False)
        codes2 = make_codes(len(coords2))
        coords = np.concatenate((coords, coords2))
        codes = np.concatenate((codes, codes2))
    path = Path(coords, codes)
    patch = patches.PathPatch(path, facecolor=color)
    plt.axes().add_patch(patch)
```

> **Outer clockwise path**
>
> **Inner counterclockwise paths**
>
> **Concatenate paths**

```
ds = ogr.Open(r'D:\osgeopy-data\global\ne_110m_admin_0_countries.shp')
lyr = ds.GetLayer(0)
for row in lyr:
    geom = row.geometry()
    if geom.GetGeometryType() == ogr.wkbPolygon:
        plot_polygon_patch(geom, 'yellow')
    elif geom.GetGeometryType() == ogr.wkbMultiPolygon:
        for i in range(geom.GetGeometryCount()):
            plot_polygon_patch(geom.GetGeometryRef(i), 'yellow')
plt.axis('equal')
plt.show()
```

This listing contains a couple of useful functions. The first, order_coords, checks if coordinates are in the order requested and reorders them if not. Most of the code in the function implements an algorithm for determining order. Once the order is determined, it's compared to the requested order, and if they differ, the coordinates are reversed.

Also, a simple function called make_codes creates a list of LINETO codes of the appropriate length, with the first one changed to MOVETO so a new path can be started.

The last function plots polygons as patches. The first thing this function does is create a list of the outer ring coordinates in clockwise order, along with a corresponding code list. Then it loops through any inner rings that might exist, and for each one creates a list of coordinates in counterclockwise order and a list of codes. Then it appends the coordinates and codes for the inner ring to the end of the master lists. Once all rings have been processed, it creates a patch and adds it to the plot.

The main part of the code simply loops through the features in a shapefile and calls the `plot_polygon_patch` function on each polygon, including those inside multipolygons. Don't forget to set the axis to `equal` before drawing the plot, because otherwise the x and y axis will probably only range from 0 to 1, and you'll end up staring at a blank plot.

ANIMATION

You can have even more fun by animating your plots. To see how it's done, you'll animate the movements of one of the albatrosses from chapter 7. Let's start by configuring the plot's extent based on the GPS data:

```
ds = ogr.Open(r'D:\osgeopy-data\Galapagos')
gps_lyr = ds.GetLayerByName('albatross_lambert')
extent = gps_lyr.GetExtent()
fig = plt.figure()
plt.axis('equal')
plt.xlim(extent[0] - 1000, extent[1] + 1000)
plt.ylim(extent[2] - 1000, extent[3] + 1000)
plt.gca().get_xaxis().set_ticks([])
plt.gca().get_yaxis().set_ticks([])
```

You get the extent of the GPS data layer and then use it to set the x and y limits for the plot, except that you add 1,000 meters in every direction to add a little buffer around the data you want to show. You also turn the tick marks off. You probably want to add the landmasses to your plot because the GPS locations aren't too interesting without context. You can use your `plot_polygon` function to do this:

```
land_lyr = ds.GetLayerByName('land_lambert')
row = next(land_lyr)
geom = row.geometry()
for i in range(geom.GetGeometryCount()):
    plot_polygon(geom.GetGeometryRef(i))
```

Now you're ready to add the animated data, but you need to store it somewhere so it's accessible to the animation routines. You have many ways you could set this up, but for this example you'll store the x,y coordinate pairs in a list, with the corresponding timestamps in another list:

```
timestamps, coordinates = [], []
gps_lyr.SetAttributeFilter("tag_id = '2131-2131'")
for row in gps_lyr:
    timestamps.append(row.GetField('timestamp'))
    coordinates.append((row.geometry().GetX(), row.geometry().GetY()))
```

You iterate through all of the features for the animal with tag '2131-2131' and add the timestamp to one list and a tuple containing the coordinates to another list. You'll use the coordinates to animate a point and the timestamps to show the current time. You need to initialize both the point and the timestamp annotation, so let's do that:

```
point = plt.plot(None, None, 'o')[0]
label = plt.gca().annotate('', (0.25, 0.95), xycoords='axes fraction')
label.set_animated(True)
```

Here you initialize the point by plotting it with no coordinates. The plot function returns a list of objects, but in this case you have only one item in the list because you only plotted one point. You grab that point graphic out of the list and store it in your point variable. Then you create an empty annotation object on the current axes (gca is short for "get current axes"). Setting the optional xycoords parameter to 'axes fraction' lets you specify the annotation's location using percentages rather than pixels or map coordinates. The annotation will be a quarter of the way across the axes (0.25) and close to the top (0.95). You also tell the annotation that it's going to be animated, which will make the text change much more smoothly.

Now you need to write a simple function that tells the animation what items are going to change, namely, your point and label. If you don't set the point coordinates to None in this function, then there is always a point at the first location in the animation, even while another point is moving around.

```
def init():
    point.set_data(None, None)
    return point, label
```

One last function you need to write is the one that moves the point and changes the label. The first parameter to this function is a counter that gets passed to it automatically, specifying which iteration of the animation is currently being processed. The rest of the parameters are up to you. It needs to accept the objects that will change and any data needed to change them. Like the init function, this function must return the objects that change.

```
def update(i, point, label, timestamps, coordinates):
    label.set_text(timestamps[i])
    point.set_data(coordinates[i][0], coordinates[i][1])
    return point, label
```

The function uses the counter variable, i, to pull the correct timestamps and coordinates out of the lists. It changes the label's text to the timestamp, and sets the point's coordinates to the values you saved from the shapefile. Then it returns the point and the label because they've changed.

Now let's run the animation using the FuncAnimation function in matplotlib. The two required parameters are the matplotlib figure object that the animation will run on and your function that tells things how to animate. The frames parameter is the counter variable, which can be a list of values, or as in this case, the number of times you want the animation to run. The init_func parameter is the initialization function that you wrote. If you don't provide this, then the first result from the animation will be used for initialization, and it will stay there throughout the animation. If your animation function requires parameters other than the counter, you need to provide them using the `fargs` argument to `FuncAnimation`. If `blit` is `True`, then only the parts of the plot that have changed will be redrawn, which will speed things up. The `interval` parameter is the number of milliseconds between frames, and `repeat` tells it whether to repeat the animation or stop after one time.

```
import matplotlib.animation as animation
a = animation.FuncAnimation(
    fig, update, frames=len(timestamps), init_func=init,
    fargs=(point, label, timestamps, coordinates),
    interval=25, blit=True, repeat=False)
plt.show()
```

It would be nice if the animation could be embedded in paper, but it can't, so you'll have to run the code yourself to see it in action. One thing you should notice is that nothing in this code will force the elapsed time to stay at a constant speed. If two consecutive GPS fixes are three days apart, they'll be treated the same as two that are only an hour apart. One way to fix that is to round the timestamps to the nearest hour and make sure entries are in the `timestamps` and `coordinates` lists for every hour. If there aren't coordinates corresponding to a specific time, then put a bogus value in the list. When you update the animation, only update the point location if the coordinates are valid. Here's a function that rounds timestamps:

```
from datetime import datetime, timedelta
def round_timestamp(ts, minutes=60):
    ts += timedelta(minutes=minutes/2.0)
    ts -= timedelta(
        minutes=ts.minute % minutes, seconds=ts.second,
        microseconds=ts.microsecond)
    return ts
```

If you use the default value of 60 for the `minutes` parameter, the function rounds to the nearest hour. In this case it adds 30 minutes to the timestamp, so if the original was 11:27:14.01, the new time is 11:57:14.01. Then it calculates the remainder of dividing the timestamp's minutes value by the number of minutes you want to round to. In this case, that value is 57 because 57 goes into 60 zero times and the entire value is the remainder. Then the numbers of seconds and microseconds from the timestamp are added to this value, so you have 57:14.01, and the result is subtracted from the timestamp. Now the timestamp is 11:00 even, which is the closest hour to 11:27:14.01.

Now that you can round timestamps, let's initialize the `timestamps` and `coordinates` lists with the first values from the dataset:

```
gps_lyr.SetAttributeFilter("tag_id = '2131-2131'")
time_format = '%Y-%m-%d %H:%M:%S.%f'
row = next(gps_lyr)
timestamp = datetime.strptime(row.GetField('timestamp'), time_format)
timestamp = round_timestamp(timestamp)
timestamps = [timestamp]
coordinates = [(row.geometry().GetX(), row.geometry().GetY())]
```

Now you can loop through the rest of the features and fill in your lists. Get the timestamp for each row and compare it to the last one in the timestamps list. Keep adding new timestamps to the list until the last one is equal to the one from the feature, and while you're at it, append a bogus set of coordinates to that list, too. The loop will stop when the last timestamp in the list is equal to the row's timestamp, so you can overwrite the last set of bogus coordinates with the feature's coordinates and they'll match up with the correct timestamp.

```
hour = timedelta(hours=1)
for row in gps_lyr:
    timestamp = datetime.strptime(row.GetField('timestamp'), time_format)
    timestamp = round_timestamp(timestamp)
    while timestamps[-1] < timestamp:
        timestamps.append(timestamps[-1] + hour)
        coordinates.append((None, None))
    coordinates[-1] = (row.geometry().GetX(), row.geometry().GetY())
```

The only other thing you need to do is change your update function so that it only moves the point if there are valid coordinates. If you don't do this, the point will disappear when there aren't coordinates for a specific time because they'll be set to None.

```
def update(i, point, label, timestamps, coordinates):
    label.set_text(timestamps[i])
    if coordinates[i][0] is not None:
        point.set_data(coordinates[i][0], coordinates[i][1])
    return point, label
```

Now you can run the animation as before, but the time increments will be constant, which makes much more sense.

If you have appropriate software installed, you can also save the animation as a video file. For example, I have FFmpeg (www.ffmpeg.org) installed, so as long as ffmpeg is in my PATH environment variable, I can save the animation like this:

```
a.save('d:/temp/albatross.mp4', 'ffmpeg')
```

If you don't have the software to save it yourself but would still like to see the results, there's a saved version in the Galapagos data folder.

13.1.2 *Plotting raster data*

You can also use matplotlib to draw raster data. Making a simple raster plot is extremely easy because you have no coordinates to worry about, and there happens to be a function for displaying data contained in a NumPy array. Let's start with a small image and draw it using the default color ramp, as shown in figure 13.10A.

```
ds = gdal.Open(r'D:\osgeopy-data\Washington\dem\sthelens_utm.tif')
data = ds.GetRasterBand(1).ReadAsArray()
plt.imshow(data)
plt.show()
```

A. Default color ramp

B. Gray color ramp

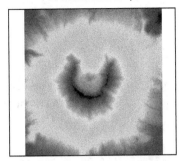 

Figure 13.10 Two plots of the same digital elevation model of Mount St. Helens. Plot A uses the default color ramp (which morphs from blue to red), and plot B uses a grayscale color ramp.

As you can see, all you have to do is read the raster data into a NumPy array as you've done many times before, and then pass that array to the `imshow` function, and you have yourself a plot. You might not like the default color ramp, but you can probably find a built-in one that you like. If not, you can create your own, although you won't learn how to do that here. To use a colormap, pass its name to `imshow` as the `cmap` parameter, like this (figure 13.10B):

```
plt.imshow(data, cmap='gray')
```

TIP As of this writing, you can see a list of matplotlib colormaps at http://wiki.scipy.org/Cookbook/Matplotlib/Show_colormaps.

If you want to plot a large image, you shouldn't read the entire band in and try to plot it. You're much better off using one of the pyramid layers because they take up much less memory and will plot considerably faster. You need to choose the appropriate overview level so that you have the resolution that you need without degrading performance. Here's a function that retrieves overview data from an image, although it doesn't check to make sure that the user requests a valid overview level.

Listing 13.5 Function to retrieve overview data

```
def get_overview_data(fn, band_index=1, level=-1):
    """Returns an array containing data from an overview.

    fn          - path to raster file
    band_index  - band number to get overview for
    level       - overview level, where 1 is the highest resolution;
                  the coarsest can be retrieved with -1
    """
    ds = gdal.Open(fn)
    band = ds.GetRasterBand(band_index)
```

```
if level > 0:
    ov_band = band.GetOverview(level)
else:
    num_ov = band.GetOverviewCount()
    ov_band = band.GetOverview(num_ov + level)
return ov_band.ReadAsArray()
```

The function requires that the user provide the path to a raster file, and optionally, a band number and overview level. If the optional parameters aren't provided, it will return the coarsest overview for the first band. Try using this function to plot the lowest resolution overview for a Landsat band:

```
fn = r'D:\osgeopy-data\Landsat\Washington\p047r027_7t20000730_z10_nn10.tif'
data = get_overview_data(fn)
data = np.ma.masked_equal(data, 0)
plt.imshow(data, cmap='gray')        ◄──┐  Mask out black edges
plt.show()
```

A. Default

B. Stretched

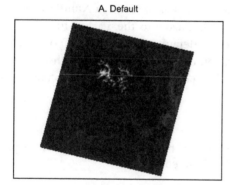 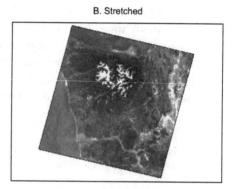

Figure 13.11 **Two plots of the same Landsat band. Plot A uses default settings, but plot B uses stretched data for much better contrast.**

As you can see from figure 13.11A, this results in an extremely dark image and in this case, at least, it's difficult if not impossible to differentiate much at all. It might even seem worse if you hadn't masked out the pixels that were equal to 0. Without that step, you'd see a rectangle with all of the outside pixels that weren't part of the satellite imagery drawn as black.

Because of the lack of contrast in figure 13.11A, this is a perfect time to stretch the data to make it look better. A *standard deviation stretch*, which is a common method, keeps pixel values that are within one or more standard deviations (usually two) from the mean, and sets everything outside that range to the minimum or maximum included values, as shown in figure 13.12. The values are then stretched between 0 and 1 for drawing, because that's what matplotlib wants.

To implement this, figure out the minimum and maximum cutoffs that are the desired number of standard deviations from the mean and then pass them as the vmin

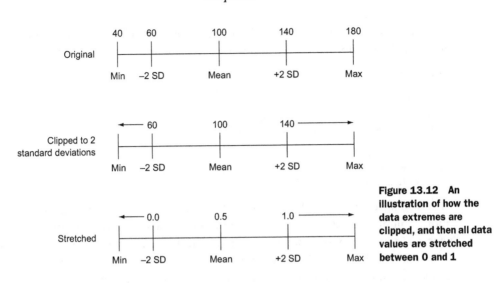

Figure 13.12 An illustration of how the data extremes are clipped, and then all data values are stretched between 0 and 1

and `vmax` parameters to `imshow`, respectively. The data will automatically be stretched for you, but you need to provide these clip values, like this:

```
mean = np.mean(data)
std_range = np.std(data) * 2
plt.imshow(data, cmap='gray', vmin=mean-std_range, vmax=mean+std_range)
```

Figure 13.11B is stretched in this way, and it's obviously a better visualization of the data than the nonstretched version.

You can also plot three bands as red, green, and blue, with an optional fourth alpha band. In this case you need to stack the bands into a three-dimensional array and pass that to `imshow`. Unlike with single bands, using masked arrays to filter out the zeros

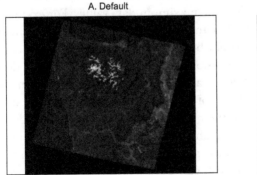
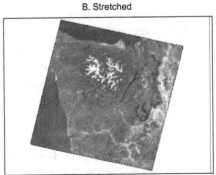

Figure 13.13 Two plots of the same three-band Landsat image. Plot A uses default settings, but plot B uses stretched data for considerably better contrast.

around the edges doesn't work in this case, so you're stuck with the black edges for the moment. The following code snippet uses three bands to create a figure like 13.13A:

```
os.chdir(r'D:\osgeopy-data\Landsat\Washington')
red_fn = 'p047r027_7t20000730_z10_nn30.tif'
green_fn = 'p047r027_7t20000730_z10_nn20.tif'
blue_fn = 'p047r027_7t20000730_z10_nn10.tif'
red_data = get_overview_data(red_fn)
green_data = get_overview_data(green_fn)
blue_data = get_overview_data(blue_fn)
data = np.dstack((red_data, green_data, blue_data))
plt.imshow(data)
```

Again, that image is too dark to be useful. Unfortunately, stretching the data is a bit more complicated if you're plotting multiple bands because the automatic scaling with vmin and vmax only works for single bands. You'll need to normalize the data yourself. The following function performs a standard deviation stretch on the data contained in a NumPy array and then scales the results between 0 and 1.

> **Listing 13.6   Function to stretch and scale data**

```
def stretch_data(data, num_stddev):
    """Returns the data with a standard deviation stretch applied.

    data       - array containing data to stretch
    num_stddev - number of standard deviations to use
    """
    mean = np.mean(data)
    std_range = np.std(data) * 2
    new_min = max(mean - std_range, np.min(data))
    new_max = min(mean + std_range, np.max(data))
    clipped_data = np.clip(data, new_min, new_max)
    return clipped_data / (new_max - new_min)
```

Instead of finding the appropriate distance from the mean, based on the desired number of standard deviations, this function makes sure that the values used aren't less than the minimum or greater than the maximum data values. For example, if you have 8-bit data that ranges from 0 to 255, the mean value is 43, and the standard deviation is 24, then the lower bound would be -5 if you subtracted two standard deviations from the mean. The minimum possible value is 0, however, and you don't want to normalize your data using impossible values, so that's why the function checks to make sure that the bounds don't fall out of the range of potential values. After determining the bounds, they're used with the np.clip function, which replaces all values that are less than new_min with new_min, and replaces all values that are greater than new_max with new_max, like what was illustrated back in figure 13.12. Then the resulting data are scaled from 0 to 1. Now you can use this function to scale each of the three bands appropriately.

Because you're scaling these data yourself, you can take advantage of the alpha channel to get rid of the black around the edges. For this particular image, you can assume that if all three bands contain 0, then the pixel is an outside edge. The alpha band should also contain 0 for these pixels, meaning it's fully transparent. Other pixels should have a 1 in the alpha band so that they'll be drawn at full opacity. Add this alpha band to your three-dimensional stack, as shown in the following snippet, and when you plot it the results will be similar to figure 13.13B.

```
red_data = stretch_data(get_overview_data(red_fn), 2)
green_data = stretch_data(get_overview_data(green_fn), 2)
blue_data = stretch_data(get_overview_data(blue_fn), 2)
alpha = np.where(red_data + green_data + blue_data > 0, 1, 0)
data = np.dstack((red_data, green_data, blue_data, alpha))
plt.imshow(data)
```

13.1.3 Plotting 3D data

You can even plot three-dimensional data, such as a digital elevation model. To do this, you need the array containing elevation data, and two other arrays of the same size containing x and y coordinates for each pixel. These latter two arrays can be created by passing arrays containing the possible x and y values to np.meshgrid, which results in data like that shown in figure 13.14. Each pixel in the x array contains a value indicating which row it's in, and each pixel in the y array indicates the column. If your pixels are square and you don't need georeferencing information in your plot, you can use arange to get the input lists for meshgrid, so getting your two-dimensional x and y arrays is as easy as this:

```
x, y = np.meshgrid(np.arange(band.XSize), np.arange(band.YSize))
```

In other cases, you can use the geotransform to compute the required information so that the x and y arrays contain real-world coordinates instead of pixel coordinates like

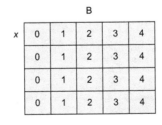

x, y = np.meshgrid(range(5), range(4))

Figure 13.14 An illustration of meshgrid output. Part A shows the x,y coordinate pair for each cell in the array. The output is two arrays, one of which contains x coordinates (part B) and the other contains y coordinates (part C).

those in figure 13.14. The following listing shows the steps to do this using a DEM of Mount St. Helens, and then it plots the data in 3D to get figure 13.15A.

Listing 13.7   Using `meshgrid` to get map coordinates

```
import numpy as np
import matplotlib.pyplot as plt
from mpl_toolkits.mplot3d import Axes3D
from osgeo import gdal
ds = gdal.Open(r'D:\osgeopy-data\Washington\dem\sthelens_utm.tif')
band = ds.GetRasterBand(1)
ov_band = band.GetOverview(band.GetOverviewCount() - 3)      ◄── Get appropriate
data = ov_band.ReadAsArray()                                      overview level
geotransform = ds.GetGeoTransform()
minx = geotransform[0]
maxy = geotransform[3]
maxx = minx + ov_band.XSize * geotransform[1]                 Calculate bounding
miny = maxy + ov_band.YSize * geotransform[5]                 coordinates
x = np.arange(minx, maxx, geotransform[1])
y = np.arange(maxy, miny, geotransform[5])
x, y = np.meshgrid(x[:ov_band.XSize], y[:ov_band.YSize])      Get x and y arrays

fig = plt.figure()
ax = fig.gca(projection='3d')
ax.plot_surface(x, y, data, cmap='gist_earth', lw=0)          Make the 3D plot
plt.axis('equal')
plt.show()
```

The first part of this listing reads overview data into memory and uses the geotransform to calculate the bounding coordinates for the DEM. These coordinates are then used in conjunction with `meshgrid` to create the x and y arrays needed for the plot.

To create the plot, you first create a matplotlib figure object and then grab its axes object. You tell the axes to use 3D, and then you call its `plot_surface` method in order to make the plot. This function requires the x and y arrays and the array

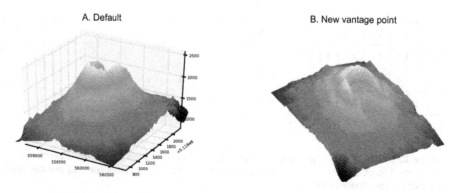

A. Default   B. New vantage point

Figure 13.15   3D plots of Mount St. Helens. Plot A uses default settings, while the elevation and azimuth have been changed for plot B, as well as the axis removed.

containing elevations. You use the colormap named gist_earth instead of the default, and you used lw=0 to set the line width to 0. If you don't change the line width, then each cell will have an outline around it, which doesn't look good in this case. By the way, a figure and axes were created automatically for your earlier plots, but you didn't need to worry about them. Here you do, because you need a handle to the axes to specify 3D and plot the surface.

What if you want to change the vantage point that you're viewing the 3D image from? Well, you can set an elevation between 0 and 90, where 0 is ground level and 90 is looking straight down, and you can also rotate the plot from 0 to 360 degrees. The image in figure 13.15B was obtained by setting the elevation to 55, rotating the figure 60 degrees, and turning the axis off. To do this, add these two lines before calling plt.show():

```
ax.view_init(elev=55, azim=60)
plt.axis('off')
```

You can make this even more fun by creating an animation. This is simpler than the Albatross animation from earlier because all you have to do is change the rotation factor for each iteration. Try adding this to your code before calling plt.show:

```
import matplotlib.animation as animation

def animate(i):
    ax.view_init(elev=65, azim=i)

anim = animation.FuncAnimation(
    fig, animate, frames=range(0, 360, 10), interval=100)
```

The animate function changes the vantage point that the plot is being viewed from. The call to FuncAnimation sets things up so that the animate function is called 36 times, once for each value in frames. This will cause the plot to rotate 10 degrees each time. Although the interval parameter specifies that there will be 100 milliseconds in between each frame, it will be slower if your computer can't draw it that fast. A saved version of this is in the Nepal data folder.

13.2 *Mapnik*

The plots you've been making so far work well for visualizing data, but a good chance exists that you'll need to make something that looks a little nicer, or more like a real map, at some point. One good way to do this using Python is with Mapnik, a popular cartographic library. In fact, you might have seen maps created with Mapnik without knowing it. Mapnik was designed for making tiled maps for web applications, and as far as I know it's not easy to put cartographic symbols such as North arrows on Mapnik images. You can do it with other graphics modules such as Cairo, but that's beyond the scope of this introduction. This section will walk you through the basics of drawing vector and raster data using this module, but you should visit mapnik.org if you want to learn more.

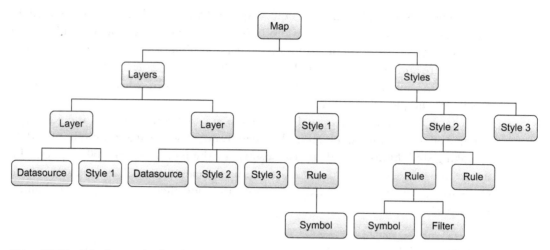

Figure 13.16 A basic organization chart of a Mapnik map. Each map has at least one layer and one style. Each layer needs to reference at least one of the styles.

Before we start drawing anything, though, let's take a quick look at the minimum requirements for a Mapnik map, as seen in figure 13.16. A map has one or more layers, as well as one or more styles. The styles are what specify how the data are to be drawn. Each style needs at least one rule, and each rule needs at least one symbol. Rules can also have filters so that they only apply to a subset of the data. Each layer needs a data source and at least one style. Layer styles aren't new style objects; they reference one of the styles that belongs to the map. You'll see how this all works in the next few examples.

13.2.1 *Drawing vector data*

Do you remember the New Orleans data from an earlier chapter? If not, you're about to be reminded, because you'll use it in the next few examples. The following listing starts by drawing the TIGER water layer from the US Census Bureau.

Listing 13.8 Creating a simple Mapnik map

```
import mapnik

srs = "+proj=longlat +ellps=GRS80 +datum=NAD83 +no_defs"
m = mapnik.Map(800, 600, srs)
m.zoom_to_box(mapnik.Box2d(-90.3, 29.7, -89.5, 30.3))          Create a mapnik map

tiger_fn = r'D:\osgeopy-data\Louisiana\tiger_la_water_CENSUS_2006'
tiger_shp = mapnik.Shapefile(file=tiger_fn)
tiger_lyr = mapnik.Layer('Tiger')                              Create a layer
tiger_lyr.datasource = tiger_shp                               from a shapefile
```

```
water_color = mapnik.Color(165, 191, 221)
water_fill_sym = mapnik.PolygonSymbolizer(water_color)
```

Create a polygon fill symbol

```
tiger_rule = mapnik.Rule()
tiger_rule.symbols.append(water_fill_sym)
tiger_style = mapnik.Style()
tiger_style.rules.append(tiger_rule)
m.append_style('tiger', tiger_style)
```

Create a symbology style and add to layer

```
tiger_lyr.styles.append('tiger')
m.layers.append(tiger_lyr)
```

Add the style and layer to map

Save the map to file

```
mapnik.render_to_file(m, r'd:\temp\nola.png')
```

The first step is to create a Mapnik map object, but you call it m instead of map because map is a reserved word in Python. You need to provide a size for the map when you create it, so this map will be 800 pixels wide and 600 tall. You can optionally provide a spatial reference in the form of a Proj.4 string or EPSG code; if you don't provide this, then it will default to WGS84 lat/lon. Because most of the New Orleans data uses NAD83 lat/lon, that's what you decide to use here. You also set a bounding box in the form of (min_x, min_y, max_x, max_y). If you don't set the bounding box, you'll end up with an empty map.

To add a layer to a map, you need to create a layer object and give it a data source. Several types of data sources exist for different data formats, such as shapefiles, GeoJSON, and PostGIS. Here you create a shapefile data source and add it to a layer that you name 'Tiger'.

Adding a layer to a map isn't enough, however. If you want the layer to be drawn, you also need to provide information about how to symbolize it. You start this off by creating a Mapnik color object (water_color) from RGB values that specify a light blue, and then used that to create a polygon symbolizer for drawing water layers. Polygons drawn with this symbolizer will be filled with the blue color defined by the RGB values.

Once you have a symbolizer, you create a symbology style. A style needs at least one rule that defines how to draw something. This particular style is simple and only contains one rule, which in turn only contains your polygon symbolizer. Then you add the style to the map so that layers could use it. Notice that you provide a name for the style at the same time you add it to the map; this is important later.

You want the Tiger layer to use the style you create, so then you add the style to the layer as well, making sure to use the same name for the style that you used when adding it to the map. The style must be added to both the layer and the map or it won't work. In addition, the style must be added to the layer before the layer is added to the map, which is what you do next.

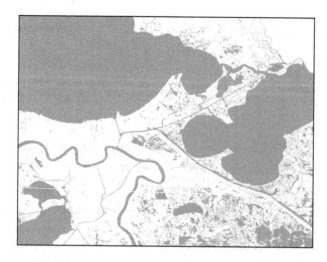

Figure 13.17   A simple plot of hydrographic data using a single layer and style rule

Finally, after everything is added in the appropriate places, you save the map to a file. If all goes well, you'll have an image like figure 13.17.

As pretty as that figure is, you want more than water bodies, so now try adding marshland, too. These data come from a national hydrography dataset that includes open water, glaciers, marshes, dry lakes, canals, and other features. In fact, including the canals and lakes from this dataset will make your map look a little better, so you'll include them as well. This next listing shows how to add this new layer with more-complicated styling to the map. This code would be added in before saving the map to an image file.

Listing 13.9   Using multiple rules in a style

```
atlas_lyr = mapnik.Layer('National Atlas')
atlas_shp = mapnik.Shapefile(file=r'D:\osgeopy-data\US\wtrbdyp010')
atlas_lyr.datasource = atlas_shp

water_rule = mapnik.Rule()
water_rule.filter = mapnik.Expression(          Create open
    "[Feature]='Canal' or [Feature]='Lake'")    water rule
water_rule.symbols.append(water_fill_sym)

marsh_color = mapnik.Color('#66AA66')                        Create marsh fill and
marsh_fill_sym = mapnik.PolygonSymbolizer(marsh_color)       outline symbolizers
marsh_line_sym = mapnik.LineSymbolizer(marsh_color, 2)

marsh_rule = mapnik.Rule()
marsh_rule.filter = mapnik.Expression(
    "[Feature]='Swamp or Marsh'")              Create marsh rule
marsh_rule.symbols.append(marsh_fill_sym)
marsh_rule.symbols.append(marsh_line_sym)

atlas_style = mapnik.Style()
atlas_style.rules.append(water_rule)        Create style
atlas_style.rules.append(marsh_rule)        and add rules
```

```
m.append_style('atlas', atlas_style)
atlas_lyr.styles.append('atlas')
m.layers.append(atlas_lyr)
```

The methods for creating this layer and adding it and its style to the map are the same as before, but this time the style is more complicated. For starters, you add two rules to this style instead of one. Let's look at the first of these, called water_rule. You use a filter to apply this rule to only those features where the "Feature" attribute column is equal to either 'Canal' or 'Lake'. Filter expressions in Mapnik are similar to the OGR filter expressions that you've already used, but attribute names must be surrounded by brackets. You use the same water polygon symbolizer for this rule that you used for the tiger data.

Before creating the second rule, you construct new symbolizers that use a green color. This time you define the color using hex notation to prove that you can, but you could use RGB again if you want. The color is then used to create another polygon fill symbol and also a line symbol that's 2 pixels wide. This line symbolizer will be used to outline the polygons with the same color that they're filled with. The reason you use the outline here is because the datasets have slight gaps between shapes that are obvious without an outline filling them up.

Now that you have your symbolizers, you create the marsh rule for this layer. First, you use a filter to make this rule apply only to features where the "Feature" attribute column is equal to the string 'Swamp or Marsh'. Then you add the green fill and outline symbols that you created previously.

After creating the rules, you create a new style and add both rules to it. Then you add the style to the map and the layer, and add the layer to the map. After rendering this map to a file, you end up with a graphic like figure 13.18.

If you compare figures 13.17 and 13.18, you might wonder where all of the little water bodies disappeared to. The layers are drawn in the same order that you add them to the map, so the marshes were drawn on top of those little water bodies. For

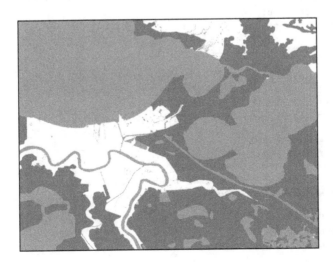

Figure 13.18 Another layer added, this time using two rules to specify that marshes and open water in the same dataset are drawn differently

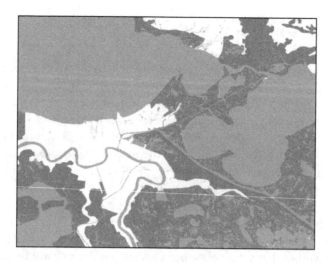

Figure 13.19   The same data as figure 13.18, but the order of the layers reversed

this reason, you need to think about which of your layers should not be covered up and plan accordingly. To get a graphic like figure 13.19 instead, move the code that appends the layers to the map down to the end of the script and then reverse the order that the layers are added, like this:

```
m.layers.append(atlas_lyr)
m.layers.append(tiger_lyr)
```

Your map is still not complete, however, because you want some roads and the New Orleans city boundary. The following listing shows the code to add these.

Listing 13.10   Adding the roads and city outline

```
roads_lyr = mapnik.Layer('Roads', "+init=epsg:4326")        ◄─── Specify the layer's SRS
road_shp = mapnik.Shapefile(
    file=r'D:\osgeopy-data\Louisiana\roads')
roads_lyr.datasource = road_shp

roads_color = mapnik.Color(170, 170, 127)

roads_primary_rule = mapnik.Rule()
roads_primary_rule.filter = mapnik.Expression("[fclass]='primary'")
roads_primary_sym = mapnik.LineSymbolizer(roads_color, 1.5)
roads_primary_rule.symbols.append(roads_primary_sym)

roads_secondary_rule = mapnik.Rule()
roads_secondary_rule.filter = mapnik.Expression(
    "[fclass]='secondary' or [fclass]='tertiary'")
roads_secondary_sym = mapnik.LineSymbolizer(roads_color, 0.5)
roads_secondary_rule.symbols.append(roads_secondary_sym)

roads_style = mapnik.Style()
roads_style.rules.append(roads_primary_rule)
roads_style.rules.append(roads_secondary_rule)
```

```
m.append_style('roads style', roads_style)
roads_lyr.styles.append('roads style')

city_lyr = mapnik.Layer('City Outline')
city_shp = mapnik.Shapefile(file=r'D:\osgeopy-data\Louisiana\NOLA')
city_lyr.datasource = city_shp

city_color = mapnik.Color('black')                    ◄─── Create a color by name
city_sym = mapnik.LineSymbolizer(city_color, 2)
city_sym.stroke.add_dash(4, 2)                        ◄─── Make a dashed line
city_rule = mapnik.Rule()
city_rule.symbols.append(city_sym)
city_style = mapnik.Style()
city_style.rules.append(city_rule)

m.append_style('city style', city_style)
city_lyr.styles.append('city style')

m.layers.append(atlas_lyr)
m.layers.append(tiger_lyr)
m.layers.append(roads_lyr)
m.layers.append(city_lyr)
```

Only a few new things were added in this example. This first is that you specify the spatial reference when creating the roads layer. This is necessary because this particular shapefile uses WGS84 instead of NAD83. You could use a Proj.4 string, as you did with the map spatial reference information, but you opt for an EPSG code instead. Notice that you use two rules for the roads style so that you can draw primary roads a little fatter than secondary and tertiary roads.

The second new concept is that you can create color objects using HTML named colors, as well. This is the technique you use to create the black line for the city outline. But you also want the city outline to be dashed instead of solid, so you edit the line's stroke property to make it dashed. The first parameter to add_dash is the length of the dash in pixels, and the second is the length of the gap between the dashes.

The result of adding all of this code to your script is shown in figure 13.20.

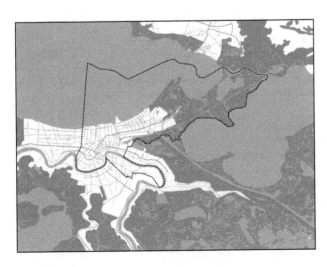

Figure 13.20   Line styles added in order to draw roads and the city outline

13.2.2 *Storing information as XML*

If you use certain styles or layers often, you can store the relevant information in XML files that can be loaded from your script. You can also store entire maps this way, meaning that you can create a map using XML and then render it with Mapnik. If you'd like to see what one of these files looks like, add this line of code to the end of your script:

```
mapnik.save_map(m, 'nola_map.xml')
```

To render the map described in this XML file to an image, write a script that imports Mapnik and then loads the XML and saves the output like this:

```
m = mapnik.Map(400, 300)
m.zoom_to_box(mapnik.Box2d(-90.3, 29.7, -89.5, 30.3))
mapnik.load_map(m, r'd:\temp\nola.xml')
mapnik.render_to_file(m, r'd:\temp\nola.png')
```

That is pretty much the entire script. You do still have to create a map object with the desired size and bounding box, but the layers and styles are pulled from the XML file.

You aren't stuck using only the information contained in the XML, however, so you can use this technique to store commonly used layers or styles. For example, if you use the hydrography dataset from the National Atlas often, you can store its information in an XML file and load it in your scripts. Pull the code pertaining to the atlas layer out of your earlier script and use it to create a new script that saves the necessary XML. The following listing shows what you need.

Listing 13.11   Create XML to describe the National Atlas hydrography layer

```
import mapnik

m = mapnik.Map(0, 0)

water_rule = mapnik.Rule()
water_rule.filter = mapnik.Expression(
    "[Feature]='Canal' or [Feature]='Lake'")
water_rule.symbols.append(
    mapnik.PolygonSymbolizer(mapnik.Color(165, 191, 221)))

marsh_rule = mapnik.Rule()
marsh_rule.filter = mapnik.Expression("[Feature]='Swamp or Marsh'")
marsh_color = mapnik.Color('#66AA66')
marsh_rule.symbols.append(mapnik.PolygonSymbolizer(marsh_color))
marsh_rule.symbols.append(mapnik.LineSymbolizer(marsh_color, 2))

atlas_style = mapnik.Style()
atlas_style.rules.append(water_rule)
atlas_style.rules.append(marsh_rule)
m.append_style('atlas', atlas_style)

lyr = mapnik.Layer('National Atlas Hydro',
                   "+proj=longlat +ellps=GRS80 +datum=NAD83 +no_defs")
lyr.datasource = mapnik.Shapefile(file=r'D:\osgeopy-data\US\wtrbdyp010')
lyr.styles.append('atlas')
m.layers.append(lyr)

mapnik.save_map(m, r'd:\temp\national_atlas_hydro.xml')
```

This script creates the styles used by the National Atlas layer, including the filters that are specific to that layer's attribute table. It also creates the layer and appends the style to it. The SRS is added to the layer, too, because your scripts that load this file may not use the same SRS as this particular layer. The style and layer are both added to a dummy map object that's used to save the information. The size of the map doesn't matter because that will be determined by the script that loads the XML.

The resulting XML looks like the following listing.

Listing 13.12 XML describing the National Atlas hydrography layer

```xml
<?xml version="1.0" encoding="utf-8"?>
<Map srs="+proj=longlat +ellps=WGS84 +datum=WGS84 +no_defs">
    <Style name="atlas">
        <Rule>
            <Filter>
                ((([Feature]=&apos;Canal&apos;) or
                ➥ ([Feature]=&apos;Lake&apos;))
            </Filter>
            <PolygonSymbolizer fill="rgb(165,191,221)"/>
        </Rule>
        <Rule>
            <Filter>([Feature]=&apos;Swamp or Marsh&apos;)</Filter>
            <PolygonSymbolizer fill="rgb(102,170,102)"/>
            <LineSymbolizer stroke="rgb(102,170,102)" stroke-width="2"/>
        </Rule>
    </Style>
    <Layer name="National Atlas Hydro"
            srs="+proj=longlat +ellps=GRS80 +datum=NAD83 +no_defs">
        <StyleName>atlas</StyleName>
        <Datasource>
            <Parameter name="file">D:\osgeopy-data\US\wtrbdyp010</Parameter>
            <Parameter name="type">shape</Parameter>
        </Datasource>
    </Layer>
</Map>
```

As you can see, the XML is straightforward, so you might even want to define your layers this way from the beginning instead of writing code. Either way, once you have this file, you can delete all of the code from listing 13.9 that creates the atlas layer and style (that's more than 20 lines) and then replace this

```
m.layers.append(atlas_lyr)
```

with this:

```
mapnik.load_map(m, r'd:\temp\national_atlas_hydro.xml')
```

Obviously, this technique will simplify your life if you use the same layers in multiple maps and is worth looking into.

13.2.3 *Drawing raster data*

Now that you know the basics of drawing vector data with Mapnik, it's time to create a simple graphic using raster data. The following listing creates an image that displays a topo map for Mount St. Helens.

Listing 13.13   Drawing a raster

```
import mapnik
srs = '+proj=utm +zone=10 +ellps=GRS80 +datum=NAD83 +units=m +no_defs'
m = mapnik.Map(1200, 1200, srs)
m.zoom_to_box(mapnik.Box2d(558800, 5112200, 566600, 5120500))

topo_lyr = mapnik.Layer('Topo', srs)
topo_raster = mapnik.Gdal(                                          Add a GDAL data
    file=r'D:\osgeopy-data\Washington\dem\st_helens.tif')          source to the layer
topo_lyr.datasource = topo_raster

topo_sym = mapnik.RasterSymbolizer()          ◄──── Use a RasterSymbolizer
topo_rule = mapnik.Rule()
topo_rule.symbols.append(topo_sym)
topo_style = mapnik.Style()
topo_style.rules.append(topo_rule)

m.append_style('topo', topo_style)
topo_lyr.styles.append('topo')

m.layers.append(topo_lyr)
mapnik.render_to_file(m, r'd:\temp\helens.png')
```

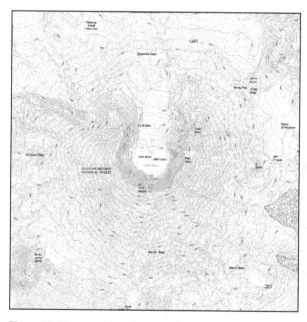

Figure 13.21   A raster plot of a topo map

Much of this example should look familiar, because it's similar to the vector example. The main differences are that you use a GDAL data source instead of a shapefile and you use a simple raster symbolizer with no options. Unlike the shapefile examples, though, you do have to specify an SRS for the raster data source even if it matches the map's SRS. Other than that, the process of creating rules, styles, and layers is still the same. The output graphic looks like figure 13.21.

This image could use a little help, though. One common technique for making something like this more aesthetically pleasing is to overlay it on a *hillshade* dataset to give it depth. A hillshade is created by assuming a height

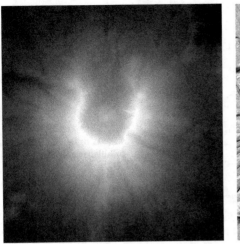

Figure 13.22   A digital elevation model of Mount St. Helens on the left, and a hillshade derived from the DEM on the right

and angle for a light source, and determining where the shadows would fall based on a digital elevation model (figure 13.22). The next listing shows how to put a hillshade derived from the Mount St. Helens DEM underneath this topo map to get a figure like 13.23.

Listing 13.14   Using a hillshade

```
import mapnik
srs = '+proj=utm +zone=10 +ellps=GRS80 +datum=NAD83 +units=m +no_defs'
m = mapnik.Map(1200, 1200, srs)
m.zoom_to_box(mapnik.Box2d(558800, 5112200, 566600, 5120500))

hillshade_lyr = mapnik.Layer('Hillshade', srs)
hillshade_raster = mapnik.Gdal(
    file=r'D:\osgeopy-data\Washington\dem\sthelens_hillshade.tif')
hillshade_lyr.datasource = hillshade_raster

hillshade_rule = mapnik.Rule()
hillshade_rule.symbols.append(mapnik.RasterSymbolizer())
hillshade_style = mapnik.Style()
hillshade_style.rules.append(hillshade_rule)

m.append_style('hillshade', hillshade_style)
hillshade_lyr.styles.append('hillshade')

topo_lyr = mapnik.Layer('Topo', srs)
topo_raster = mapnik.Gdal(
    file=r'D:\osgeopy-data\Washington\dem\st_helens.tif')
topo_lyr.datasource = topo_raster
```

```
topo_sym = mapnik.RasterSymbolizer()
topo_sym.opacity = 0.6
topo_rule = mapnik.Rule()
topo_rule.symbols.append(topo_sym)
topo_style = mapnik.Style()
topo_style.rules.append(topo_rule)

m.append_style('topo', topo_style)
topo_lyr.styles.append('topo')

m.layers.append(hillshade_lyr)
m.layers.append(topo_lyr)
mapnik.render_to_file(m, r'd:\temp\helens2.png')
```

Make topo layer semitransparent

Topo is drawn on top of hillshade

Figure 13.23   A topo raster drawn partly transparent so that an underlying hillshade layer provides shadows

In this example, the hillshade layer is added exactly the same way as the topo layer was added previously, but this time you make one change to the topo layer's symbolizer. Because you want the topo layer to be semitransparent to let the hillshade layer show through, you change the opacity property to a value of 0.6. A value of 1.0 (the default) makes the layer fully opaque, so the hillshade layer might as well not even be there. A value of 0 is fully transparent, so you'd only see the hillshade. You can play with this value to see what level of transparency you like best, but figure 13.23 shows what effect a value of 0.6 has.

13.3   Summary

- The matplotlib module is a general-purpose plotting module for Python and works well for quickly visualizing data.
- You can use the matplotlib interactive mode to see immediately what effect something has.
- Use the Mapnik module if you want prettier maps and images than what you can easily get with matplotlib.
- You can store Mapnik styles and layers in XML files to make them easily reusable.

appendix A
Installation

You'll need to install several components in order to work through this book. The most obvious is Python itself, but the basic Python distribution doesn't come with the geoprocessing modules bundled with it. Some third-party Python distributions do include these libraries, or at least some of them, and you're welcome to use them if you'd like. Although I haven't tested the examples in this book with it, I've had success teaching classes using Anaconda Python, which is available for Windows, OS X, and Linux (https://www.continuum.io/downloads). Three other examples that I'm aware of, but haven't tested things with, are OSGeo4W, Enthought Canopy, and Python(x,y). If you want to use one of these distributions, make sure that you check the package list first to make sure it includes the Python modules that you're interested in. The instructions provided here will show you how to obtain Python and the required modules without using one of these distributions, as well as how to get them using Anaconda. Unfortunately, everybody's system is different, even if they're using the same operating system, so these instructions can in no way cover all cases.

Here's the list of modules that we'll work with in this book:

- Python itself: www.python.org
- GDAL/OGR, for reading and writing geospatial data: www.gdal.org
- NumPy, the basic Python array-processing module: www.numpy.org
- Matplotlib, for plotting data graphically: www.matplotlib.org
- SciPy, a scientific computing module: www.scipy.org
- Pyproj, a Python wrapper for the PROJ.4 Cartographic Projections library: https://code.google.com/p/pyproj/
- Folium, for making Leaflet.js maps using Python: https://github.com/python-visualization/folium
- Spectral Python, for processing hyperspectral image data: http://www.spectralpython.net/

- scikit-learn, for data analysis: http://scikit-learn.org/stable/
- Mapnik, for making beautiful maps: http://mapnik.org/

There are two main versions of Python, 2.x and 3.x, and there are a few significant differences between them, so they aren't completely interchangeable. A lot of code will run in both, however, and I've tried to write the examples in this book so that they'll work with either one. The latest version of the 2.x branch is 2.7, and there will be no more major releases in this branch. The 3.x branch is being actively developed, and if you have no specific version requirements, I would suggest going with this because as the Python website says, it's "the present and the future of the language." However, you may be forced to use an older version of Python if you need to use a third party module that has not been updated to work with Python 3.x.

For example, I use both Python 2.7 and 3.3, but at work it's usually 2.7 because ArcGIS software is used extensively at the university, and it requires Python 2.7. Because my coworkers and students almost always have ArcGIS installed, they already have Python 2.7 even if they don't realize it. It makes sense to help them install open source tools to work with the Python version they already have, and this way I can take advantage of both GDAL and ArcGIS in the same script if I want, and teach them to do the same.

It's possible to have multiple versions of Python installed on one computer, so you can always pick and choose which version to use for which project. It's also possible to have different environments for one version of Python, meaning that you can have different workspaces, each with different modules installed. This allows you to have different configurations for different applications built with Python. Although that won't be covered in this book, please see www.virtualenv.org if you're interested in how to do it.

Python comes with a command-line utility called pip, which you'll become familiar with if you install many extra modules for Python because it's usually the easiest way to do it. The pip utility lives in the scripts folder inside your Python installation directory. Because this is a command-line tool, you need to use it from a terminal window or command prompt. To install a module using pip, you do something like this:

```
pip install module_name
```

To see a list of modules in the default pip repository, see https://pypi.python.org/pypi. You can install modules from other locations, but if you're using a new version of pip, you might get an error about the repository not being trusted. The error message will tell you what to add to the command to override this, but you should only do it if you trust the location you're trying to download the module from. (But you don't ever download something from a source you don't trust, right?) For more information on using pip, see https://pypi.python.org/pypi/pip.

There's more information about installing GDAL available online at http://trac .osgeo.org/gdal/wiki/DownloadingGdalBinaries, but I'm providing some information for you here.

A.1 Anaconda

Anaconda (https://www.continuum.io/downloads) includes Python and a large number of modules designed for scientific computing. Some are installed by default, and many more can be installed with a command-line tool that comes with Anaconda. Unfortunately, Anaconda quit supporting GDAL with version 2.5, but you can still download older versions that have GDAL. This is definitely the easiest way to get set up, although not all of the modules used in this book are included, and it doesn't have all possible capabilities compiled into GDAL. When you download and install an older version of Anaconda Python from https://repo.continuum.io/archive/index.html, you'll automatically have Python, NumPy, SciPy, matplotlib, and scikit-learn. GDAL/OGR and pyproj are optional installs, and you can get them by opening up the Anaconda command prompt and typing the following command:

```
conda install gdal pyproj
```

More information about using the conda utility to install and manage packages can be found at http://conda.pydata.org/docs/using/pkgs.html. You can use conda to install packages that aren't part of Anaconda, too. To find out how to do it, go to http://pypi.anaconda.org/ and search for the package you're interested in. Or you can use pip, and that's what we'll do here to install folium and spectral:

```
pip install folium spectral
```

See the information for your operating system to get hints for how to get Mapnik and set it up to work with Python. As of this writing, version 2.2 is the latest with downloadable binaries from the Mapnik website, but that's fine and is what I used for the examples.

A.2 Nonbundled installations

If you don't want to use a set of prebundled modules such as Anaconda, you can install everything individually. Every system is different and has its own little idiosyncrasies, however, so these are just general guidelines.

A.2.1 Linux

I can't provide information for all flavors of Linux here, but these Ubuntu directions might be enough to get you started. This example uses the standard Ubuntu repository, but more recent builds for these packages may be available from www.launchpad.net/~ubuntugis.

You can use apt-get to install GDAL and its dependencies:

```
sudo apt-get install gdal-bin libgdal-dev python-gdal
```

The easiest way to install most of the other needed packages is also to use apt-get. I think this should do it:

```
sudo apt-get install gdal-bin libgdal-dev python-scipy \
python-matplotlib python-pyproj python-scikits-learn libmapnik2.2 \
    libmapnik2-dev mapnik-utils python-mapnik2 qgis
```

You can install spectral python using pip. If you don't already have the pip tool for installing Python modules, you should install that now:

```
sudo apt-get install python-pip
```

And then to install Spectral Python:

```
sudo pip install spectral
```

A.2.2   Mac OS X

Although I haven't used a Mac in several years, back when I had one, the easiest way I found to install GDAL and several other geospatial packages was to take advantage of the great KyngChaos Wiki resource maintained by William Kyngesburye at www.kyngchaos.com/software:frameworks. This site contains prebuilt frameworks that are designed to work with the system version of Python on newer versions of OS X (Lion, Mountain Lion, and Mavericks, as of this writing). They don't support other versions of Python, including ones you download and install from www.python.org.

You can get everything GDAL, including some of the Python modules, by installing the GDAL Complete framework from the KyngChaos Wiki. This includes everything required by GDAL/OGR, but it doesn't include the optional plugins such as drivers for the Esri FileGDB and MrSID file formats. If you want them, they're available for download in the GDAL section farther down on the same download page.

The GDAL Complete framework also includes the NumPy module, but it doesn't include many of the other modules discussed in this book. If you follow the link to the Python Modules section of the wiki (www.kyngchaos.com/software/python), you'll see downloads for SciPy and matplotlib, among other useful modules not discussed here. While at this website, you might as well follow the link for QGIS (www.kyngchaos.com/software/qgis) and install that as well.

You should now be able to use pip from your terminal window to install the remaining Python packages:

```
pip install folium spectral scikit-learn
```

The remaining package is Mapnik. See http://mapnik.org/pages/downloads.html to get a precompiled binary (although not for the latest version as of this writing), and then see https://pypi.python.org/pypi/mapnik2 to see how to get the correct set of Python bindings. If I were to install this today, the version of Mapnik would be 2.2 and I would install the Python bindings like this:

```
easy_install -U mapnik2==2.2.0
```

A.2.3   Windows

If you're using Windows, the first thing you need is a copy of Python, because unlike many other operating systems, Windows doesn't come with Python already installed. The easiest way to get everything working on Windows is to download an official copy of Python from www.python.org. Several versions are available there, but as I said earlier, I'd suggest the latest one unless you have other requirements. If you have a 64-bit operating system, which is likely these days, you'll probably want to get one of the 64-bit versions of Python (please note, however, that I am not aware of a 64-bit version of Mapnik for Windows, so you'll need a 32-bit version if you want to use that in the last chapter). A 32-bit version will run on a 64-bit operating system, but the performance won't be as good. A 64-bit version of Python will definitely not run on a 32-bit operating system, however.

Assuming you're using an official Python distribution from www.python.org, you can get all of the other modules used in this book, except for folium and mapnik, from Christoph Gohlke's excellent resource at www.lfd.uci.edu/~gohlke/pythonlibs, and that's how I'd suggest doing it. The Spectral Python module is listed in the Misc section at the bottom of the page, as "spectral." When downloading packages from his site, be sure to download the ones corresponding to whatever version of Python you installed. If you installed a 64-bit version of Python, make sure you download the 64-bit versions of all packages. The same goes for 32-bit.

You can install folium using pip:

```
pip install folium
```

To install Mapnik, download and extract the Mapnik zip file from http://mapnik.org/pages/downloads.html. As of this writing, the latest version with a Windows binary provided is 2.2, and then only for 32-bit (it won't work with a 64-bit version of Python). After extracting the archive, set the following environment variables, assuming you extracted the files to C:\mapnik-v2.2.0:

- Add C:\mapnik-v2.2.0\bin and C:\mapnik-v2.2.0\lib to PATH
- Add C:\mapnik-v2.2.0\python\2.7\site-packages to PYTHONPATH (you'll need to create this environment variable if it doesn't already exist)

Another option for installing GDAL is to download the latest version from http://www.gisinternals.com/. If you do this, you need to set the following environment variables, assuming the installation folder is C:\Program Files\GDAL (the default location for 64-bit):

- Add C:\Program Files\GDAL to PATH
- GDAL_DATA = C:\Program Files\GDAL\gdal-data
- GDAL_DRIVER_PATH = C:\Program Files\GDAL\gdalplugins
- PROJ_LIB = C:\Program Files\GDAL\projlib

I've used this method to tell Anaconda where to find GDAL instead of using the version of GDAL provided by Anaconda because this one has more options precompiled. I've also used it with the new Anaconda Python suite that doesn't include GDAL.

A.3 *Environment variables*

Because several of these Python modules need external libraries to work, you need to make sure that Python can find them. This is where environment variables come in. For example, GDAL isn't really a Python program, and you need the GDAL program itself before the GDAL Python bindings will work. The Python GDAL module just allows you to use the real GDAL program through Python, but you can't do this if the real GDAL libraries can't be found.

I'm guessing that those of you on Linux who used apt-get will have everything in the correct place, so you won't have issues. I'm not sure about OS X, although I wouldn't be surprised if there were problems with Mapnik, if nothing else. The Windows packages from Gohlke's website tend to put the required binaries in the same folder as the Python bindings, so you probably won't have problems there. But again, Mapnik will be an issue.

If you do have problems, here are some environment variables that might help (if you don't know how to set environment variables, a quick web search will tell you how for your operating system):

- PATH: Should include the install folders for GDAL and Mapnik; for example:

```
C:\mapnik-v2.2.0\bin;C:\mapnik-v2.2.0\lib;C:\Python33\Lib\site-
packages\osgeo;C:\Program Files\<the rest of your PATH>
```

- GDAL_DATA: Set this to the folder in the GDAL installation directory that contains a bunch of .csv and .wkt files. It's usually called data or gdal-data; for example:

```
C:\Python33\Lib\site-packages\osgeo\data\gdal
```

- GDAL_DRIVER_PATH: Set this to the folder containing optional GDAL drivers, if you installed some; for example:

```
C:\Python33\Lib\site-packages\osgeo
```

- PROJ_LIB: This should be set to the folder in the pyproj installation that contains a large collection of files, most of them without extensions (one will be "epsg"); for example:

```
C:\Python33\Lib\site-packages\pyproj\data
```

- PYTHONPATH: This includes folders that Python will search in when looking for modules. If you've put modules in nonstandard locations, those need to be specified in this variable. If you install Mapnik on Windows, for example, the Python module isn't moved into a standard Python location, so you need to point Python to the Mapnik site-packages subfolder; for example:

```
C:\mapnik-v2.2.0\python\2.7\site-packages
```

A.4 *Source code and data*

The source code for the examples in the book is available from the Manning website at https://www.manning.com/books/geoprocessing-with-python and from GitHub at https://github.com/cgarrard/osgeopy-code. There's also a custom Python module for the book that contains some convenience functions and some crude tools for viewing data. This is contained in a file called ospybook-latest.zip in the code download. You can use pip to install it like this (assuming it's in my C:/temp folder):

```
pip install c:/temp/ospybook-latest.zip
```

If you download the book's source code, you'll have all of the code for this module. Using pip puts it in a standard location so that Python can find it. You can download the data used in the examples from www.manning.com/books/geoprocessing-with-python, and from https://app.box.com/osgeopy.

A.5   *Development environments*

Python comes with an interactive environment that you can use from a terminal window or command prompt, but most people aren't huge fans of this (I use it a lot for playing with short snippets of code). Python also comes with a graphical interface called IDLE. This includes an interactive environment, similar to the command line, but with syntax highlighting and code completion, meaning you can start typing and then press TAB to see a list of options for completing your code, such as function or variable names. It also has a text editor in which you can edit files containing Python code and run them from inside IDLE.

If you want something nicer, however, you have a lot of options. Two examples, both of which happen to come with Anaconda Python, are IPython and Spyder. IPython is an interactive shell, but is more functional than the default Python interactive environment. It has syntax highlighting, tab completion, system shell access, aliases in the form of "magic commands," macros, and much more. You can learn about IPython at http://ipython.org/. Another nice feature of IPython is its notebook support. You can embed text along with Python code and output into a notebook that you can share with others or convert to another format such as HTML. More information about Jupyter notebooks is available at http://jupyter.org/. Anaconda also installs Spyder, which is an interactive development environment (IDE) for Python. It uses IPython and puts a code editor, interactive shell, output windows, variable lists, and other information all in one package. See https://pythonhosted.org/spyder/ for more information.

I have to admit that I use a text editor instead of an IDE much of the time. This has its downsides, such as the fact that I don't have a code editor and interactive shell that are linked together. With Spyder, for example, you can run a script from a file, and the variables set in the script are then available from the interactive shell. This makes it really easy to play with and explore your data. The lack of integration can also be a good thing, however, because my scripts always start with a blank slate when I run them from a text editor. More than once I've seen students get tripped up when they accidentally broke their code by deleting a line that set a variable, but the IDE remembered the variable and their code still ran. Well, it ran until they restarted the IDE, and then things no longer worked.

Another advantage to IDEs is that they make it easier to walk through and debug your code. Python has a debugger called pdb built in, but you might find using an IDE to be easier. A debugger allows you to set breakpoints on lines of your code, and then if you run your script it will run until it hits the breakpoint. You can inspect the current values of your variables at that point, and you can also step through your code line by line and watch how your variables change or see if the code is executing the way you intended it to (in other words, you can check your logic). So if you do use an IDE, you'd be wise to read the help documentation about its debugger, or else just write some code, hit the debug button, and see what happens. Playing is the best way to learn, after all.

There's a long list of Python IDEs available online at https://wiki.python.org/moin/IntegratedDevelopmentEnvironments.

appendix B
References

Data used in figures

The numbers in parentheses correspond to items in the second section of this appendix, "Data References."

Chapter 1

1.1: Library of Congress (7)
1.2–1.4: Natural Earth (9)
1.5: USDA NAIP (16)
1.6: PRISM (12)
1.8: Grand Canyon NP (6), USGS Grand Canyon (19)
1.9: Snow (13)
1.11, 1.13: USGS Small-scale data (24)

Chapter 3

3.1: USGS Small-scale data (24), GSHHG (10), OpenStreetMap (11)
3.2: OpenStreetMap (11), USGS NHD (22)
3.3: Natural Earth (9)
3.4: Natural Earth (9), USGS Small-scale data (24)
3.5: USCB TIGER (15), USGS Small-scale data (24), City of New Orleans (2), OpenStreetMap (11)
3.10, 3.13: Natural Earth (9)
3.14: USGS Small-scale data (24)
3.15, 3.17: Natural Earth (9)

Chapter 4

4.2: Natural Earth (9)
4.4: USGS Small-scale data (24), NWS (8)
4.5: USGS Small-scale data (24)

4.7: NWS (8), OpenStreetMap (11)
4.8: NWS (8), Stamen (14)

Chapter 5

5.1–5.9: Natural Earth (9)
5.11: USGS Small-scale data (24)

Chapter 6

6.7: Natural Earth (9)

Chapter 7

7.4: City of New Orleans (2), USGS Small-scale data (24), USCB TIGER (15), OpenStreetMap (11)
7.5, 7.6: USGS Small-scale data (24), City of New Orleans (2)
7.8: NREL Wind data (29), US Census Tract Data (28), USGS Small-scale data (24)
7.9: US Census tract data (28), USGS small-scale data (24)
7.10, 7.11: NREL Wind data (29)
7.12–7.16: Natural Earth (9), Env-DATA (3 and 4)

Chapter 8

8.1–8.4: Natural Earth (9)
8.5: Utah AGRC (27)
8.7: USGS Small-scale data (24)
8.8–8.10: Natural Earth (9)
8.12: USDA NAIP (16)
8.13: Natural Earth (9)

Chapter 9

9.1: USGS GAP (26)
9.2, 9.4: USDA NAIP (16)
9.7, 9.11, 9.13: USGS Landsat (20)
9.16: USGS Ortho (30)

Chapter 10

10.1: USGS GAP (26)
10.2: USGS DOQ (17), USGS TOPO (25)
10.6: USGS DOQ (17)
10.7, 10.8: USGS GTOPO30 (18)
10.12, 10.13, 10.15: USGS Landsat (20)

Chapter 11

11.2: USDA NAIP (16)
11.7, 11.9: USGS GTOPO30 (18)
11.12: USGS GAP (26), EPA (5)
11.14: USGS Roads (23), BLM Wilderness areas (1)
11.16: USGS Roads (23)

Chapter 12

12.1: USGS GAP (26), USGS Landsat (20)

12.2: USGS GAP (26)

12.4: USGS GAP (26), USGS Landsat (20)

12.5: USGS GAP (26)

Chapter 13

13.4–13.7, 13.9: Natural Earth (9)

13.10: USGS NED (21)

13.11, 13.13: USGS Landsat (20)

13.15: USGS NED (21)

13.16: USCB TIGER (15)

13.18, 13.19: USCB TIGER (15), USGS Small-scale data (24)

13.20: USCB TIGER (15), USGS Small-scale data (24), NOLA (2), OpenStreetMap (11)

13.21: USGS TOPO (25)

13.22: USGS NED (21)

13.23: USGS NED (21), USGS TOPO (25)

Data references

1   Bureau of Land Management and Forest Service. Wilderness areas. http://research.idwr .idaho.gov/index.html

2   City of New Orleans—Office of Information Technology & Innovation, Enterprise Information Team. https://data.nola.gov/Geographic-Base-Layers/NOLA-Boundary/2b2j-u6kh

3   Cruz, S.; Proaño, C.B.; Anderson, D.; Huyvaert, K.; Wikelski, M. (2013). Data from: The Environmental-Data Automated Track Annotation (Env-DATA) System: Linking animal tracks with environmental data. Movebank Data Repository. DOI:10.5441/001/1.3hp3s250.

4   Dodge, S.; Bohrer, G.; Weinzierl, R.; Davidson, S.C.; Kays, R.; Douglas, D.; Cruz, S.; Han, J.; Brandes, D.; Wikelski, M. 2013. The Environmental-Data Automated Track Annotation (Env-DATA) System—linking animal tracks with environmental data. Movement Ecology 1:3. DOI:10.1186/2051-3933-1-3.

5   Environmental Protection Agency. Level III ecoregions of North America. http://archive .epa.gov/wed/ecoregions/web/html/na_eco.html

6   Grand Canyon National Park—Science Center GIS. Roads, Routes, Streets and Trails of Grand Canyon National Park and Arizona. https://catalog.data.gov/dataset/roads-routes-streets-and-trails-of-grand-canyon-national-park-and-arizona

7   Library of Congress, Geography and Map Division. http://www.loc.gov/item/2005634035/

8   National Weather Service. http://gis.srh.noaa.gov/arcgis/rest/services

9   Natural Earth. http://www.naturalearthdata.com/

10 NOAA National Geophysical Data Center, GSHHG, October 2015, http://www.ngdc.noaa.gov/ mgg/shorelines/shorelines.html

11 OpenStreetMap. Map tiles by OpenStreetMap, under CC BY 2.0. Data by OpenStreetMap, under ODbL: http://www.openstreetmap.org/copyright

12 PRISM Climate Group, Oregon State University, http://prism.oregonstate.edu. Map created 31 May 2015. Copyright © 2015.

13 Snow, John. https://en.wikipedia.org/wiki/File:Snow-cholera-map-1.jpg

14 Stamen Design. Map tiles by Stamen Design, under CC BY 3.0. Data by OpenStreetMap, under ODbL: http://maps.stamen.com/#toner

15 US Census Bureau. 2006 TIGER/Line shapefiles. https://www.census.gov/geo/maps-data/data/tiger-line.html

16 USDA FSA Aerial Photography Field Office. NAIP Imagery. http://earthexplorer.usgs.gov/

17 US Geological Survey. Digital orthophoto quadrangle (DOQs). http://earthexplorer.usgs.gov/

18 US Geological Survey. Global 30 arc-second elevation (GTOPO30). http://earthexplorer.usgs.gov/

19 US Geological Survey. Grand Canyon data repository. http://pubs.usgs.gov/ds/121/grand/grand.html

20 US Geological Survey. Landsat imagery. http://earthexplorer.usgs.gov/

21 US Geological Survey. National elevation dataset (NED). http://nationalmap.gov/elevation.html

22 US Geological Survey, 2013, National Hydrography Geodatabase (NHD). http://viewer.nationalmap.gov/viewer/nhd.html?p=nhd

23 US Geological Survey. Roads. http://research.idwr.idaho.gov/index.html

24 US Geological Survey. Small-scale data. http://nationalmap.gov/small_scale/atlasftp.html

25 US Geological Survey. US topo quadrangles (TOPO). http://nationalmap.gov/ustopo/

26 USGS National Gap Analysis Program. 2004. Provisional Digital Land Cover Map for the Southwestern United States. Version 1.0. RS/GIS Laboratory, College of Natural Resources, Utah State University. http://earth.gis.usu.edu/swgap/landcover.html

27 Utah Automated Geographic Reference Center. http://gis.utah.gov/data/

28 US Department of Commerce, U.S. Census Bureau, Geography Division/Cartographic Products Branch. 2010. 2010 Cartographic Boundary File, State-County-Census Tract for California, 1:500,000. http://www2.census.gov/geo/tiger/GENZ2010/

29 US National Renewable Energy Laboratory. Wind data. http://www.nrel.gov/gis/data_wind.html

30 USGS EROS 1-foot orthoimagery from the National Map. http://raster.nationalmap.gov/arcgis/rest/services/Orthoimagery/USGS_EROS_Ortho_1Foot/ImageServer

index